AF443398

# Programming and Mathematical Method

# NATO ASI Series

## Advanced Science Institutes Series

*A series presenting the results of activities sponsored by the NATO Science Committee, which aims at the dissemination of advanced scientific and technological knowledge, with a view to strengthening links between scientific communities.*

The Series is published by an international board of publishers in conjunction with the NATO Scientific Affairs Division

| | | |
|---|---|---|
| A | Life Sciences | Plenum Publishing Corporation |
| B | Physics | London and New York |
| C | Mathematical and Physical Sciences | Kluwer Academic Publishers Dordrecht, Boston and London |
| D | Behavioural and Social Sciences | |
| E | Applied Sciences | |
| F | Computer and Systems Sciences | Springer-Verlag Berlin Heidelberg New York |
| G | Ecological Sciences | London Paris Tokyo Hong Kong |
| H | Cell Biology | Barcelona Budapest |
| I | Global Environmental Change | |

## NATO-PCO DATABASE

The electronic index to the NATO ASI Series provides full bibliographical references (with keywords and/or abstracts) to more than 30 000 contributions from international scientists published in all sections of the NATO ASI Series. Access to the NATO-PCO DATABASE compiled by the NATO Publication Coordination Office is possible in two ways:

- via online FILE 128 (NATO-PCO DATABASE) hosted by ESRIN, Via Galileo Galilei, I-00044 Frascati, Italy.

- via CD-ROM "NATO-PCO DATABASE" with user-friendly retrieval software in English, French and German (© WTV GmbH and DATAWARE Technologies Inc. 1989).

The CD-ROM can be ordered through any member of the Board of Publishers or through NATO-PCO, Overijse, Belgium.

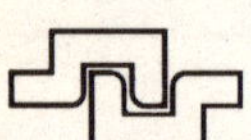

Series F: Computer and Systems Sciences Vol. 88

# Programming
# and Mathematical Method

International Summer School directed by
F. L. Bauer, M. Broy, E. W. Dijkstra, C. A. R. Hoare

Edited by

## Manfred Broy

Institut für Informatik
Technische Universität München
Postfach 20 24 20
W-8000 München 2, FRG

Springer-Verlag
Berlin Heidelberg New York London Paris Tokyo
Hong Kong Barcelona Budapest
Published in cooperation with NATO Scientific Affairs Division

Proceedings of the NATO Advanced Study Institute on Programming and
Mathematical Method, held at Marktoberdorf, FRG, July 24- August 5, 1990.

ISBN 3-540-55558-7 Springer-Verlag Berlin Heidelberg New York
ISBN 0-387-55558-7 Springer-Verlag New York Berlin Heidelberg

Library of Congress Cataloging-in-Publication Data
Programming and mathematical method / edited by Manfred Broy.  p.  cm. -- (Computer and systems
sciences ; sers. F88)
ISBN 0-387-55558-7
1. Electronic digital computers--Programming. 2. Programming (Mathematics) I. Broy, M., 1949-  . II. Series:
NATO ASI series. Series F, Computer and system sciences ; no. 88.  QA76.6P75137  1992  005.1'01'5113--
dc20  92-26379



© Springer-Verlag Berlin Heidelberg 1992
Printed in Germany

Typesetting: Camera ready by authors
45/3140 - 5 4 3 2 1 0 - Printed on acid-free paper

# Preface

The Summer School in Marktoberdorf 1990 had as its overall theme the development of programs as an activity that can be carried out based on and supported by a mathematical method. In particular mathematical methods for the development of programs as parts of distributed systems were included. Mathematical programming methods are a very important topic for which a lot of research in recent years has been carried out.

In the Marktoberdorf Summer School outstanding scientists lectured on mathematical programming methods. The lectures centred around logical and functional calculi for the

- specification,
- refinement,
- verification

of programs and program systems.

Some extremely remarkable examples were given. Looking at these examples it becomes clear that proper research and teaching in the area of program methodology should always show its value by being applied at least to small examples or case studies.

It is one of the problems of computing science that examples and case studies have to be short and small to be presentable in lectures and papers of moderate size. However, even small examples can tell a lot about the tractability and adequacy of methods and being able to treat small examples does at least prove that the method can be applied in modest ways. Furthermore it demonstrates to some extent the notational and calculational overhead of applying formal methods.

The Marktoberdorf Summer School 1990 proved to be – like the previous summer schools – a highlight in my academic life. It was not only the well-prepared lectures that proved to be a stimulating contribution, it was also the participants of the summer school who again provided one of the most stimulating environments one can think of.

It's a pleasure for me to thank all the people that helped to make the summer school into a success. My special thanks go to Birgit Schieder who did a very careful job in assisting me to edit this book and to Hans Wössner from Springer-Verlag for the trouble he has taken.

München, July 1991 Manfred Broy

# Table of Contents

# Chapter 1

# Examples of Derivations

The presentation of formal proofs and derivations is for computing science even more important than for mathematical sciences. Adequate formalization and structuring of proofs and derivations, the choice of adequate concepts are one of the most important tools for getting proofs simple and understandable such that they can actually be used as a foundation for understanding and reasoning about programs, algorithms, and systems.

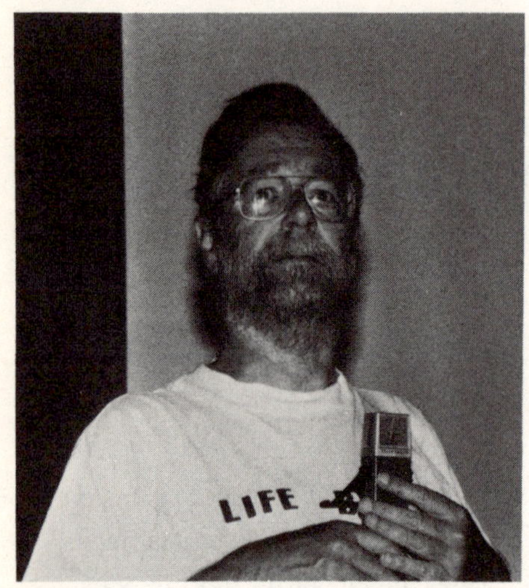

E.W. Dijkstra

W.H.J. Feijen

# On the design of a simple proof for Morley's Theorem

Edsger W. Dijkstra
Department of Computer Sciences
The University of Texas at Austin
Austin, TX 78712 - 1188
USA

The general opinion - with which I concur - is that Frank Morley's Theorem about the angle trisectors of a triangle is a geometrical curiosity that is of historical interest at best. Independently of the (in)significance of the theorem proved by it, a proof may deserve our attention, for instance, by virtue of its structure, its simplicity, or its brevity. For the proof to be shown below, such a claim can be made: requiring no auxiliary lines or points, it is so simple that the theorem's late discovery (1899) and the elapsed decade before the first proofs were published (1909) become the more striking. When I found this proof years ago, I was only too willing to ascribe that discovery to my great ingenuity and all that. The purpose of this note, however, is to show how, in the mean time, the art and science of proof design have advanced to a stage in which the design of such proofs has almost become a routine exercise, requiring the usual care in arrangement and notation, but a minimum of invention.

Consider a plane figure - consisting of 6 distinct points and 12 straight line segments - of the following configuration -

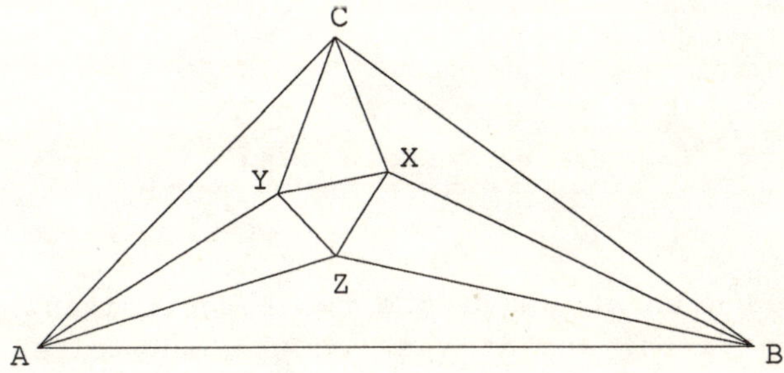

Morley's Theorem states that $\triangle XYZ$ is equilateral if, in this figure, the angles of $\triangle ABC$ are trisected.

Note that there are all sorts of nonredundant ways of defining that a triangle is equilateral, such as

(0) its edges are of equal length

(1) its angles are of equal size

(2) it is isosceles, and its top angle equals 60°.

Definition (1) makes it clear that Morley's Theorem can be formulated in terms of angles only, i.e. that in the above figure we can isolate size from shape.

Let us call the above configuration with the angles of $\triangle ABC$ trisected a Morley Figure. The simplest way of giving the shape of a Morley Figure independently of its size is to give the angles of $\triangle ABC$. The simplest way of giving the size of a Morley Figure independently of its shape is to give the size of $\triangle XYZ$.

This is a very nice disentanglement. To do justice to it, we propose to prove Morley's Theorem by showing

For any triple $(\alpha,\beta,\gamma)$ of positive angles whose sum equals 60° and for any equilateral triangle, one can erect on the three sides of the equilateral triangle triangles with top angles $\alpha,\beta,\gamma$ respectively so that the angles of the triangle formed by their tops are trisected.

The existence of these three triangles will be demonstrated by showing how to erect them. To this end we shall derive how to erect them. We exploit the cyclic symmetry (which we have been careful not to destroy) in order to avoid repetition, and focus our attention on side AB.

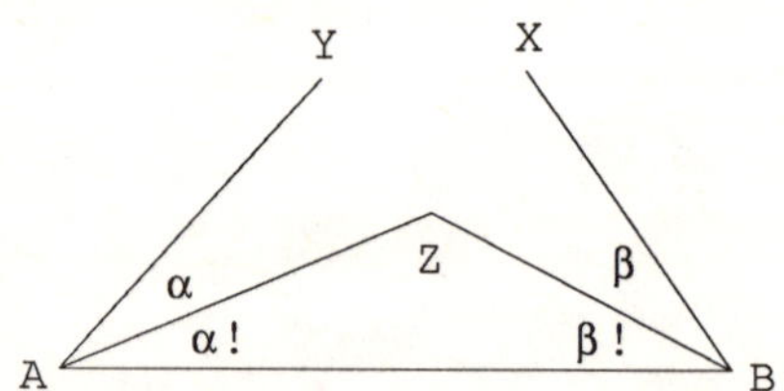


We have marked with $\alpha!$ and $\beta!$ the two target sizes, with $\alpha$ and $\beta$ the two given sizes, and in view of the many alternative definitions we have left open how we are going to take into account that $\triangle XYZ$ is equilateral. In the above picture, we can draw one immediate conclusion from the target sizes: $\angle AZB = 180° - \alpha - \beta$. Because the remainder of the picture has to be taken into account, we have to conclude that its full complement equals $180°+\alpha+\beta$. Now it stands to reason to draw XZ and YZ and to take into account that $\angle XZY = 60°$.

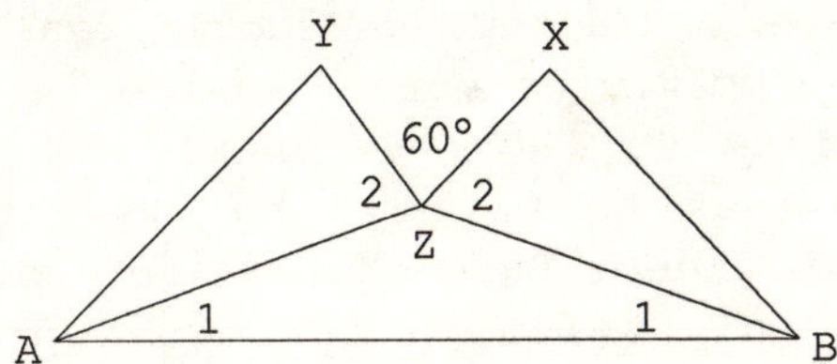

From $\angle XZY = 60°$ we thus conclude

$\quad$ ($\alpha+\beta$ = the sum of angles marked 1) $\equiv$

$\quad$ ($120°+\alpha+\beta$ = the sum of the angles marked 2).

A prescribed sum of two values gives a one-dimensional degree of freedom, for which we now have to find the most suitable parameterization. For the angles marked 1, the simplest parameterization is

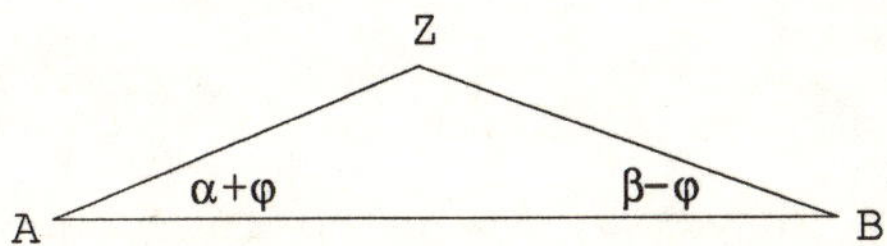

since that translates our target into $\varphi = 0$. For reasons of symmetry it stands to reason to split $120°+\alpha+\beta$ into $60°+\alpha+\psi$ and $60°+\beta-\psi$, but which is which? Because of $\alpha+\beta+\gamma = 60°$, we get the nicest fit if we allocate $60°+\alpha+\psi$ in the triangle with top $\beta$. Thus we find ourselves contemplating the following parameterized figure:

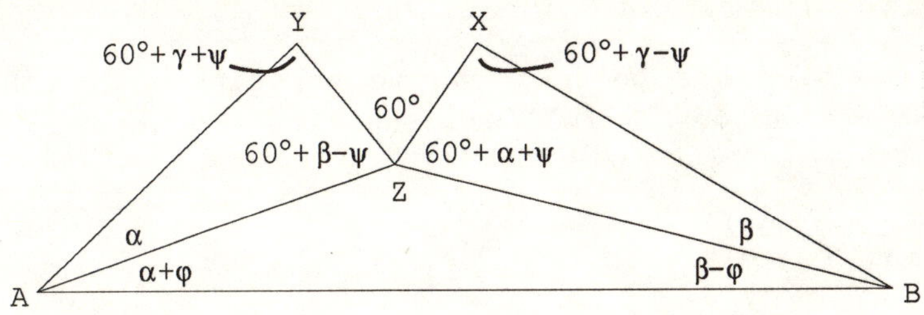

Were we to complete it, we would find in clockwise direction $60°+\alpha+\psi$, $60°+\beta-\psi$, $60°+\gamma+\psi$, $60°+\alpha-\psi$, $60°+\beta+\psi$, $60°+\gamma-\psi$. Our remaining duty is to show that $\psi$ can be chosen in such a way that $\varphi = 0$ follows.

Looking at our last diagram, we should realize that, so far, we have used from geometry no more than the fact that the angles of a triangle add up to $180°$. We should be willing to use a little bit more so as to be able to exploit that the three triangles share two sides. Parameter $\varphi$ (for which $= 0$ has to be concluded) occurs only twice in the above picture, more precisely, in $\Delta AZB$ opposite to the shared sides. So we are looking for a fact from geometry that links angles of a triangle to their opposite sides. Here the Rule of Sines is supposed to come to the reader's mind. Furthermore, of the fact that $\Delta XYZ$ is equilateral, so far we used only $\angle XZY = 60°$ and thus can expect to have to use $XZ = YZ$.

So, let us start with the expression in $\varphi$ that we can manipulate with the Rule of Sines

$$\frac{\sin(\alpha+\varphi)}{\sin(\beta-\varphi)}$$

$= \{$Rule of Sines in $\Delta AZB\}$

$$\frac{BZ}{AZ}$$

$= \{$Rule of Sines in $\Delta BZX/\Delta AZY$ so as to introduce $XZ/YZ\}$

$$\frac{\sin(60°+\gamma-\psi)\cdot XZ/\sin(\beta)}{\sin(60°+\gamma+\psi)\cdot YZ/\sin(\alpha)}$$

$= \{XZ = YZ\}$

$$\frac{\sin(60°+\gamma-\psi)\cdot\sin(\alpha)}{\sin(60°+\gamma+\psi)\cdot\sin(\beta)} \quad ,$$

in summary

$$(3) \quad \frac{\sin(\alpha+\varphi)}{\sin(\beta-\varphi)} = \frac{\sin(60°+\gamma-\psi)}{\sin(60°+\gamma+\psi)} \cdot \frac{\sin(\alpha)}{\sin(\beta)}$$

Now the question has become: can we substitute in (3) such a value for $\psi$ that $\varphi = 0$ follows? Substituting into (3) $\varphi := 0$ yields

$$(4) \quad \frac{\sin(60°+\gamma-\psi)}{\sin(60°+\gamma+\psi)} = 1 \quad ,$$

a solvable equation (e.g. $\psi = 0$). Substituting (4) into (3), we get

$$\sin(\beta)\cdot\sin(\alpha+\varphi) = \sin(\alpha)\cdot\sin(\beta-\varphi)$$

= {addition formula for sin}

$$\sin(\beta)\cdot(\sin(\alpha)\cdot\cos(\varphi)+\cos(\alpha)\cdot\sin(\varphi)) =$$

$$\sin(\alpha)\cdot(\sin(\beta)\cdot\cos(\varphi)-\cos(\beta)\cdot\sin(\varphi))$$

= {algebra}

$$(\sin(\beta)\cdot\cos(\alpha)+\sin(\alpha)\cdot\cos(\beta))\cdot\sin(\varphi) = 0$$

= {addition formula for sin}

$$\sin(\alpha+\beta)\cdot\sin(\varphi) = 0$$

= {$0 < \alpha+\beta < 60°$}

$$\sin(\varphi) = 0$$

= {$\alpha+\varphi \geq 0 \Rightarrow \varphi > -60°$; $\beta-\varphi \geq 0 \Rightarrow 60° > \varphi$}

$$\varphi = 0 \quad ,$$

which concludes the proof. (To solve the last equation with the addition formula for the sine, rather than via a monotonicity argument, was suggested to me by Ambuj Kumar Singh.)

*              *

*

For a traditional presentation of a geometric proof, see [0] - Coxeter amazingly contrasts "a trigonometrical proof" with "an elementary proof" -; for a compact presentation of the above trigonometrical proof, see [1] (where the proof is reduced to almost a one-liner). This note is a critical comment on such proof presentations, which are correct, but rather uninstructive.

Coxeter starts his proof with "Much trouble is experienced if we try a direct approach, but the difficulties disappear if we work backwards, beginning with an equilateral triangle and building up a general triangle which is afterwards identified with the given triangle ABC.". Coxeter at least mentions the "trouble" - be it as a fact learned from unanalysed experience gained from trying to prove Morley's Theorem -; Dijkstra does not even mention the trouble and just announces "We start in our proof not with the arbitrary triangle, but with the equilateral one.".

They both pull the same big rabbit out of their mathematical hat. Above we showed how this rabbit can be avoided. In order not to interrupt the flow of the presentation, we mentioned the catchword "disentanglement" - see [2] -; here is the place to point out that it refers to a general principle. The idea is to write in particular the information to be used as a conjunction of as many independent terms as possible because this is a precondition for subsequently being able to indicate precisely what is appealed to where. (If a datum can be factored differently - such as "$\Delta XYZ$ is equilateral" - the advice is to list the options and to postpone the choice until it emerges which one is appropriate.) The decision not to fix the size of the Morley Figure by, say, the length of AB rests on the principle - see [2] - not to destroy symmetry lightly.

The second big rabbit that [0] and [1] share is the choice of very special angles that new lines make with the edges of the equilateral triangle. The above analysis, based on angles alone, shows that we are essentially left with one degree of freedom, viz. the relative orientation of the two triangles; hence $\psi$. The magician's proof immediately substitutes for $\psi$ its proper value (so that $\psi$ need not be mentioned at all), thus suggesting that without such spark of genius one cannot find such proofs. It is good to know that a whole class of such sparks are superfluous: in case of freedom, capture that freedom by the introduction of as yet undetermined parameters, which, if not truly arbitrary, will be later constrained by the needs of the subsequent proof.

The third big rabbit in [0] and [1] is the absence of any motivation for the choice of mathematical fact to exploit. In [0], out of the blue Coxeter refers to a theorem about the incenter of a triangle that I was not aware of. In [1], Dijkstra writes equally unmotivated "Using the rule of sines three times (in triangles AXB, BXZ, and AXY) we deduce [...]", leaving the reader in awe for the miracle how subsequently all nasty terms cancel out. If we wish to teach our students how to design proofs we should teach them how to let their logical needs guide them through their mathematical knowledge.

Finally I would like to point out that the above design has been driven by different concerns. Using that AZ is a shared edge of two triangles was a logical necessity, our choice of parameters was a matter of expedience. Instead of angles $\alpha+\varphi$ and $\beta-\varphi$ we could have introduced $\varphi'$ and $\alpha+\beta-\varphi'$; it would have uglified the later development.

[0] Coxeter, H.S.M., "Introduction to Geometry", John Wiley &
    Sons, Inc., New York, 1969, pp. 23-25.

[1] Dijkstra, Edsger W., "Selected Writings on Computing: A
    Personal Perspective", Springer-Verlag New York Inc., 1982,
    pp. 182-183.

[2] Gasteren, A.J.M. van, "On the shape of mathematical argu-
    ments", Ph.D. Thesis, 20 December 1988, Technical University
    Eindhoven.


Prof. Dr. Edsger W. Dijkstra                    Austin, 23 April 1989

# Well-foundedness and the transitive closure


A.J.M. van Gasteren
Edelweislaan 20
5582 BW WAALRE
The Netherlands

Edsger W. Dijkstra
Department of Computer Sciences
The University of Texas at Austin
Austin, TX 78712-1188
USA


The transitive closure of relation R is defined as the <u>strongest</u> relation S that satisfies for all x,y (in the domain of R)

$$(0) \quad xSy \equiv xRy \lor (\underline{E}z:\ zRy:\ xSz)\ .$$

A corollary of the following theorem is that the restriction to the strongest relation can be omitted in the case of well-founded R.

<u>Theorem 0</u> For well-founded R, at most 1 relation S satisfies (0).

<u>Proof</u> Well-foundedness of R means that for any predicate P (on the domain of R)

$$(1) \quad (\underline{A}y::\ P.y) \Leftarrow (\underline{A}y::\ P.y \Leftarrow (\underline{A}x:\ xRy:\ P.x)),$$

a formula that captures the notion of proof by mathematical induction over the domain of R.

Let S and S' satisfy (0). We prove our demonstrandum S = S', that is

$$(\underline{A}x::\ (\underline{A}y::\ xSy \equiv xS'y))\ ,$$

by proving for any x

$$(\underline{A}y::\ xSy \equiv xS'y)$$

by mathematical induction over y. According to (1) our proof obligation is to show for any y

$$(xSy \equiv xS'y) \Leftarrow (\underline{A}z:\ zRy:\ xSz \equiv xS'z)\ .$$

Starting with the consequent, we observe

$$xSy \equiv xS'y$$

$$= \quad \{S \text{ and } S' \text{ satisfy } (0)\}$$

$$xRy \lor (\underline{E}z: zRy: xSz) \equiv$$

$$xRy \lor (\underline{E}z: zRy: xS'z)$$

$$\Leftarrow \quad \{\text{Leibniz}\}$$

$$(\underline{E}z: zRy: xSz) \equiv (\underline{E}z: zRy: xS'z)$$

$$\Leftarrow \quad \{\text{predicate calculus}\}$$

$$(\underline{A}z: zRy: xSz \equiv xS'z) \quad .$$

(End of Proof.)

<u>Examples</u> Let R be a relation on the natural numbers. With $xRy \equiv x+1=y$, the only S satisfying (0) is $xSy \equiv x<y$. With $xRy \equiv x=y \lor x+1=y$, the strongest S satisfying (0) is $xSy \equiv x{\leq}y$; in this case, however, (0) is also satisfied by $xSy \equiv$ true. Our first R is well-founded, our second R, being reflexive, is not.

(End of Examples.)

The significance of Theorem 0 is that, for a well-founded relation, its transitive closure need not be handled as extreme solution. This is illustrated in the proof of our next theorem.

<u>Theorem 1</u> The transitive closure of a well-founded relation is well-founded.

<u>Proof</u> With R and S satisfying (0) and (1) we have to prove for any P (1) with R := S, i.e. - naming the most complicated sub-expression -

$$(3) \quad (\underline{A}y:: P.y) \Leftarrow (\underline{A}y:: P.y \Leftarrow Q.y) \quad \text{with}$$

$$(4) \quad Q.y \equiv (\underline{A}x: xSy: P.x) \quad .$$

Since the occurrence of S is confined to Q, we start manipulating Q.y:

$$Q.y$$

$$= \quad \{(4)\}$$

$$(\underline{A}x: xSy: P.x)$$

$$= \quad \{(0)\}$$

$$(Ax: xRy \lor (Ez: zRy: xSz): P.x)$$

$=$  {splitting the range}

$$(Ax: xRy: P.x) \land (Ax: (Ez: zRy: xSz): P.x)$$

$=$  {trading and $\lor$ distributes over $A$}

$$(Ax: xRy: P.x) \land (Ax:: (Az: zRy \land xSz: P.x))$$

$=$  {renaming; interchange of quantifications}

$$(Az: zRy: P.z) \land (Az: zRy: (Ax: xSz: P.x))$$

$=$  {conjoining the terms; (4) with $y := z$}

$$(Az: zRy: P.z \land Q.z) \ .$$

Thus we have established – from (0) –

(5)   $Q.y \equiv (Az: zRy: P.z \land Q.z)$ .

Our gain from the above use of (0) is that, whereas (4) expressed Q in terms of S, (5) expresses Q in terms of R, which is well-founded.

To prove (3) we now observe

$$(Ay:: P.y \Leftarrow Q.y)$$

$=$  {(5)}

$$(Ay:: P.y \Leftarrow (Az: zRy: P.z \land Q.z))$$

$=$  {(5) and pred.calc., preparing for (1)}

$$(Ay:: P.y \land Q.y \Leftarrow (Az: zRy: P.z \land Q.z))$$

$\Rightarrow$  {R is well-founded: (1) with $P.y := P.y \land Q.y$}

$$(Ay:: P.y \land Q.y)$$

$\Rightarrow$  {pred.calc.}

$$(Ay:: P.y) \ .$$

(End of Proof.)

We would like to stress that from the point of view of manipulation, the structure of the above proof is very nice. The first part uses (0) – which connects R and S – to eliminate S

14

and to introduce R. The second part the uses (1), which states R's relevant property. We also think we greatly benefitted from the introduction of the identifier Q. The frequency with which it occurs gives an idea of how much Q has shortened the text. Moreover it highlights that in a number of steps the internal structure of Q – and a fortiori the choice between (4) and (5) – is totally irrelevant. In the second part of the proof, we first replaced the antecedent and then conjoined the consequent with Q.y to show the heuristics for the latter operation. Starting with

$$(\underline{A}y:: P.y \Leftarrow Q.y)$$

$$= \quad \{\text{pred. calc.}\}$$

$$(\underline{A}y:: P.y \wedge Q.y \Leftarrow Q.y)$$

would have reduced the number of appeals to (5), but the above first step is more of a rabbit.

Finally we prove

<u>Theorem 2</u> If the transitive closure of a relation is well-founded, so is the relation itself.

<u>Proof</u> We have to prove (1), given (0) and the fact that S is well-founded. To this end we observe for any P

$$(\underline{A}y:: P.y \Leftarrow (\underline{A}x: xRy: P.x))$$

$$\Rightarrow \quad \{\text{from (0): } xRy \Rightarrow xSy, \text{ antimonotonicity twice}\}$$

$$(\underline{A}y:: P.y \Leftarrow (\underline{A}x: xSy: P.x))$$

$$\Rightarrow \quad \{S \text{ is well-founded}\}$$

$$(\underline{A}y:: P.y) \hspace{4cm} \text{(End of Proof.)}$$

We point out that one can base the proofs of theorems 1 and 2 on an alternative definition of well-foundedness, viz. that all decreasing chains are of finite length. (In that setting, theorem 1 is simpler to prove than theorem 2!) One may be under the impression that the alternative proofs are shorter and simpler than the ones developed here, but that impression could very well be wrong. For a fair comparison, the alternative proofs should be as explicit as ours on what they use; we mention arithmetic, the notion of infinity, and the connection between decreasing chains for the original relation and its transitive closure.

Finally, please note that we have established the equivalence between the classical mathematical induction with base and step, and "course of values induction": with $xRy \equiv x+1=y$, the transitive closure S is $xSy \equiv x<y$.

Austin, 28 April 1990

Dr. A.J.M. van Gasteren      Prof. Dr. Edsger W. Dijkstra

# Designing the proof of Vizing's Theorem

Josyula R. Rao
Edsger W. Dijkstra
Department of Computer Sciences
The University of Texas at Austin
Austin, TX 78712 - 1188
USA

We shall design the proof of the following theorem, due to V.G. Vizing.

<u>Theorem</u> For a finite undirected graph without autoloops and without multiple edges, at any vertex of which fewer than N edges meet, N colours suffice for an edge colouring such that edges incident on the same vertex are of different colour.

$$*\qquad\qquad\qquad *$$

$$*$$

The graph being <u>acceptably coloured</u> is expressed by

(0) $(\underline{A}V:: acc.V)$

where predicate acc on vertices is given by

$acc.V \equiv$ (no two edges incident on vertex
V have the same colour).

We shall prove Vizing's Theorem constructively by showing how the entire graph can be acceptably coloured. To this end it suffices to design a procedure that, given an acceptably coloured graph with one uncoloured edge, constructs an acceptable colouring in which this last edge is coloured as well. Repeated application of this procedure, starting with a graph consisting of vertices only (whose colouring is trivially acceptable) and adding one uncoloured edge at a time, colours the whole graph acceptably. We shall design the procedure for colouring the next uncoloured edge.

In the rest of this note, we confine ourselves to acceptably coloured graphs. In order not to interrupt the subsequent development of the procedure too much, we first introduce a concept, the need of which will emerge, viz. the <u>alternating path</u>. An alternating path is a maximal path of at most two given colours. Because the colouring of the graph is acceptable, those two colours alternate along such a path; hence the name. One can show the following:

(i) an alternating path is either a cycle (i.e. without end points) or a simple path with two end points. (In order to avoid case analysis, we allow the end points to coincide, in which case the alternating path has length 0.)

(ii) a pair of colours and a vertex determine a unique alternating path of those colours and through that vertex. (We allow the two given colours to be equal, in which case the length of the alternating path is at most 1 edge.)

(iii) swapping the colours of the edges of an alternating path leaves the colouring acceptable (because alternating paths are maximal, i.e. have no end points where they can be extended); its significance is that it provides a way for changing the sets of colours at the end points while leaving the colouring acceptable.

Before the procedure starts, we define for each vertex V a colour c.V, in terms of which we define predicate f given by

$$f.V \equiv (\text{no edge incident on V has colour c.V}).$$

The fact that the number of colours is higher than the number of edges meeting at any vertex is exploited by choosing c in such a way that initially ($\underline{A}$V:: f.V) holds.

In the following algorithm for the procedure, X, Y, and Z are variables of type vertex. The procedure's functional specification in terms of pre- and postcondition is given in full; intermediate assertions are only named and will be determined later. Explanation and missing definitions will be given afterwards, when we design the algorithm. (We give this text now for future reference. At this stage, the reader should not try to understand this program fragment.)

```
{(A̲V:: acc.V) ∧ (A̲V:: f.V) ∧

  (XY is the only uncoloured edge)}
{P0: invariant}
do the c.Y-path ends in Y → {P1}

      determine Z so that edge

      XZ has colour c.Y {P2}
```

```
    ; give edge XY colour c.Y and

      uncolour edge XZ {P3}

    ; Y := Z {P0}

  od {P4}

  ; swap the colours along the c.Y-path {P5}

  ; give edge XY colour c.Y

    {(AV:: acc.V) ∧ (all edges are coloured)}   .
```

The term

    (0)  (AV:: acc.V)

is a conjunct of all intermediate assertions; the term

    (1)  (XY is the only uncoloured edge)

is a conjunct of all, exept P3. Note that, in view of the absence of autoloops, (1) implies that X and Y are different vertices.

We shall now show how the above algorithm can be designed. The leading principle is to keep things as simple as possible and not to introduce complications unless forced to do so. In particular: the simplest way, if possible, of asserting a term that has been asserted before is to maintain it in between.

We now start the design of the algorithm, beginning with its last statement. Because (0) - i.e. (AV:: acc.V) - occurs in both pre- and postcondition, we decide - see previous paragraph - to maintain (0) all through the algorithm. Consequently, the final act of the algorithm is to colour the uncoloured edge with an acceptable colour. Let, as initially, the uncoloured edge be XY; colouring it while maintaining (0) means that it has to be coloured with a colour that is incident on neither X nor Y. Initially, since f.X ∧ f.Y holds, c.X is not incident on X and c.Y is not incident on Y. Somewhat asymmetrically we decide to give edge XY the colour c.Y. This choice of last statement yields - note that in view of the definition of f

    f.Y ≡ (c.Y not incident on Y) -

for its precondition P5:

P5: (0) ∧ (1) ∧ (c.Y not incident on X) ∧ f.Y   .

We now consider P5 as the postcondition to be established. Because (0) ∧ (1) ∧ f.Y is implied by the precondition, we concentrate our attention on establishing P5's conjunct

(c.Y not incident on X)

while maintaining its three other conjuncts. If an edge incident on X has colour c.Y, its colour has to be replaced by a colour not incident on X. Since initially f.X holds, (i.e. colour c.X not incident on X), we propose, in view of (iii), to swap the colours c.X and c.Y along the alternating path through X that has those colours. This ensures that c.Y is no longer incident on X.

Since we need the concept a number of times, we define, for graphs satisfying (A̲V:: acc.V) ∧ f.X and for any colour p, the _p-path_ to be the alternating path through X with colours p and c.X. Due to f.X, i.e. c.X not incident on X,

• the p-path with p = c.X is empty

• a non-empty p-path starts at X with an edge of colour p and

  ends at a vertex different from X.

In the above terminology, we propose to etablish (c.Y not incident on X) by swapping the colours along the c.Y-path; this maintains (0) ∧ (1), but maintains f.Y _only_ provided the c.Y-path does _not_ end in Y. (Because of f.Y, a c.Y-path ending at Y does so with an edge of colour c.X; changing that colour into c.Y would falsify f.Y.) Collecting our requirements, we get as precondition P4 for the colour swap along the c.Y-path

P4: (0) ∧ (1) ∧ f.X ∧ f.Y ∧

(the c.Y-path does not end in Y)   .

We now consider P4 as the postcondition to be established. Observing that the first four conjuncts of P4 are implied by the precondition, we apply the same heuristic principle as before and decide to look for a statement that establishes the last conjunct of P4 while maintaining the others. The following methodological interlude explains why we look for a statement of the form of a repetition.

_Methodological Interlude_ Suppose we look for a program S that satisfies for some P, Q

{P} S {P ∧ Q}   .

In those initial states in which Q holds, S may act as a skip. Sometimes this requires no explicit measures. (We saw an example of this in the statement we derived earlier: swapping the colours along an alternating path automatically reduces to a skip if the path is empty.) Otherwise we need an explicit guard in order to distinguish between the cases Q and $\neg$Q. Unless an S of the form

    **if** Q $\rightarrow$ skip $[\!]$ $\neg$Q $\rightarrow$ D **fi**

obviously does the job, it is wiser to begin looking for an S of the form

    **do** $\neg$Q $\rightarrow$ D **od**    ,

since this allows us to separate the concerns for partial correctness and termination. If, while studying termination, we learn that D is executed at most once because it establishes Q, we are still free to replace the repetition by the alternative construct. (End of Methodological Interlude.)

The repeatable statement we are looking for should have the potential of falsifying the guard

    (the c.Y-path ends in Y)    .

Because of an example in which it is impossible to falsify this guard by only changing c and not changing the colouring of the edges, and because, for simplicity's sake, we would not like to change both, we decide to try to keep c constant and let the repeatable statement affect the edge colouring only.

The next constraint we adopt is that the repeatable statement is confined to the combination of colouring the uncoloured edge and uncolouring a coloured one. This is inspired by the observation that this is the simplest transformation whose repeated application can transform a colouring with one uncoloured edge into any other such colouring. It is a constraint - and therefore not adopted without optimism - because, in addition, we intend to maintain at each step

    (0) $\wedge$ (1) $\wedge$ f.X $\wedge$ f.Y    .

As asymmetrically as before, we decide to give the uncoloured edge the colour c.Y; on account of the initial validity of f.Y, this colouring act maintains acc.Y. In order to maintain acc.X as well, the first edge of the c.Y-path is uncoloured. Thus we have designed the repeatable statement given above. (The variable Z has primarily been introduced to ease the discussion.) Our gamble is that we can prove invariance and termination. Let us concentrate on the invariance first.

As we shall see shortly, all is available for the invariance of (0) $\wedge$ (1) $\wedge$ f.X; the invariance of f.Y, however, poses a problem. The invariance of f.Y requires the conjunct f.Z in P3, and the simplest way of justifying it there is by requiring f.Z as conjunct in P2. (Besides this being the simplest way, our termination argument, as we shall see later, relies on the fact that no false f-value is truthified.) In order to justify f.Z in P2 the invariant has to be strengthened with a new conjunct; ($\underline{A}$V:: f.V) would justify f.Z but is too strong to be maintained. We could also justify f.Z from

for any colour p such that the p-path

is not empty:

$$f.S \quad ,$$

where S is the second vertex on the p-path.

This, however, is also too strong: the repeatable statement transforms the p-path XZ...Y into the p-path XY...Z with XY of colour c.Y, so $\neg$f.S for p = c.Y. The following slightly weaker conjunct does the job:

(2) for any colour p such that the p-path is not empty:

f.S $\vee$ c.L $\neq$ p ,

where S is the second vertex of the p-path

and L its last vertex.

Now the time has come to list and justify assertions P0 through P3.

P0: (0) $\wedge$ (1) $\wedge$ (2) $\wedge$ f.X $\wedge$ f.Y $\wedge$ X $\neq$ Y

Assertion P0 is implied by the precondition of the procedure: its first two conjuncts occur in the precondition, the next three follow from ($\underline{A}$V:: f.V), and X $\neq$ Y follows from (1) and the absence of autoloops. We return later to the re-establishment of P0 at the end of the repeatable statement.

P1: P0 $\wedge$ (the c.Y-path ends in Y) $\wedge$

(the c.Y-path starts at X with an edge of colour c.Y)

The first two conjuncts follow from the topology of the program; the last conjunct is a property of nonempty p-paths and the c.Y-path is nonempty because it connects the different vertices X and Y.

P2: P1 $\wedge$ (XZ has colour c.Y) $\wedge$ X $\neq$ Z $\wedge$

    f.Z $\wedge$ c.Z $\neq$ c.Y $\wedge$ Y $\neq$ Z    .

The first conjunct P1, which does not refer to Z, is maintained and implies the existence of a unique vertex Z such that XZ has colour c.Y; XZ is an edge and not an autoloop, hence X $\neq$ Z; the conjunct f.Z follows from (2) with the instantiation p,S,L := c.Y,Z,Y; from f.Z and the fact that XZ has colour c.Y, we conclude c.Z $\neq$ c.Y, from which, with Leibniz's Principle, Y $\neq$ Z follows.

P3: (XZ is the only uncoloured edge) $\wedge$

    (XY has colour c.Y) $\wedge$ (0) $\wedge$

    f.X $\wedge$ $\neg$f.Y $\wedge$ f.Z $\wedge$ (2)    .

Since X,Y,Z are three distinct vertices, XY and XZ are distinct edges and the (un)colouring statement is well-defined; with (1) it establishes the first two conjuncts. For (0) and the next three conjuncts we need only consider vertices X,Y,Z since they are the only vertices with incident edges changing colour; acc.X and f.X are maintained because the bag of colours incident on X remains the same; to the bag of colours incident on Y the colour c.Y is added and, because of f.Y in P2, this maintains acc.Y and falsifies f.Y; acc.Z and f.Z are maintained because the bag of colours incident on Z decreases. For the invariance of (2), we distinguish two cases.

In the case p $\neq$ c.Y, we observe that the (un)colouring statement leaves the p-path unchanged; in particular, if the p-path is not empty, its vertices S and L remain the same. The constancy of L ensures the constancy of c.L $\neq$ p; for the constancy of f.S we observe that Y is the only vertex whose f-value changes, while S $\neq$ Y because XS has colour p and XY was uncoloured. (Here we have used the absence of multiple edges.)

In the case p = c.Y we observe that the p-path XZ...Y is replaced by the p-path XY...Z, i.e. we have to demonstrate (2) for the instantiation p,S,L := c.Y,Y,Z, i.e.

f.Y $\vee$ c.Z $\neq$ c.Y    ,

which follows from P2. Hence P3 is justified.

The axiom of assignment tells us that P3 is for Y := Z a precondition strong enough for the re-establishment of P0.

For the termination argument we observe that, the function c not being changed, f-values can only be changed by changing edge colours, i.e. by the (un)colouring statement. Comparison of P2 and P3 shows that it decreases (NV:: f.V) by 1 (by falsifying f.Y); hence the repetition terminates.

And this concludes our design of the proof of Vizing's Theorem.

<u>Acknowledgements</u> We thank R.E. Tarjan for having drawn our attention to the ugliness of the existing proofs; for constructive criticism we thank the ATAC - in particular K.L. Calvert and J. Misra - and the ETAC - in particular W.H.J. Feijen -; for their active assistance we thank A.J.M. van Gasteren and David Gries.


## In retrospect


The above is our nth explanation of the second algorithm we designed to prove Vizing's Theorem. (This algorithm is much better than our first one, which was probably still too much influenced by the published proofs.)

We are very pleased by the modesty of what we needed: the function c, the predicates acc and f, the notion of the p-path, and three variables X,Y,Z of type vertex (of which X is a constant and Z really only a local variable of the repeatable statement).

We also have reservations, for the algorithm is probably more subtle than our heuristic explanation suggests. The conclusion that in the repeatable statement the uncoloured edge should be allowed to migrate over the graph is <u>not</u> surprising; the decision to restrict that migration by keeping X constant (and confining Y to range over the neighbours of X) is much less obvious. (Note that at the beginning of the repeatable statement, the situation is symmetric in X and Y: they are connected by a path of even length along which the edge colours c.Y and c.X alternate.) The only justification we can think of for the optimistic decision to keep X constant is that thus, <u>independently of the size of the graph</u>, the number of possible values of XY is bounded by N. The underlying hope was obviously to prevent the shape of the graph from generating a case analysis.

Moreover it is only fair to say that we worked very hard on our definitions. The a priori restriction to acceptably coloured graphs simplified the introduction of the alternating path; the restriction to graphs with f.X as well simplified the introduction of the p-path; the formulation of (2) in terms of universal quantification over the dummy p of type colour simplified its proof of invariance. (It is straightforward predicate calculus to eliminate p from (2): "For each L such that the nonempty c.L-path ends at L, etc."; in the case p = c.Y we have to compare XZ...Y with XY...Z: the end points differ, the colour that alternates with c.X remains the same. Hence the colour p, and not the end point L, should be the dummy in (2).)

We are pleased that we needed only one case analysis and no pictures; we are glad that we could avoid notations such as $\mu_{\beta\gamma_3}[a,z]$ - used by Claude Berge -, but we would like to master a more calculational style of reasoning about graphs.


                                        Austin, 22 September 1990

Josyula R. Rao                          Prof. Dr. Edsger W. Dijkstra

# Phase Synchronization for two machines


Wim H.J. Feijen
Dept. of Mathematics & Computing Science
Eindhoven University of Techn.
P.O. Box 513/Den Dolech 2
NL-5600 MB Eindhoven
Tel.: (040)479111
Telex 51163


This is an exercise in formally deriving little multiprograms with the theory of Owicki and Gries and the predicate calculus as our only tools for reasoning. But before we embark on the selected example, we will first explain the Owicki-Gries theory in a very rudimentary fashion.

A multiprogram is a set of ordinary sequential programs (also called "component programs" or "components" or "machines"). Each sequential program is built out of atomic statements, and it may be annotated in such a way that each assertion is a postcondition of an atomic statement. For such an entourage the Owicki-Gries theory prescribes the following two proof obligations for any assertion:

- the assertion is "locally correct", i.e. it is established by the component in which it occurs. More precisely, for atomic statement $S$ with postassertion $P$ we have to show that the precondition of $S$ implies wlp.$S.P$. Note that the fulfillment of this proof obligation may necessitate the introduction of a preassertion of $S$.

- the assertion is "globally correct", i.e. it is maintained by any atomic statement of any other component program. More precisely, for assertion $P$ in some component program to be globally correct we have to show that the precondition of any atomic statement $S$ taken from a different component implies $P \Rightarrow$ wlp.$S.P$. Note that the fulfillment of this proof obligation may necessitate the introduction of a preassertion of $S$, perhaps for many $S$'s. (It may even necessitate a change of $P$.)

So much for the rudiments of the Owicki-Gries theory.

*       *

*

The example to be tackled is a drastic simplification of one that we owe to Jayadev Misra (Notes on UNITY: 12–90). We consider a multiprogram with two components, $p$ and $q$. Program $p$ is given by the straight-line text

$$\text{Prog.}p : \qquad\qquad S0.p\,;\, S1.p\,,$$

and Prog.$q$ is Prog.$p$ with the rôles of $p$ and $q$ interchanged. Our task is to synchronize the components in such a way that

(0)     the one component does not start on the execution of its $S1$ before the other component has completed its $S0$, and

(1)    the synchronization is correct irrespective of the initial values of the (shared) variables to be introduced.

Furthermore, the ultimate multiprogram has to be expressed in terms of atomic statements of the traditional type "at most one reference to at most one shared variable".

*        *

*

Our first step in developing the algorithm is to formalize the problem statement. On account of (0), we must ensure that the precondition of $S1.p$ satisfies

$$\text{``}S0.q \quad \text{has been completed''} .$$

In order to get rid of this phrase, we introduce a boolean $x.q$ with the intent that the precondition of $S1.p$ satisfy $x.q$ and

$$R.q: \qquad\qquad x.q \;\Rightarrow\; \text{``}S0.q \text{ has been completed''} .$$

On account of (1), nothing may be assumed about the initial value of $x.q$, and therefore we must see to it that Prog.$p$ establishes $R.q$ all by itself. There is, however, only one way to enforce the local correctness of $R.q$, viz. by an assignment $x.q := \text{false}$. In order to ensure $R.q$'s global correctness as well we will see to it that $x.q := \text{true}$ will only occur in Prog.$q$ if "$S0.q$ has been completed". Having thus ensured the correctness of $R.q$ as a precondition of $S1.p$ we are left with the obligation to guarantee the correctness of assertion $x.q$ in what is our first approximation of

Prog.$p$:
$$
\begin{aligned}
&S0.p \\
&; \; x.q := \text{false} \\
&; \; \{x.q\} \\
&\quad S1.p .
\end{aligned}
$$

As for the local correctness of assertion $x.q$ there is not much more that we can do than to accomplish it by the test  **if** $x.q \;\rightarrow\;$ skip **fi**  (which is a synonym for  **do** $\neg x.q \;\rightarrow\;$ skip **od** ).  The global correctness offers no problem whatsoever since no atomic statement in Prog.$q$ can falsify $x.q$, even not if we use the freedom to insert $x.q := \text{true}$ in Prog.$q$. With the introduction of the if-statement, that freedom will be needed badly, lest the multiprogram exhibit the danger of deadlock. In plugging in the statements $x := \text{true}$ in the two components there is only one reasonable option, viz. this:

Prog.$p$:
$$
\begin{aligned}
&S0.p \\
&; \; x.q := \text{false} \\
&; \; x.p := \text{true} \\
&; \; \textbf{if } x.q \;\rightarrow\; \text{skip } \textbf{fi} \\
&\quad S0.q .
\end{aligned}
$$

Our next step is to prove the absence of deadlock. We do this by constructing preconditions of the two if-statements such that their conjunction implies the disjunction of the guards. For that purpose we modify the components with operations on thought booleans $h.p$ and $h.q$ as follows (while omitting the now superfluous $S0$'s and $S1$'s):

Prog.$p$:
$$x.q, h.p := \text{false}, \text{true}$$
$$; \ x.p, h.p := \text{true}, \text{false}$$
$$; \ \{(x.p \lor h.q \lor x.q) \land \neg h.p\}$$
$$\textbf{if } x.q \ \rightarrow \ \text{skip } \textbf{fi}$$

(The multiple assignments are supposed to be atomic.) The assertion is correctly established by Prog.$p$, and the twin assertion

$$(x.q \lor h.p \lor x.p) \land \neg h.q$$

is maintained by Prog.$p$. Hence the two preassertions of the if-statements are correct, and their conjunction implies $x.q \lor x.p$ — the disjunction of the guards — as demanded. Hence, there is no danger of both components getting blocked simultaneously.

For later use we wish to mention that further insertions of statements $x.p := \text{true}$ in Prog.$p$ and $x.q := \text{true}$ in Prog.$q$ are harmless to the two assertions, and will therefore not introduce deadlock.

**Remark**  The preassertion in Prog.$p$,

$$(x.p \lor h.q \lor x.q) \land \neg h.p \ ,$$

has not simply been pulled out of a magic hat, but is the result of a construction process. That construction process begins with $x.p$ as a first approximation of the preassertion, which is, however, falsified by statement $x.p := \text{false}$ in Prog.$q$. We remedy the situation by changing that statement into $x.p, h.q := \text{false}, \text{true}$, yielding $x.p \lor h.q$ as a second approximation. Now the reader may complete the construction, guided by the requirement that the conjunction of the preassertions imply the disjunction of the guards.

(**End** of Remark.)

Our last step is concerned with proving that there is even no danger of just one component getting blocked. Following a suggestion of Rob Hoogerwoord's, we do this by showing that the precondition of $S1.p$ implies $x.p$, i.e. the guard of the other component. For that purpose, we introduce fresh thought booleans $h.p$ and $h.q$, with initial values false, and modify the components as follows:

Prog.$p$:
$$\{\neg h.p\}$$
$$x.q := \text{false}$$
$$; \ x.p := \text{true}$$
$$; \ \{Q.p\}$$

$$\textbf{if } x.q \;\rightarrow\; \textbf{skip } \textbf{fi}$$
$$;\; \{x.p \wedge h.q,\text{ hence } x.p\}$$
$$S1.p\;.$$

The only reason for us to introduce the booleans $h$ is that we have trouble in demonstrating the global correctness of target assertion $x.p$ in Prog.$p$. Indeed, $x.p := $ false in Prog.$q$ does falsify it. In order to remove the trouble we strengthen the target assertion into $x.p \wedge h.q$, while at the same time requiring that the precondition of $x.p := $ false in Prog.$q$ imply $\neg h.q$. But now the strengthened assertion $x.p \wedge h.q$ is no longer vacuously established locally in Prog.$p$. For this to be the case we must ensure that the precondition $Q.p$ of the if-statement implies

$$\text{wlp.}(\textbf{if } x.q \rightarrow \textbf{skip } \textbf{fi}).(x.p \wedge h.q)\;,$$

which is

$$(*) \hspace{4cm} \neg x.q \;\vee\; (x.p \wedge h.q)\;.$$

Unfortunately, condition $(*)$ is troublesome in almost any respect. Statement $x.p := $ false in Prog.$q$ may falsify $(*)$ in an almost incurable fashion, as does statement $x.q := $ true. It is the occurrence of subexpression $x.p$ in $(*)$ that is the source of all the complication.

**Remark** The reader who meanwhile has had an operational flirt with the multiprogram as is, will have discovered that there is a danger of one component getting blocked forever.

(**End** of Remark.)

How did subexpression $x.p$ enter $(*)$ in the first place? Because it was a postcondition of the if-statement. How can we remove it there? The simplest possibility is by postfixing the if-statement with an assignment $x.p := $ true. (Recall that insertion of such an assignment in Prog.$p$ does not introduce the danger of both components getting blocked.) Following this proposal we obtain, using the same thought booleans $h$,

Prog.$p$:
$$\{\neg h.p\}$$
$$x.q := \text{false}$$
$$;\; x.p, h.p := \text{true}, \text{true}$$
$$\{\neg x.q \vee h.q\}$$
$$;\; \textbf{if } x.q \;\rightarrow\; \textbf{skip } \textbf{fi}$$
$$\{h.q\}$$
$$;\; x.p := \text{true}$$
$$;\; \{x.p \wedge h.q,\text{ hence } x.p\}$$
$$S1.p\;.$$

The verification of the annotation's correctness is purely mechanical and offers no problems. And hence, we are done.

This completes the derivation of the algorithm, of which the raw code runs

Prog.$p$:

$$S0.p$$
$$;\ x.q := \text{false}$$
$$;\ x.p := \text{true}$$
$$;\ \textbf{if}\ x.q\ \rightarrow\ \text{skip}\ \textbf{fi}$$
$$;\ x.p := \text{true}$$
$$;\ S1.p \ .$$


## Final Remarks

- I would like to confess that I find the ultimate multiprogram a pretty nice one: it is short and not completely obvious from an operational point of view. The textually second occurrence of $x.p :=$ true introduces a communication pattern that I was not really aware of.

- I would like to admit that the mode of derivation in which the foregoing presentation has been rendered is not nearly as calculational as I would like it to be.

- It has been argued that the sequential programs form a very compact record of our intellectual labour. We should be prepared, I think, to accept that in the case of multiprograms, the record is even more compact.

- I would like to thank Huub Schols for his careful listening to my explanation of this derivation.

(**End** of Final Remarks.)

W.H.J. Feijen,
27 November 1990,
Eindhoven.

# The Lexicographic Minimum of a Cyclic Array


Wim H.J. Feijen
Dept. of Mathematics & Computing Science
Eindhoven University of Techn.
P.O. Box 513/Den Dolech 2
NL-5600 MB Eindhoven
Tel.: (040)479111
Telex 51163


We consider a cyclic array $b\,(i:\ 0 \le i \wedge i < N)$ of integers, $1 \le N$. It being cyclic means that for any $i$ we have $b.i = b.(i+N)$. In terms of it we define, for any $h$, sequence $X.h$ of length $N$, as follows

$$X.h = b\,(i:\ \ h \le i \wedge i < h + N)\,.$$

As a result we have $X.h = X.(h+N)$, so that there are at most $N$ different sequences $X$. Our purpose is to identify a lexicographically minimal one. More precisely, with sequence $M$ given by

$$(0) \qquad\qquad (\mathbf{E}h:\ 0 \le h \wedge h < N:\ M = X.h)$$

$$(1) \qquad\qquad (\mathbf{A}h:\ 0 \le h \wedge h < N:\ M \le X.h)\,,$$

we wish to construct a program that, for some variable $r$, has

$$R: \qquad\qquad\qquad M = X.r$$

as a postcondition. Moreover, we aim at a program with a "time complexity" that is linear in $N$.

*      *

*

Our target program will contain a repetitive construct and we therefore must propose an invariant. A number of candidate invariants readily present themselves. One of them is

$$(\mathbf{A}h:\ 0 \le h \wedge h < n:\ X.r \le X.h)\,.$$

It is easily initialized, and conjoined with $n = N$ — the negation of an acceptable guard — it implies $R$.

Another possible invariant is

$$(\mathbf{E}h:\ r \le h \wedge h \le s:\ M = X.h)\,;$$

together with $r = s$ it implies $R$. Now someone may come up with a third or a fourth proposal, but meanwhile we get saddled with the problem of which invariant to choose.

In view of the relatively high demand on the required efficiency of the program, we do not want to choose an invariant too lightheartedly, because the price to be paid for making too airy a choice could become pretty high. Not only could it hamper us in attaining the efficiency goal, but — much worse — it could prevent us from keeping the design simple. Therefore, we rather analyze our programming problem a little more cautiously, and then *design* our invariant from our findings.

$$*  \qquad  *$$

$$*$$

Our first observation is that, whatever algorithm we may compose, carrying out a lexicographic comparison between two distinct sequences $X.p$ and $X.q$ seems to be an indispensable ingredient. Since such sequences are to be compared element-wise and "from left to right", we will somehow need to investigate invariant relation

$$P: \qquad 0 \leq d \wedge (\mathbf{A}i:\ 0 \leq i \wedge i < d:\ X.p.i = X.q.i)\,,$$

for $p \neq q$. Two obvious ways of handling it are by

$$d := 0\,,$$

which establishes $P$, and by

$$X.p.d = X.q.d \quad \longrightarrow \quad d := d + 1\,,$$

which maintains $P$. The latter immediately raises the question of what we can say in the case $X.p.d \neq X.q.d$

Here, the notion of lexicographic order forces us to distinguish between the cases

$$(2) \qquad\qquad X.p.d < X.q.d$$

and $X.q.d < X.p.d$. Fortunately, relation $P$ is symmetric in $p$ and $q$ so that the case analysis does not hurt.

By the definition of lexicographic order, $P$ conjoined with (2) implies that sequence $X.p$ lexicographically precedes sequence $X.q$, i.e.

$$X.p < X.q\,.$$

Can this be of any significance for our purpose of identifying sequence $M$? From (1) — one of the two things we know about $M$ — we have $M \leq X.p$, so that together with $X.p < X.q$ we conclude

$$M \neq X.q\,.$$

As a result, we have identified one "non-$M$" sequence.

But the result is not nearly strong enough, since we have identified only one "non-$M$" sequence at the expense of $d+1$ element comparisons, viz. those recorded in $P$ and in (2). If we cannot perform substantially better, we will simply arrive at an algorithm that is quadratic in $N$. Perhaps, this is the best we can do if nothing more specific is known about the sequences $X$. But in our example they are tightly related.

Relation $P$ attracts our attention to common prefixes of $X.p$ and $X.q$. Let $H$ be such a prefix, and let $h$ be its length. From $P$ we then conclude that for any $H$ such that $0 \leq h \wedge h < d+1$,

$$\text{(i)} \qquad X.p = H \circ Y \qquad \text{and}$$

$$\text{(ii)} \qquad X.q = H \circ Z \;,$$

for some $Y$ and $Z$. (Symbol $\circ$ stands for concatenation.) Now let us see how we can manipulate $X.p < X.q$ (, which was the conclusion that we drew from $P$ and (2)):

$$X.p < X.q$$
$$= \quad \{\text{(i) and (ii)}\}$$
$$H \circ Y < H \circ Z$$
$$= \quad \{\text{property of lexicographic order}\}$$
$$Y < Z$$
$$= \quad \{\text{property of lexicographic order}\}$$
$$Y \circ H < Z \circ H \;.$$

It is here where the tight relationship between the sequences $X$ enters the picture. From (i) and the definition of $X$ we can derive

$$\text{(iii)} \qquad X.(p + h) = Y \circ H \;,$$

and with the aid of (ii)

$$\text{(iv)} \qquad X.(q + h) = Z \circ H \;.$$

Now we continue the above calculation:

$$Y \circ H < Z \circ H$$
$$= \quad \{\text{(iii) and (iv)}\}$$
$$X.(p + h) < X.(q + h)$$
$$\Rightarrow \quad \{\text{with (1)}\}$$
$$M \neq X.(q + h) \;.$$

In summary, we have derived

$$P \wedge X.p < X.q \qquad \Longrightarrow \qquad (\mathbf{A}h:\ 0 \le h \wedge h < d+1:\ M \ne X.(q+h))\,,$$

or, by dummy transformation $h := h - q$,

$$(3) \quad P \wedge X.p < X.q \qquad \Longrightarrow \qquad (\mathbf{A}h:\ q \le h \wedge h < q+d+1:\ M \ne X.h)\,.$$

As a result, we have identified $d+1$ "non-$M$" sequences at the expence of $d+1$ element comparisons. And this looks much more promising!

Finally, we take advantage of the symmetry between $p$ and $q$ to conclude

$$(4) \quad P \wedge X.q < X.p \qquad \Longrightarrow \qquad (\mathbf{A}h:\ p \le h \wedge h < p+d+1:\ M \ne X.h)\,.$$

This concludes our exploration of a lexicographic comparison between two sequences $X.p$ and $X.q$, $p \ne q$, be it that we have not yet addressed the remaining case $X.p = X.q$. We return to that in a moment.

$$*\qquad *$$

$$*$$

The consequents of the lemmata (4) and (3) are — by their very shapes — calling for invariants of the form

$$Qp:\qquad\qquad (\mathbf{A}h:\ p_0 \le h \wedge h < p:\ M \ne X.h)\qquad \text{and}$$

$$Qq:\qquad\qquad (\mathbf{A}h:\ q_0 \le h \wedge h < q:\ M \ne X.h)\,,$$

and the question that then remains is how to choose $p_0$ and $q_0$ — which is our only freedom left — .

Because $X$ is periodic — recall, $X.h = X.(h+N)$ — there is no restriction involved in choosing $p_0 = 0$. At the same time we adopt invariant

$$Z:\qquad\qquad\qquad\qquad p < q$$

in order to ensure $p \ne q$. For deciding on $q_0$ we now turn our attention to the as yet undiscussed case $X.p = X.q$.

The analysis given so far has only yielded "non-$M$" sequences. In view of our ultimate goal to isolate a sequence $M$, the analysis of the case $X.p = X.q$ will (have to) be driven by the desire to find a sequence $M$. From $X.p = X.q$ we conclude that $X$ has $q - p$ as a period. (This may be shown by showing that for any $h$, $X.p = X.q \Rightarrow X.(p+h) = X.(q+h)$, which follows in a straightforward manner from the connections (i)–(iii) and (ii)–(iv), given before.) In conjunction with (0) — which has not been used yet — we then obtain

$$(\mathbf{E}h:\ p \le h \wedge h < q:\ M = X.h)\,.$$

Now we confront this with invariant $Qq$

$$(\mathbf{A}h: \quad q_0 \leq h \wedge h < q: \quad M \neq X.h) \,,$$

and here we see that with the choice $q_0 = p+1$ the conjunction of these two conditions implies $M = X.p$. Adopting that choice we therefore can rely on

$$(5) \qquad\qquad Qq \wedge X.p = X.q$$
$$\Longrightarrow$$
$$M = X.p \,.$$

$$* \qquad\qquad *$$
$$*$$

This completes our analysis of the problem and our design of a hopefully suitable invariant for the repetitive construct. In summary, it runs

$$P: \qquad\qquad 0 \leq d \;\wedge\; (\mathbf{A}i: \; 0 \leq i \wedge i < d: \; X.p.i = X.q.i)$$
$$Z: \qquad\qquad p < q$$
$$Qp: \qquad\qquad (\mathbf{A}h: \; 0 \leq h \wedge h < p: \; M \neq X.h)$$
$$Qq: \qquad\qquad (\mathbf{A}h: \; p+1 \leq h \wedge h < q: \; M \neq X.h) \,,$$

with as a little accompanying theory the lemmata (3), (4), and (5). This and the givens (0) and (1) is, in fact, all we need to know for carrying out the forthcoming construction of a program text establishing postcondition $M = X.p$

$$* \qquad\qquad *$$
$$*$$

After the above has been said and done, the construction of the program text offers no further surprises, except for the guard of the repetitive construct which we shall discuss in isolation. Here is the text, lavishly annotated at some crucial places.

$$p, q, d := 0, 1, 0$$

$$\{\text{Inv:} \quad P \wedge Z \wedge Qp \wedge Qq\}$$

$$; \mathbf{do} \; \ldots \qquad \longrightarrow$$

$$\qquad \mathbf{if} \;\; X.p.d = X.q.d \qquad \longrightarrow \qquad d := d+1 \qquad \{\text{Inv}\}$$

$$\qquad [\!] \;\; X.p.d < X.q.d \qquad \longrightarrow$$

$\qquad\qquad$ {By $P$, we have $X.p < X.q$, and hence, with the aid of lemma (3), $(\mathbf{A}h: \; q \leq h \wedge h < q+d+1: \; M \neq X.h)$. Conjoined with $Qq$ this yields $(\mathbf{A}h: \; p+1 \leq h \wedge h < q+d+1: \; M \neq X.h)$, so that }

$$\qquad q, d := q+d+1, 0$$

$\qquad\qquad$ {maintains $Qq$, and of course the rest of Inv. }

▯ $\quad X.q.d < X.p.d \qquad \longrightarrow$

> { By $P$, we have $X.q < X.p$, and with the aid of lemma (4) and $Qp$,
> $(\mathbf{A}h: \ 0 \leq h \land h < p + d + 1: \ M \neq X.h)$. Conjoined with $Qq$ — using
> $0 \leq d$ — this yields $(\mathbf{A}h: \ 0 \leq h \land h < (p + d + 1)\,\mathbf{max}\,q: \ M \neq X.h)$,
> so that }

$\qquad p, d := (p + d + 1)\,\mathbf{max}\,q, 0$

$\qquad$ {maintains $Qp$.}

$\qquad ; q := p + 1 \qquad$ {Inv}

$\quad$ **fi**

**od** .

What remains to be done is to construct a guard and to find a suitable variant function. As for the latter, choice $p + q + d$ is not too far-fetched, since we may observe that the first two alternatives each increase $p + q + d$ by exactly 1, and the third alternative by at least 1 (even by at least 2). So, this is okay. But is $p + q + d$ bounded from above?

From $Qp$ and datum (0),

$$(0) \qquad\qquad (\mathbf{E}h: \ 0 \leq h \land h < N: \ M = X.h) ,$$

we conclude that $p < N$ is an invariant of the repetition. This gives us an upperbound on $p$ that is linear in $N$. We now could proceed finding or imposing linear upperbounds on $q$ and on $d$, but we can do better in one fell swoop.

The invariance of $p < N$ implies that the precondition of statement

$$p := (p + d + 1)\,\mathbf{max}\,q$$

in one way or another has to imply

$$(p + d + 1)\,\mathrm{max}\,q < N ,$$

or, equivalently,

$$(6) \qquad\qquad p + d + 1 < N \land q < N .$$

This precondition does not simply follow from the program text as is, unless (6) can be taken as a guard of the repetition. If we can take it as a guard, the precondition of the step implies

$$p + q + d \leq 2 * N - 3 ,$$

and as a result — $p + q + d$ being equal to 1 initially — the number of steps of the repetitive construct will be at most $2 * N - 3$. This, then, nicely complies with our efficiency requirements.

Next we shall show that (6) can be taken as a guard indeed, by demonstrating that its negation conjoined with the invariant implies $M = X.p$. We argue by case analysis.

**Case** $N \leq q$

$$N \leq q$$
$$\Rightarrow \quad \{\text{using } Qp \text{ and } Qq\}$$
$$(\mathbf{A}h: \ 0 \leq h \wedge h < p: \ M \neq X.h)$$
$$\wedge \ (\mathbf{A}h: \ p+1 \leq h \wedge h < N: \ M \neq X.h)$$
$$\Rightarrow \quad \{\text{using } (0)\}$$
$$M = X.p$$

**Case** $N \leq p + d + 1$

*Subcase* $X.p < X.q$

$$X.p < X.q$$
$$\Rightarrow \quad \{\text{using } P \text{ and } (3)\}$$
$$(\mathbf{A}h: \ q \leq h \wedge h < q+d+1: \ M \neq X.h)$$
$$\Rightarrow \quad \{\text{using } Z, \text{ i.e. } p < q\}$$
$$(\mathbf{A}h: \ q \leq h \wedge h < p+d+1: \ M \neq X.h)$$
$$\Rightarrow \quad \{\text{using } N \leq p+d+1\}$$
$$(\mathbf{A}h: \ q \leq h \wedge h < N: \ M \neq X.h)$$
$$\Rightarrow \quad \{\text{conjoining with } Qp \text{ and } Qq\}$$
$$(\mathbf{A}h: \ 0 \leq h \wedge h < p: \ M \neq X.h)$$
$$\wedge \ (\mathbf{A}h: \ p+1 \leq h \wedge h < N: \ M \neq X.h)$$
$$\Rightarrow \quad \{\text{using } (0)\}$$
$$M = X.p$$

*Subcase $X.q < X.p$*

$$X.q < X.p$$

$\Rightarrow$ {using $P$ and (4)}

$$(\mathbf{A}h: \ p \leq h \wedge h < p + d + 1: \ M \neq X.h)$$

$\Rightarrow$ {conjoining with $Qp$}

$$(\mathbf{A}h: \ 0 \leq h \wedge h < p + d + 1: \ M \neq X.h)$$

$\Rightarrow$ {using $N \leq p + d + 1$ and (0)}

false

$\Rightarrow$ {pred. calc.}

$$M = X.p$$

*Subcase $X.p = X.q$*

$$X.p = X.q$$

$\Rightarrow$ {using $Qq$ and (5)}

$$M = X.p$$

(**End** of Cases.)

So in any case we have $M = X.p$ as a postcondition of the repetition.

We conclude this derivation by depicting the raw code, phrased in terms of the original cyclic array $b \ (i: \ 0 \leq i \wedge i < N)$.

$$p, q, d := 0, 1, 0$$
$$; \mathbf{do} \ \ p + d + 1 < N \ \wedge \ q < N$$
$$\longrightarrow \quad \mathbf{if} \ \ b.(p + d) = b.(q + d)$$
$$\longrightarrow \quad d := d + 1$$
$$[\!] \ \ b.(p + d) < b.(q + d)$$
$$\longrightarrow \quad q, d := q + d + 1, 0$$
$$[\!] \ \ b.(q + d) < b.(p + d)$$
$$\longrightarrow \quad p, d := (p + d + 1) \, \mathbf{max} \, q, 0$$

$$; \ q := p + 1$$

**fi**

**od**

$$\{M = X.p\}$$

*　　*

*

## Final Remarks

a)   There are some technicalities concerning the above algorithm that the reader may wish to know, to check, to appreciate, or to ignore.

- From the guard $p + d + 1 < N$ and the invariance of $0 \leq p$, it follows that $d + 1 < N$ is a valid precondition of the step of the repetition. Hence, $d < N$ is an invariant as well. In the light of the invariance of $P$ this means that no execution of the program will ever "find" $X.p = X.q$. If this sounds inconceivable, the conclusion must be that operational reasoning is impossible to comprehend.

- The repetitive construct can do with the much simpler and weaker guard

$$d < N \ \wedge \ q < N \ ,$$

  but the price to be paid is a nonnegligible loss of efficiency, viz. from at most $2 * N - 3$ to at most $3 * N - 4$ element comparisons. For us, the reason to "tolerate" the stronger guard is not only a matter of efficiency, but that its incorporation requires no extensive justification and that we can handle it in terms of the theory already developed.

- The latter remark is related to the experience that all sorts of efforts to improve on the program's efficiency were all attended by a serious complification of the argument. In this respect Yossi Shiloach has pushed the exercise to a limit — see his incomprehensible "Fast Canonization of Circular Strings" in the Journal of Algorithms 2 (1981) — .

- The weaker guard

$$d < N \ \wedge \ q < N$$

  is attractive however, if the algorithm is to deliver the (least) period of $X$ as well, because this can then be achieved by postfixing the repetitive construct with

$$
\begin{array}{llll}
\textbf{if} & N \leq d & \longrightarrow & period := q - p \\
\square & N \leq q & \longrightarrow & period := N \\
\textbf{fi} \ .
\end{array}
$$

b)   Our main purpose in the foregoing presentation has been to get the design process across. In general, the corner stones of a design process are formed by a number of outstanding design decisions (which usually are prompted by heuristic considerations and by the desire to keep a design simple). It is hoped and believed that each time a design decision has been taken, a more or less exhaustive exploration of its consequences will lead to a point where the need for a next design decision presents itself. In our present example the corner stones are formed by

- the mere decision to attach the name $M$ to the final answer

- the decision to focus on a lexicographic comparison between two sequences

- the adoption of the invariants $Qp$ and $Qq$, but more in particular the choice of $q_0$.

When armed with such a limited number of corner stones, we should be able to reconstruct the design without much effort, not only today or tomorrow but even next year. (There is something completely wrong if a University Professor of Programming needs to bring notes to guide his lectures.) In this respect, I was pleasantly surprised when some time ago Richard Bird sent me a note containing the algorithm that we just developed. The surprise was that the development had been explained to him by Netty van Gasteren *without* the use of pencil and paper (during a rail journey). I am very grateful to them for having done that experiment.

c) I am also indebted to Edsger W. Dijkstra and Carel S. Scholten for a considerable embellishment, back in 1979, of a first draft describing the development of this program.

W.H.J. Feijen,
29 November 1990,
Sterksel.

# Chapter 2
# Rules of Programming

In the verification and the development of programs like in all other calculi rules are needed. Finding an adequate collection of rules that are powerful enough to express all the forms of reasoning needed to verify and develop a program is one of the key issues of programming methods. The simpler and more tractable the systems of rules are the better they can be used and supported.

R.L. Constable

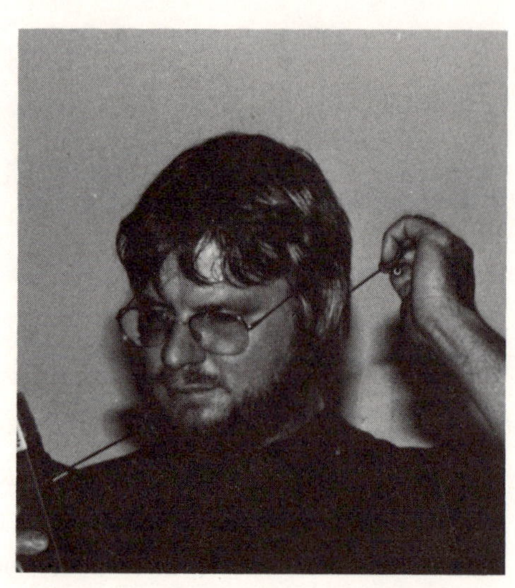

B. Möller

C.A.R. Hoare

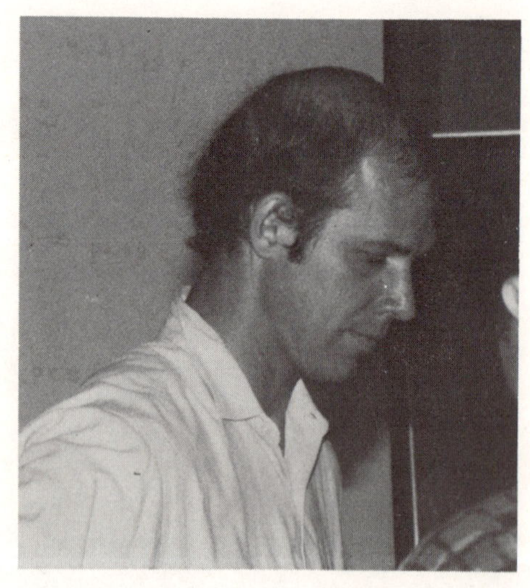

G. Nelson

# Metalevel Programming in Constructive Type Theory


Robert L. Constable*
Department of Computer Science
Cornell University
Ithaca, NY 14853
rc@cs.cornell.edu



**Abstract**

This paper describes a particular version of constructive type theory, a subset of
the one underlying the Nuprl proof development system, and it examines its role
as a programming environment. Special attention is given to the property of the
theory to talk about itself. This reflective capability is the basis for automating
proof development and for metalevel programming.


## 1 Introduction

A mature theory of computing will be a basis for a rigorous approach to programming. Already we see this in practice; the algorithms we use to solve problems are
often those developed in the mathematical theory of algorithms, and they are chosen
based on properties proven to hold for them, such as economy of space or time or
robustness or generality. Moreover the algorithms are often expressed in a programming notation that is typed, and the programming environment will include a type
checker to insure that the types are sensible. The type checker and the program
evaluator are parts of the environment that were developed in the mathematical
theory of semantics.

It may soon be the case that most parts of the environment will be based on systematic theoretical results. We see more concepts from logic appearing such as
formal specification languages, and it is now a smaller step from these disconnected
parts, programming notation, evaluator, type checker, specification language, to a
full-blown theory whose axioms and rules of inference are as explicit and rigorous


*This research supported by NSF grant CCR86-16552 and ONR grant N00014-88K-0490.

as the other components and yet is an integral part of the programming environment. In such a setting more results of computing theory can be *directly* applied in programming. It seems that constructive type theory is one candidate for such a unified computing theory, perhaps the first candidate. In these notes we will examine certain aspects of programming in this setting.

It is interesting to note that such a theory might serve as a foundations for computer science in the sense that logicians searched to find such a foundational theory for mathematics. Many people are happy with Set Theory in that role. But for computer science the formal theory should offer a programming notation as well since computing involves programming in a central way; and a formal notation is needed for that. If the programming notation were independent of the foundational theory, then we could not directly use the theoretical insights in programming and vise versa. Once we are committed to a formalized theory for computing, we need one that can express arguments about algorithms and systems actually built, that is, it should be a programming logic as well as a foundational theory.

These two requirements place a large burden on such a theory. Simply adding a programming notation to Set Theory for example results in a costly and cumbersome solution. Constructive type theories have arisen as a more elegant and lean approach. In this article we will work in one such theory, but the ideas are applicable in most others being developed.

This article begins with remarks about the variety of type theories. Then it defines a small subset of the Nuprl type theory, which is an instance in the class of Martin-Löf type theories. The treatment is intuitive and informal since there reader can consult other sources for a complete and rigorous account [15,31], and the book of Nordstrom *et al* [43] covers similar ground. After this there is a presentation of the syntax and deductive machinery of the theory in itself, following exactly the method in [2]. This serves as a basis for the applications of metaprogramming. The main example is a derivation of a term matching algorithm critical in theorem proving and metamathematics. This example comes from [16] and was implemented by Howe in Nuprl. It is the example used in the Marktoberdorf course. The article ends with a discussion of *logic variables* which was not part of the Marktoberdorf course in 1990 but is germane to the topic and is now an active concern of type theory research.

## 2   Choosing a Type Theory

I do not intend to survey all of the candidate type theories, but there are at least three main species: Martin-Löf's predicative type theories and their extensions [35,36,15, 43,30,38,40,14], Girard's impredicative theories and their extensions [23,24,17,34] [4], and Fefeman's theories of functions and classes [21,28,10]. In some ways these are converging in that they are treating the same class of concepts, and none is clearly right at the expense of the others. Indeed all are not right, but they represent the best we can do at present. They all offer the following advantages over taking the union of Set Theory and Programming Logic (say over ones favorite programming language).

## 1. Simple connections between programs and specifications

These type theories all use a functional programming notation and assign types essentially as for the typed lambda calculus. The connection to logic in the case of the ML and Girard theories is via the propositions-as-types principle. These two mechanisms combine elegantly, and the connection is robust, supporting many extensions to both type and programs. It offers also a richer type system than any in existing functional programming language, so in principle the result is a richer programming notation.

## 2. Rich specification languages

These type theories can express a great deal of mathematics. One reason for this is that they allow quantification over propositions. This is possible in Martin-Löf style theories because the notion *Type* presented above is refined into a hierarchy of types called *universes*. A universe is a type closed under all the type forming operations, so it looks like the collection of all types. But it is itself a type and is not contained in itself. The first of these is called $U1$. There is a larger universe $U2$ which contains all of the types from $U1$ plus $U1$ itself. Indeed there is a sequence of universes, $Ui$ with $Ui$ contained in $Ui + 1$. (We will not be concerned with universes here and simplify the whole account by taking *Type* to be $U1$ and omitting mention of $U2$ and higher universes for the most part.)

These universes allow the user to define recursive types, and with recursive types one can define the natural numbers and lists and trees and other inductively defined data types. This is how we proceed here, but in Nuprl the integers are taken as primitive since we want a very efficient implementation, including a fast decision procedure for a subset of quantifier free arithmetic over plus,minus and times. In Nuprl the list forming operator is also primitive, but this is for historical reasons.

With recursive types one can even define infinitary versions of the logical connectives and obtain a version of infinitary logic. It is also possible to represent dynamic logic and other programming logics. In the case of the type theories with partial objects [48,14] [4] this can be done directly taking programs as partial functions on state.

Of course Set Theory allows all of these concepts as well and can represent higher-order logic by taking propositions to be boolean valued functions. It is simple to define sets inductively and to provide a denotational semantics for dynamic logic and programming logics. However, the encoding of these ideas in Set Theory does not help us keep track of what can be computed and what can't be. So typically although we can for example define a model of a programming language, the partial functions used in it will not be computable, so we cannot actually execute the semantic definition. Also the definition of a mathematical concept in Set Theory is often not the one we want to compute with. Let us examine two examples of this phenomenon, first consider the definition of *multi-sets*, say as in [Manna&Waldinger/89book].

A multi-set is defined axiomatically as a set with the operations of adding an element, say $add(a, S)$. There is an empty mulit-set, say []. There is an axiom about equality, saying that $add(x, add(y, S)) = add(y, add(x, S))$, and there is a principle of induction for proving things about such sets. It has the form that if we can prove $P$ on [] and if for every multi-set $S$ and any element $a$ we can prove that $P$ holding on $S$ implies that it holds on $add(a, S)$, then we know $P$ on any multi-set. From this basis Manna&Waldinger show that we can define a choice function, *choose*, from any nonempty multi-set $S$ to an element, $choose(S)$. They also show how to define the function $count(a, S)$ which denotes the number of times that the element a appears in $S$. Now a constructive account of multi-sets requires a more delicate approach. A complete account occurs in [5]. First these sets must be defined over some underlying type, say $A$. Then some concrete representation must be defined, say the sets of Lists over $A$. An equality relation is imposed on these representations, the same as above, and an induction combinator must be defined. It is like the induction principle stated above except that one must explicitly say that the method of proving $P$ on $add(x, S)$ from $P$ on $S$ must respect the defined equality. It turns out that this restriction prevents us from defining the choice function, $choose(x, S)$, which is a good thing since it is not in general computable for multi-sets over any base $A$.

As a second example, consider the definition of the real numbers in Set Theory. There one can choose from among many equivalent definitions, say Dedekind cuts versus Cauchy sequences versus nested intervals and so forth. But one cannot compute with any of these definitions, and some of them are difficult to modify to make them computable. Dedekind cuts do not lead to a notion of computable reals in a natural way whereas the Cauchy definition does. Namely, following Bishop, [7], we say that a constructive real number is a sequence of rational numbers, $x_1, x_2, \ldots$ such that $|x_i - x_j| < 1/i + 1/j$. Two of them , $x_1, x_2, \ldots$ and $y_1, y_2, \ldots$ are equal iff for all $i$, $|x_i - y_i| < 2/i$. These are examples of two very basic mathematical concepts which arise frequently in computer science and for which the set theoretic definitions do not necessarily lead to good data structures for practical computation. So new computational definitions must be introduced and related to the other definitions. In most cases this process requires an elaborate definition of computability prior to the definitions. In type theory we can start with the computationally sensible definition right at the start because the right computing concepts are part of the basic theory.

## 3. A range of programming styles

**functional programming**  The terms of constructive type theories include those that constitute a functional programming language, so they support directly the functional programming style. In fact in a system like Nuprl we expect to be able to achieve the performance of a language like Standard ML on the subset of programs common to both languages, which is essentially the functional subset of Standard

ML. This is sufficient performance for a wide range of algorithms and systems, including by the way the implementation of Nuprl itself [41].

**object oriented programming**   The richness of these type theories allows them to express directly concepts such as modules and algebraic data types, ADT's. So it is possible to program in the style of the so-called object oriented languages to some extent.  For instance in [5] there are many examples including lists, multi-sets, formulas and proofs all formulated as ADT's using the constructors *fun* and *prod*. However, no one has explored in detail the adequacy with which type theory can represent the various operations on algebraic structures that make this kind of programming so attractive.  I would conjecture that many of these operations will have a pleasant formulation.

**logic programming**   In a system like Nuprl it is possible to use the metalanguage to enable a logic programming style. Lipton [33] has actually built a Prolog engine in Nuprl in just this way. But we can see the main idea by looking at an example of a Prolog style computation done using just the basic types of a Martin-Löf theory plus the Nuprl-style tactics.  This will be done in section 3 where we also discuss logic variables.

**imperative programming**   There is good evidence to believe that the mechanisms being used in functional programming to attain the performance levels of imperative programming languages will be available in these type theories, either in the implementation or in some cases in the logic as well.  For example the work of Griffin and Murthy [27,41] shows how to type the control operator call/cc ( the "go to's" of functional programming). Moreover this work reveals a deep connection between these evaluation strategies in functional programming and certain transformations important in proof theory.

## 4. Expressing basic ideas from computer science

The constructive type theories are able to explain and generalize some of the concepts basic to computer science. In the first place they provide a rigorous and general notion of data type which generalizes many of the earlier notions. For example, variant records become dependent products, function spaces can be seen as special cases of dependent function spaces, set types can be defined over any type, modules are instances of parameterized dependent products (as are abstract types) and recursive types can be generalized to include the constructive ordinals. In addition new relationships are revealed by the definitions as in the case of lazy data types being the co-inductive types [37].

The underlying deductive mechanism of a system such as Nuprl can be used to unify many important basic concepts. For instance there is one uniform notion of binding and scope which applies to the entire system. Also the notion of a transformation tactic gives as well the concept of a program transformation. The use of tactics to define the idea of a high-level proof applies as well to defining high-level programs.

One of the most important unifications provided by constructive type theories occurs at the interface between mathematics and computer science. Here we see that many of the concepts from constructive mathematics are exactly what is needed in computing, e.g. the real numbers, and conversely many of the algorithms needed in constructive mathematics can be developed by techniques from computer science. An illustration of this later point occurs in Howe's library of basic real analysis, [29], the proof of the Intermediate Value Theorem is based on a simple binary search algorithm whose structure is directly apparent from the proof. In contrast the same proof in Bishop&Bridges hides the algorithm. In another field, the constructive proof of Higman's lemma provided by Murthy and Russell [42] gives an algorithm based on regular expressions which is much simpler to understand than the one which can be obtained from the classical proofs, [41].

### 5. Naturally automated

One of the most successful and interesting approaches to automating reasoning is the method of tactics pioneered by the LCF system [26]. This approach fits very well with type theory because it involves using a richly typed programming language as a metalanguage for proof construction. These type theories can use themselves as programming notations for automating reasoning in them. So for example it is possible to write and prove correct a theorem proving procedure for first-order matching, in the programming styles suggested above and then use this for facilitating formal reasoning in the type theory. An interesting example of this is the match theorem presented in this article.

# 3  A Simple Type Theory

Here is a simple type theory that illustrates the basic concepts needed in the rest of the paper. The basic syntactic concept is a *term*.

**Definition 1** *The terms are divided into two groups, the* canonical *ones and the* noncanonical *ones. Here are all of the canonical terms:* $prod(A; x.B)$, $fun(A; x.B)$, $union(A; B)$, $void$, $rec(x.b)$, $Type$, $pair(a; b)$, $\lambda(x.b)$, $inl(a)$, $inr(b)$, $any(b)$, $eq(t; e1; e2)$, $axiom$.

*Here are the noncanonical ones:* $spread(p; u, v.t)$, $ap(f; a)$, $decide(d; u.t1; v.t2)$, $rec - ind(t; h, x.t)$.

The notions of *binding variable*, *bound variable*, *free variable*, *scope* and *closed term* are carried over to these terms from the lambda calculus in a natural way. For instance for the lambda term $\lambda(x.b)$ we say that $x$ before the dot is a binding occurrence of $x$ whose scope is $b$. This occurrence of $x$ along with the dot forms a binding operator which binds all free occurrences of $x$ in $b$. A variable is free in an expression if it does not occur in the scope of a finding operator. These concepts are defined precisely for the reflected concept of terms, denoted *Term*, to come later.

There are *computation rules* telling how to *reduce* the noncanonical terms; the canonical ones reduce to themselves. In these rules the notation $exp[a/x]$ denotes the term obtained from $exp$ by substituting $a$ for each occurrence of $x$ in $exp$.

**Definition 2** *To reduce $spread(p; u, v.t)$, first reduce $p$. If the result is $pair(a; b)$, then continue by reducing $t[a/u, b/v]$.*

*To reduce $ap(f; a)$, first reduce $f$; if the result is $\lambda(x.b)$, then continue by reducing $b[a/x]$.*

*To reduce $decide(d; u.t1; v.t2)$, first reduce $d$; if the result is $inl(a)$, then continue by reducing $t1[a/u]$; if the result is $inr(b)$, then continue by reducing $t2[b/v]$.*

*To reduce $rec - ind(t; h, x.b)$, reduce $b[t/x, \lambda(w.rec - ind(w; h, x.b))/h]$.*

## 3.1   typing rules

These terms are the basis of a simple type theory. A type will be defined when we give it a canonical name and say which canonical terms are elements of it. One type is determined by the term *void*. It is a type with no members. We generally speak of the term which is the canonical name of a type as a type, so we say *void* is a type.

Another type is determined by the name *Type*.(This is the simple name we are using for the first universe, $U1$.) To define its canonical members we proceed inductively, and we introduce the auxiliary concept of a family of types over a type $A$. If $B$ is a term and $A$ is a type, and if for every element $a$ of $A$ the term $B[a/x]$ is a type, then $B$ is a family of types indexed by $A$ (using index variable $x$).

If $A$ and $C$ are types and if $B$ is a type indexed by $A$ (with index variable $x$) then $prod(A; x.B)$, $fun(A; x.B)$ and $union(A; C)$ are types. If $x$ does not occur in B, then we omitt the $x$. binding operator. If when $T$ is a type, then $F$ is, and if $T$ occurs only positively in $F$ then $rec(T.F)$ is a type. The positivity condition is given in detail in [15], but here we will use for F only terms built from *prod* and *union* which are all positive. (The essential idea is that $T$ occurs only in the range of a function type, not in its domain.)

To specify what types these are, we must define their canonical elements. The canonical elements of the types $union(A; C)$ are those terms $inl(a)$ and $inr(c)$ where $a$ is a term of type $A$ and $c$ is one of type $C$; neither $a$ nor $c$ need be canonical, so we must also to define what it means generally for a term to be of some type $T$. We do this below.

The canonical elements of $prod(A, x.B)$ are the terms $pair(a; b)$ where $a$ is of type $A$ and $b$ is of type $B[a/x]$. We know that these $B[a/x]$ are types since $prod(A; x.B)$ is.

The canonical element of $fun(A; x.B)$ are the terms $\lambda(x.b)$ such that for each $a$ of type $A$, $b[a/x]$ is of type $A$.

The canonical elements of $rec(T.F)$ are those of the type $F[rec(T.F)/T]$.

A term with no free variable (i.e. a *closed term*) is of type $A$ if it reduces to a canonical term of type $A$. Thus if $t[a/x, b/y]$ reduces to a canonical term type $T$, then $spread(pair(a; b); x, y.t)$ is of type $T$.

A term with free variables can be contingently of type $A$ depending on assumption about the types of the free variables. To define this, let $x_1, \ldots, x_n$ be the free variables of $t$ assume $A_1$, is a type, and say that $A_2$ is a type, possibly depending on $A_1$, if for each $a_1$ of type $A_1$, $A_2[a/x]$ is a type. Likewise for $A_n$ depending on $A_1, \ldots, A_{n-1}$, and $T$ depending on $x_1, \ldots, x_n$. Now $t$ is of type $T$ if for each $a_1$ of type $A_1$, $A_2$ of type $A_2[a_1/x], \ldots, a_n$ of type $A_n[a_1/x_1, \ldots, a_{nil}/x_{nil}]$, $t[a_1/x_1, \ldots, a_n/x_n]$ is of type $T[a_1/x_1, \ldots, a_n/x_n]$. For example, $inr(x)$ is of type $union(A; B)$ if $x$ is assumed to be of type $A$.

## 3.2   semantics

Just as in the lectures we adopted a semantic approach to type theory, so at this point we do not give an enumeration of the rules which embody the ideas just described concerning reduction of terms (which become *computation rules*), the formation of types (which become *formation rules*) and the conditions for membership in types (which for the canonical terms become *introduction rules* and in the case of the noncanonical terms become *elimination rules*). A complete set of rules can be found in [15].

In the lectures at Marktoberdorf the semantic methods of Stuart Allen [3] were used as a rigorous basis for this semantic approach. But due to space limitations here we merely cite Allen's work and proceed informally since there is no need to use the rigorous approach for the results stated in this article.

## 3.3   encoding constructive logic

The *propositions-as-types principle* establishes a correspondence between types and propositions based on the idea that a proposition is true if and only if the corresponding type is inhabited. For decidable atomic propositions it is easy to begin the correspondence, namely a proposition $P$ is taken to be an empty type if it is false and a type with at least one element if it is true; so *false* corresponds to the type *void*. One can use the proposition itself as a type. The correspondence is extended to compound propositions by the following definitions.

$$A \& B \text{ corresponds to } prod(A; B)$$
$$A \vee B \text{ corresponds to } union(A; B)$$
$$A \Rightarrow B \text{ corresponds to } fun(A; B)$$
$$\forall x : A.B \text{ corresponds to } fun(A; x.B)$$
$$\exists x : A.B \text{ corresponds to } prod(A; x.B)$$

Among the atomic propositions are these types $eq(A; s; t)$ provided $A$ is a type and $s,t$ are of that type. These are types which are inhabited by *axiom* precisely when $s$ and $t$ are equal elements of type $A$; otherwise the types are empty. The conditions under which equality holds are given next.

If $pair(a1; b1)$ and $pair(a2; b2)$ are in $prod(A; x.B)$, then they are equal if and only if $a1$ and $a2$ are equal in $A$, and $b1$ and $b2$ are equal in $B[a1/x]$. If $\lambda(x.b1)$ and $\lambda(y.b2)$ are in $fun(A; x.B)$, then they are equal if and only if $b1[a/x]$ equals $b2[a/x]$ in $B[a/x]$ for all $a$ in $A$. If $inl(a1)$ and $inl(a2)$ are in $union(A; B)$, then they are equal if and only if $a1$ equals $a2$ in $A$. Similarly for the right injections $inr(b1)$ and $inr(b2)$.

The equality types can be used to express type membership as well. When we know that the type $eq(A; a; a)$ is inhabited, always by *axiom* and nothing more, then we know that $a$ in fact belongs to the type $A$. We adopt the Nuprl convention that the judgement that $A$ is a type is expressed as $eq(Type; A; A)$.

This correspondence between propositions and types is justified on semantic grounds and can be seen as a *realizability* semantics for constructive logic as is by now well-known.

## 3.4   abbreviations

The uniform prefix operator notation for terms is not the most convenient for readability, especially when writing the logical operators. In Nuprl there is an extensive definition facility that allows users to write new operator names and provide *display forms* for them. We will not explain that facility but instead provide several abbreviations for this article which can easily be written with the definition facility.

First, for the operators *fun* and *prod* we allow the more conventional mathematical symbols: $fun(A; B)$ is abbreviated $A \to B$ and the dependent form, $fun(A; x.B)$ is written $x : A \to B$. The product $prod(A; B)$ is abbreviated $A \times B$ and sometimes because of historical ties to Cambridge and Edinburgh ML this was written $A \sharp B$. The dependent form, $prod(A; x.B)$ is written $x : A \times B$ or $x : A \sharp B$. The disjoint union, $union(A; B)$, is abbreviated as $A + B$ and in some Nuprl texts as $A \mid B$. The equality type $eq(A; a; b)$ is writtn also as $a = b \epsilon A$ or as $a = b\ inA$ in Nuprl. In the special case of $eq(A; a; a)$ used to express membership, we also write $a \epsilon A$. When we want to stress the logical meaning of types, we will use the notation from the correspondence listed above.

## 3.5   defining numbers and lists

**natural numbers**   We can define the natural numbers using the recursive type constructor. First define **1** to be $fun(void; void)$ and let $*$ be its element $\lambda(x.x)$. Now let $Nat$ be $rec(N.1 + N)$. Some typical elements are $inl(*)$, $inr(inl(*))$, $inr(inr(inl(*)))$, $\ldots$. Of course the first of these represents zero. Define the successor function, $succ$, from $Nat$ to $Nat$ as $\lambda(n.inr(n))$. If we can define functions by primitive recursion and prove propositions by Peano induction, then we can carry out elementary number theory in the normal way.

A *primitive recursive* function definition can be put in the form

$$f(0) = b$$
$$f(n+1) = exp(n, f(n)).$$

Suppose that $e$ is an expression in which $u$ is a variable for the argument $n$ and $v$ is one for the inductive argument $f(n)$, then a linear form of this definition is given by the *induction combinator* $ind(n; b; u, v.e)$ whose evaluation rule is that $ind(0; b; u, v.e)$ evaluates to the value of $b$, and $ind(n+1; b; u, v.e)$ evaluates to the value of $e[n/u, ind(n; b; u, v.e)/v]$. For example the curried addition function is defined as

$$add(0) = \lambda(y.y)$$
$$add(n+1) = \lambda(y.succ(add(n)(y))).$$

The linear form of this definition is just $ind(n; \lambda(y.y); u, v.\lambda(y.succ(v(y))))$. This operator is defined as $recind(n; h, x.decide(x; u.b; v.e[h(u)/v]))$.

**lists**   Lists over a type $A$ are defined as $rec(L.(1) + (A \times L))$ where $inl(*)$ now serves as $nil$, the empty list. We abbreviate these types as $Alist$. The elements include $nil$, $inr(pair(a; nil))$, $inr(pair(a; inr(pair(b; nil))))$ where $a$ and $b$ are in $A$. The *cons* function which takes an element of $A$ and a list and produces a new list with the element conjoined to the given list is $\lambda(x.\lambda(L.pair(x; L)))$. As in the case of natural numbers, we can develop the theory of lists if we have a primitive recursive function which defines new functions and builds proofs of propositions. To obtain this *list induction combinator* we build it from $recind$. The pattern of primitive recursion on lists which we capture is

$$f(nil) = b$$
$$f(cons(a)(L)) = exp(a, L, f(L)).$$

For example the function which computes the length of a list is defined as

$$len(nil) = 0$$
$$len(cons(a)(L)) = 1 + len(L).$$

The linear form of this recursion combinator is $listind(l; b; h, t, v.e)$ which reduces according to the rules that $listind(nil; b; h, t, v.e)$ reduces to $b$, and $listind(cons(a)(L); b; h, t, v.e)$ reduces to $e[a/h, L/t, listind(L; b; h, t, v.e)/v]$. This can be defined from $recind$ as $recind(L; f, x.decide(L; v.b; u.spread(u; h, t.e[f(t)/v])))$.

## 4   Basis for Metalevel Programming

In this section we discuss a particular way to provide metalevel programming facilities in the type theory we are using. The method is widely applicable as is shown in [2]. The basic idea is that we can define the term and proof structure of this type theory in itself if the right primitives are present. Then we can manipulate expressions which denote programs and proofs; this is the essence of metaprogramming. First let us remind ourselves of one of the most useful applications of metaprogramming (indeed the only one we examine in detail in this article), namely theorem proving tactics.

Formal proofs of interesting theorems tend to be large objects, so in practice some mechanism is used to shorten and abbreviate them. One of the most effective is the use of *tactics* (LCF[26], Nuprl [15], $\lambda$Prolog [39], HOL [25] and Oyster [12]). Proofs then are *tactic-trees*. These are proofs represented as trees in which certain steps are justified by programs (tactics) that compute subtrees. Tactics are metaprograms. We need to see how they relate to the ordinary kind.

In all heretofore existing systems, the tactics are written in a separate language, called the *metalanguage*, hence the origin of the acronym ML for the programming language ML. So the tactic trees use two languages, one the *object language*, the other a *metalanguage*. In LCF the object language is the polymorphic predicate calculus (PP$\lambda$), in Nuprl and Oyster it is constructive type theory and in HOL it is a simple type theory. The metalanguage in these systems is a programming language: ML in LCF, Nuprl and HOL,but Prolog in Oyster. We are now considering the situation in which the object language itself has a notion of computation adequate for expressing tactics. This raises the possibility of using a single language as both object language and metalanguage if we can manage to reflect proofs from the metalanguage into the object language. (This enterprise is especially interesting in the case of systems like Nuprl in which programs are extracted from proofs, for then the language of proofs is itself a metalanguage for proofs.)

For Nuprl we know that there is such a class of proofs. This was shown in [2] and is related to an extensive study of the problem in [32,30,32]. The work of Allen,Constable,Howe and Aitken which is summarized here uses the reflection of the proof concept in a formal theory to define the notion of proof in that theory. An interesting closure principle is revealed in these results. It is based on a fixed point construction over a class of proof-like trees which interleave objects from the object language and the metalanguage. In this context metaprograms are just ordinary programs of the object language that compute with the special internal data types of terms and proofs that represent terms of the object language.

The resulting concept of reflected proof is especially noteworthy because it is encapsulated in a single recursive type, in contrast to other approaches which result in an infinite tower of types or languages [32,50] or would result in an infinite tower if extended to provide the same closure [51].

The notion of a reflected proof turns out to be more useful than the unreflected kind. In the LICS paper [2] there are seven applications of the concept mentioned. They span the areas of theorem proving, type definition and functional program evaluation. First, it is mentioned how to shorten proofs and speed-up the generation and checking of proofs by proving that the tactics are correct. Second, a kind of polymorphism can be provided for theorems involving universe types. Third, it is mentioned that one can build evaluation mechanisms provably equivalent to the built-in evaluator yet more efficient on certain terms. Fourth, as a special case of the third application, it is claimed that expressions can be detected that can be destructively updated, thereby allowing a procedural evaluation for them. Fifth, it is said that reflection gives one approach to expressing computational complexity. Sixth,it is suggested that certain systems extensions can be explored in a mathematically controlled way by running code from incomplete proofs of metatheorems describing them. Finally, proof transformations are mentioned. One can see that these could be used to build new logics on top of the built-in logic of a theory.

## 4.1    representation

The context of this work is that we are standing outside of a formal language $L$ (traditionally called the *object language*). In this article it is the type theory we introduced. The syntax of $L$ is given by an inductively defined class of terms, as we have seen in the previous section defining the type theory. This is a type in the observer's language or *metalanguage*, *ML*. We are especially interested in those languages $L$ with a computable evaluation relation on terms [24,15,26,28,45], again of the kind given by the reduction relation we defined above on the type theory terms.

We are concerned with languages $L$ in which types can be defined. From outside of $L$ we can see a type $T$ of $L$ as the class of terms which have that type, and we require that for any term $t$, $t$ and its value $t'$ have the same types, *i.e.*, if $t \in T$ then $t' \in T$; furthermore, we require that if $t \in T$ then $t$ and $t'$ are equal as members of $T$, *i.e.*, $t = t'$ *in* $T$. To be specific, the results of [2] will apply to the *type systems* defined in [3] which includes the one defined here.

A basic concept in defining reflection is the notion that a type of the language $L$ represents a mathematical class $S$.

**Definition 3** *Given a relation $t$ reps $s$ between terms $t$ and members $s$ of a class of objects $S$, we say that a type $T$ of the language $L$ represents $S$ via reps if*

1. *for any term $t$ and any $s, s' \in S$, if $t$ reps $s$ and $t$ reps $s'$ then $s$ is $s'$.*

2. *for any $s \in S$ there is a term $t$ such that $t$ reps $s$.*

3. *for any terms $t$ and $t'$, $t = t'$ in $T$ if and only if there is an $s \in S$ such that $t$ reps $s$ and $t'$ reps $s$.*

The first two conditions are general requirements for calling something a representation relation. The third is specific to the conventional use of a type as a system of notations for objects of some class. If $t$ is a member of $T$ then $\downarrow(t)$ denotes the member of $S$ that $t$ represents. Note that a given object may have many representatives. In the representation schemes used below we pick out for each $s$ a *standard* representation denoted $\lceil s \rceil$; the function taking $s$ to $\lceil s \rceil$ will be computable.

In the case of the class of terms we have defined so far, there is a natural representation relation that can be designed into the language. Recall that the set of terms is the least set containing variables and some set of operator names, and such that if $op$ is an operator name, $t_1, \ldots, t_n$ are terms, $n \geq 0$, and $\overline{v_1}, \ldots, \overline{v_n}$ are lists of variables, then $op(\overline{v}_1.t_1; \ldots; \overline{v}_n.t_n)$ is a term. We say that $t_i$ are the *subterms*; the $\overline{v}_i$ are *binding variables* with *scope $t_i$* and they bind the variables occurring in $t_i$. The operators defined so far have at most three of these subterms, in the *decide* for example. Also we have used the convention that if there are no binding variables, then we do not write the dot, and if there are no subterms, as say in *Type*, then we do not write the parentheses either. We need now to include as well operator names among the terms in anticipation of the need for a term to represent each operator name. Suppose that we capitalize the operator for that purpose; so we add *VOID*, *FUN*, *PROD*, etc.

We also need now types *Var* and *Op* in $L$ that represent variables and operator names, respectively. Indeed, we add them to the list of canonical terms of this type theory. Then a natural representation of a term $op(\overline{v}_1.t_1; \ldots, \overline{v}_n.t_n)$ is a pair

$$\langle OP, (\langle \overline{V}_1, T_1 \rangle, \ldots, \langle \overline{V}_n, T_n \rangle) \rangle$$

where $OP$ represents $op$, $\overline{V}_i$ represents $\overline{v}_i$ and $T_i$ represents $t_i$. Using this representation scheme we can show that the type *Term* defined as

$$rec(T.\ Op\ +\ Var\ +\ Op \times ((Var\ list \times T)\ list))$$

represents the set of terms of $L$.

The first disjunct of the disjoint union in the body of *Term*, *Op*, is used to represent the names of the operator names. So for example, *VOID* is now a new operator with zero arguments. We need a term to represent it. Instead of following the infinite regress of making up new operator names for this and then new names for the names, etc., we agree that these names of operator names can represent themselves. Thus the term representing the operator *VOID* is *inl*(*VOID*), and the term representing the term *VOID* is *inr*(*inr*(*pair*(*inl*(*VOID*); *nil*))). By contrast the term representing *void* is *inr*(*inr*(*pair*(*VOID*; *nil*))).

## 4.2   representing proofs

We are now going to define the representation of proofs. But since we have not included a metalevel definition of them and have not give specific primitive rules, we proceed differently than we did for terms. First we rely on the internal definition to inform us about the external one. Second, we describe proofs schematically, leaving out the particular rule names. We will not need those names at all in this article and they only serve to clutter the presentation. But infact there are examples of Nuprl proofs in the subsequent sections which serve to make the concept sufficiently concrete.

The proofs we deal with involve *sequents*, which are objects of the form

$$x_1\!:\!A_1, \ldots, x_n\!:\!A_n \;\gg\; B$$

where each $x_i\!:\!A_i$ is a *hypothesis* consisting of a variable $x_i$ and a term $A_i$, and where $B$ is a term called the *conclusion* of the sequent. There is a natural representation of this class based on the representation of terms, namely,

$$Sequent \equiv (\,Var \times \,Term)\;list \times \,Term.$$

Reflection of terms and evaluation is studied in more depth in [1].

The design of this class of proofs is intended to exploit their representability in two ways. Tactics are internally represented and a reflection rule is included.

First consider these design constraints. We take from the Nuprl design the general form of proof as a finite tree whose nodes are labeled with a sequent and some discrete justification "text" which is intended to indicate how the inference of the sequent from the sequents labeling the offspring is justified. The sequent at the root is called the *goal* of the tree. The leaves of a proof tree that are unjustified are the *premises* of the proof. Next, we require that the sequent and justification text completely (and effectively) determine the list of immediate *subgoals*, that is, the goals of the immediate subtrees. Finally, we require that proofs be effectively recognizable. However, we do not require that proofhood be effectively decidable; this is a necessary result of allowing arbitrary computable tactics, since one cannot then decide which (untyped) programs will produce proofs. To simplify this presentation, we shall develop the basic ideas without incorporating extraction of programs from proofs.

The rules that exploit the internal proof representation are the tactic rule and the reflection rule. An application of a tactic rule can be written

$$s \quad \text{By tactic } t$$
$$s_1$$
$$\vdots$$
$$s_n$$

where $s$ is the goal and the $s_i$ are the subgoals. The justification text here contains a term $t$ which represents a proof with goal $s$ and premises $s_1, \ldots, s_n$. An application of the reflection rule can be written

$$s \quad \text{By refl } i,t$$
$$\gg provable(\lceil s \rceil, t, \lceil i \rceil)$$
$$s_1$$
$$\vdots$$
$$s_n$$

where $t$ represents the list $s_1, \ldots, s_n$, $\lceil s \rceil$ is the standard representative of sequent $s$, and $\lceil i \rceil$ is the standard representative of the number $i$. The number $i$ is called the *level* of the reflection rule. $provable(\lceil s \rceil, t, \lceil i \rceil)$ is a type representing the proposition that there is a proof with goal (root) $s$, premises $s_1, \ldots, s_n$, and in which every use of reflection is of level less than $i$.

Another kind of justification is a primitive rule. We do not need to specify the form of justification text for applications of primitive inference rules. If $g, s_1, \ldots, s_n$, $n \geq 0$, are sequents and $r$ is the justification text for an instance of a primitive rule then define $instance(r, g, s_1, \ldots, s_n)$ to be true if $g$ is the conclusion and $s_1, \ldots, s_n$ are the premises of the instance of the rule named by $r$. Below we will refer to $r$ simply as a primitive rule. More generally, we will not refer to "justification text" and instead only speak of justifications and rules.

**Definition 4** *A* justification *is either a primitive rule, a* tactic rule *consisting of a single term, a* reflection rule *consisting of a nonnegative integer and a term, or the* empty justification.

The first problem considered in [2] was to devise a (non-circular) definition of a self-referential class of proofs and to demonstrate their validity. It is easier to deal with representation before defining proofs, so following the LICS presentation exactly we introduce *preproofs*, which are trees having the same components as proof trees but without restrictions on how the components fit together.

**Definition 5** *A* preproof *is a tuple*

$$(s, r, \Delta_1, \ldots, \Delta_n)$$

*where $s$ is a sequent, $r$ is a justification, $n \geq 0$ and for each $i$, $1 \leq i \leq n$, $\Delta_i$ is a preproof.*

The *root* and *leaves* of a preproof are defined in terms of the obvious interpretation of preproofs as trees. The *premise list* of a preproof is the list of the sequents, in left-to-right order, occurring at leaves that have the empty justification. We will refer to the premise list simply as the *premises* of the preproof. The root of a preproof is also called the *goal* of the preproof.

Above we defined types for the terms that represent terms and sequents. We can similarly define types representing justifications and preproofs (we assume that the justification text for primitive rules is representable—this is easy to arrange). Details are given in the final subsection in exactly the form worked out by Stuart Allen.

The definition of reflective proofs involves a syntactic fixed-point construction, so we need to introduce a parameter on which to take the fixed point. (For those interested only in being able to treat tactics internally, this detail is not necessary.) So for any term $q$ we will define $q$-proofs; these will be just like proofs except for the reflection rule. The purpose of the reflection rule is to allow us to make an inference step by proving that there is a proof justifying the step; the sequent which expresses the existence of such a proof must involve a type representing proofs. In a $q$-proof we use the term $q$ in place of this as-yet unknown type of proofs. We will also need a particular fixed term $T$ in the definition of $q$-proofs. We will not define $T$ at this point; later we will specify the property required of $T$ for validity of proofs, and make it plausible that $T$ can be constructed (details are given at the end). The important point at this stage is that $T$ is simply a particular term; exactly what term we pick is relevant *only* to the validity argument, since it enters in the definition of proof only as essentially a piece of text in a proof tree.

**Definition 6** *Let $q$ be any term. We inductively define the set $pf(q)$ of $q$-proofs as a subset of the set of preproofs. A $q$-proof is a preproof*

$$(s, r, \Delta_1, \ldots, \Delta_n)$$

*where $s$ is a sequent, $r$ is a justification and*

1. *$\Delta_1, \ldots, \Delta_n$ are $q$-proofs,*

2. *if $r$ is the empty justification then $n = 0$,*

3. *if $r$ is a primitive rule then*

$$instance\,(r, s, g_1, \ldots g_n)$$

   *where $g_1, \ldots, g_n$ are the goals of the preproofs $\Delta_1, \ldots, \Delta_n$,*

4. *if $r$ is a tactic rule consisting of a term $t$ then $t$ represents a preproof $\Delta$ such that $\Delta$ is a $q$-proof, the goal of $\Delta$ is $s$ and the premises of $\Delta$ are the goals of $\Delta_1, \ldots, \Delta_n$, and*

5. *if $r$ is a reflection rule consisting of a number $i$ and term $t$ then $n \geq 1$, $t$ represents the sequent list consisting of the goals of $\Delta_2$, ..., $\Delta_n$, and the goal of $\Delta_1$ is the sequent*

$$\gg T(q)(\lceil s \rceil)(t)(\lceil i \rceil).$$

The *immediate subproofs* of a $q$-proof

$$(s, r, \Delta_1, \ldots, \Delta_n)$$

are $\Delta_1, \ldots, \Delta_n$ and also, in the case that $r$ is a tactic rule consisting of a term $t$, the preproof represented by $t$. A $q$-proof has *level $k$* if any number appearing in a reflection rule is less than $k$. More precisely,

$$\Delta = (s, r, \Delta_1, \ldots, \Delta_n)$$

has level $k$ if each immediate subproof of $\Delta$ has level $k$ and in the case that $r$ is a reflection rule $r$'s number is less than $k$.

We now need to argue that most of the above can be internalized in type theory. We have given a definition of a type *Term* which represents the class of terms. Given a term $t$, we can easily determine the member $\lceil t \rceil$ of *Term* which represents $t$. It is straightforward to write a function in type theory which reflects this. In particular, we can construct a term *Rep* of type *Term* $\rightarrow$ *Term* such that for every member $t$ of *Term*, the term $Rep(t)$ represents $\lceil u \rceil$ where $u$ is the term that $t$ represents.

We also need a type that represents the class of $q$-proofs. This is straightforward to construct since the forms of inductive definition we have used are directly supported in this type theory. In the next subsection we define a term *Pf* such that if $Q$ is a term representing a term $q$, then $Pf(Q)$ represents $pf(q)$.

We can now define the fixed-point exactly as it occurs in the LICS paper. Let $d$ be the term

$$\lambda x.\, Pf(Ap(x, Rep(x)))$$

where $Ap$ is a term which is a type theoretic function such that $Ap(\lceil a \rceil, \lceil b \rceil)$ evaluates to $\lceil a(b) \rceil$. Finally, let $p$ be the term $d(\lceil d \rceil)$.

**Definition 7** *The set pf of proofs is $pf(p)$.*

Note that $Pf(\lceil p \rceil)$ represents the class of all proofs. If we reduce the term $p \equiv d(d)$ we get $Pf(\lceil p \rceil)$, so in Nuprl's type theory $p$ is a type equal to $Pf(\lceil p \rceil)$. Hence $p$ is also a type representing the class of all proofs.

Before we can prove the validity of proofs, we need to assume a certain property of the term $T$ that was used in the definition of $q$-proofs. Suppose that $S$ represents a sequent $s$, $L$ represents a list $l$ of sequents, and $I$ represents a number $i$. We assume that the term $T(p)(S)(L)(I)$ is a type representing the class of all members $\Delta$ of $pf$ such that $\Delta$ has level $i$, the goal of $\Delta$ is $s$, and the premise list of $\Delta$ is $l$.

$T$ is actually quite straightforward to construct. We want $T(p)(S)(L)(I)$ to be a type consisting of all members of $p$ with level $I$, goal $S$ and premises $L$. This requires functions in type theory which compute the premises and goal of a proof, and a predicate for the level of a proof. Although it may appear that we need to know what proofs are in order to define $T$, functions in the type theory of Section reftype-definition are untyped (as they are in full Nuprl) and can be defined to work on (representations of) preproofs. For the level predicate, the preproof represented by a tactic rule must be computed, and this can be done by a simple recursive algorithm. The level predicate will be partial on preproofs but total on proofs.

A proof is *valid* if the goal is true when the premises are. The proof of validity is mostly abstract with respect to the definition of truth for sequents. We assume, of course, that the primitive rules are valid. The only property of truth we will need is that if the sequent $\gg A$ is true, then $A$ is a type that has a member.

**Theorem 1** *Proofs are valid.*

*Proof.*    We show by induction on $n$ that proofs of level $n$ are valid. For each $n$ we argue by induction on the immediate-subproof relation. Suppose, then, that

$$\Delta = (s, r, \Delta_1, \ldots, \Delta_n)$$

is a proof whose premises are true. Let $\overline{\Delta}$ denote the list $\Delta_1, \ldots, \Delta_n$. By induction the goal of each $\Delta_i$ is true. We now consider the cases for $r$.

If $r$ is the empty justification then $s$ is a premise hence true.

If $r$ is a primitive rule then $s$ is true by our assumption of the validity of primitive rules and the fact that $instance(r, s, g_1, \ldots, g_n)$ where $g_i$ is the goal of $\Delta_i$.

If $r$ is a tactic rule with term $t$ then $t$ represents a subproof $\Delta'$ of $\Delta$, so by induction $\Delta'$ is valid. The premises of $\Delta'$ are the goals of $\overline{\Delta}$, so they are true, hence the goal of $\Delta'$ is true, and this is $s$.

If $r$ is a reflection rule with integer $i$ and term $t$, then the goal of $\Delta_1$

$$\gg T(p, \lceil s \rceil, t, \lceil i \rceil)$$

is true, so the term in this sequent is a type that has a member, call it $t$. From our assumption about $T$ it follows that $t$ represents a proof $\Delta'$ of level less than $i$. By the induction hypothesis (of the induction on level) $\Delta'$ is valid. The premises of $\Delta'$ are the goals of $\Delta_2, \ldots, \Delta_n$ (by the assumption about $T$) and so are true. The goal of $\Delta'$ must then be true, and this is $s$. $\square$

Tactic rules and reflection rules are eliminable from complete proofs. A *primitive* proof of $s$ is a proof with goal $s$ where every node is either a premise or is justified by a primitive rule. A *complete* proof is one with no premises. By an inductive proof of the same structure as the one just given for validity of proofs, we can show the following

**Theorem 2** *If $s$ has a complete proof then it has a complete primitive proof.*

In fact, given the extraction property to be discussed in the next subsection, there is a simple procedure for computing the primitive proof.

## 4.3   extraction

For this type theory, we are not satisfied with the validity of proofs. We want to be able to *extract* programs from proofs. For example, if we have proven a sequent of the form

$$\gg \forall x.\, \exists y.\, R(x, y)$$

then we expect to be able to derive from the proof a program which when given an $x$ will produce a $y$ such that $R(x, y)$ is true.

We need to refine our notion of truth of sequents somewhat. For the discussion here, say that $t$ *satisfies* a sequent

$$x_1 \colon A_1, \ldots, x_n \colon A_n \gg B$$

if the free variables of $t$ are among $x_1, \ldots, x_n$ and if for all members $x_1, \ldots, x_n$ of $A_1, \ldots, A_n$ respectively, $t$ is a member of $B$.[1]

We require an automatic procedure for producing a satisfying term from a complete proof; we call this an *extraction* procedure. In analogy with our definition of validity for proofs, we say a partial term-valued function $F$ of proofs and term lists is valid if for any term list $b_1, \ldots, b_m$ and any proof $\Delta$ with $n \leq m$ premises, if $b_1, \ldots, b_n$ satisfy the premises of $\Delta$ then $F(\Delta, b_1, \ldots, b_m)$ satisfies the goal of $\Delta$. We allow the list $b_1, \ldots, b_m$ to be longer than need be in order to make the definition of extraction given below easier.

As an implicit parameter for constructing our extraction procedure we assume a procedure $prextr(r, s, b_1, \ldots, b_m)$ which is used to extract from proofs justified by a primitive rule.

We can now recursively define extraction as a partial function $extr$. To define

$$b = extr((s, r, \Delta_1, \ldots, \Delta_n), b_1, \ldots, b_m),$$

let $g(i)$ be one plus the sum of the lengths of the premise lists of $\Delta_1, \ldots, \Delta_i$ and let $h_i$ be $extr(\Delta_i, b_{g(i-1)}, \ldots, b_m)$. If $r$ is the empty justification then $b = b_1$. If $r$ is a primitive rule then $b = prextr(r, s, h_1, \ldots, h_n)$. If $r$ is a reflection rule then

$$b = extr(\downarrow(fst(extr(h_1))), h_2, \ldots, h_n)$$

where $fst(t)$ is the first component of the pair term that $t$ evaluates to. Finally, if $r$ is a tactic rule consisting of a term $t$ then

$$b = extr(\downarrow(t), h_1, \ldots, h_n).$$

---

[1]This is only an approximation of Nuprl's satisfaction relation. For more details, see [3].

## 4.4   details of the proof term

We will now present exactly Stuart Allen's construction of a proof term. In the following formalization of proofs, we use the notation $\overline{x_{i\ldots j}}$ to denote a list containing the elements $x_i$, $x_{i+1}$, ..., $x_j$. The *case* expression is used as destructor for the *Justification* type. It has the form *case $j$ of $c$* where $c$ is a sequence of clauses of the form $t_i x_i \to r_i$. If $j$ evaluates to an $n$th injection of $v$, then the value of the expression is the value of $r_n$ with $x_n$ bound to $v$. Note that $t_n$ is merely a mnemonic tag. The version of the *rec* type constructor used in the definition of *IsAPf*, $rec(z, x.\ t;\ a)$, is used to define an instance of a parameterized family of recursive types. $t$ is a term that defines the type. In it $z$ stands for the family of types being defined, and $x$ for the parameter. $a$ is the actual parameter. As a simple example of the form's use, the proposition that a function $f$ over the natural numbers eventually takes on the value 0 can be represented by the type $rec(P, x.\ f(x) = 0 \lor P(x+1); 0)$.

The only function we will define on preproofs is *level*; *premises* and *root* are similar inductions over *PPf* (the type representing preproofs). The term $T$ used in Section 4.2 is $\lambda qsli.\ Pvbl(q; i; s; l)$. The *recind* form used below is a primitive form for constructing recursive functions over members of a recursive type.

$$Term \equiv rec(t.\ Op\,|\,Var\,|\,Op \times ((\,Var\ list \times t)\ list))$$

$$Sequent \equiv Term\ list \times Term$$

$$Justification \equiv \{1\}\,|\,Prim\,|\,int \times Term\,|\,Term$$

$$PPf \equiv rec(s.\ Sequent \times s\ list \times Justification)$$

$$level(p; l) \equiv recind(p; \langle s, j, \overline{p_{1\ldots n}}\rangle, h.\ \bigwedge_{i=1}^{n} h(p_i)\ \land$$
$$\quad case\ j\ of$$
$$\qquad PREMISE\ x \to true$$
$$\qquad PRIM\ x \to true$$
$$\qquad REFL\ \langle n, t\rangle \to n < l$$
$$\qquad TACTIC\ t \to \exists p\colon PPf.\ t\ RepsPPf\ p \land h(p))$$

$$Pvbl(Q; l; g; s) \equiv \exists p\colon Q.$$
$$\quad level(p; l)\ \land$$
$$\quad root(p) = g \in Sequent\ \land$$
$$\quad premises(p) = s \in Sequent\ list$$

$$roots(x) \equiv map\ root\ x$$

$$IsAPf(p; Q) \equiv rec(z, \langle s, j, \overline{p_{1...n}} \rangle. \; \bigwedge_{i=1}^{n} z(p_i) \wedge$$

$$case \; j \; of$$

$$PREMISE \; x \rightarrow n = 0 \in int$$

$$PRIM \; r \rightarrow instance\,(r; s; roots(\overline{p_{1...n}}))$$

$$REFL \; \langle l, t \rangle \rightarrow$$

$$n > 0 \wedge$$

$$t \; RepsSequentList \; roots(\overline{p_{2...n}}) \wedge$$

$$p_1 = \langle nil, T(Q)(Repsequent(s))(t)(repint(l)) \rangle$$

$$\in Sequent$$

$$TACTIC \; t \rightarrow \exists x \colon PPf.$$

$$root(x) = s \in Sequent \wedge$$

$$premises(x) = roots(\overline{p_{1...n}}) \in Sequent \; list$$

$$t \; RepsPPf \; x \wedge z(x);$$

$$p)$$

$$Pf(Q) \equiv \{ \, p \colon PPf \mid IsAPf(p; Q) \, \}$$

The final definition of $Pf(()Q)$ is given using the *set type* of Nuprl. In the type theory of this article that type is not used, but it can be replaced by a dependent product with no loss of expressive power. So here we take $\{x \; x \colon A \mid B\}$ as $prod(A; x.B)$. This is in fact the way Martin-Löf treated sets in his type theories.


# 5 Metaprogramming

We now want to use the results of the previous section in support of metalevel programming. There is a sufficient basis for us to be able to use the type theory terms as a programming language operating on themselves and on proofs. Infact there is much more, there is a basis for proving that these programs are correct *and* using these correctness proofs to justify tactics as new rules of inference. For this later capability we need the reflection rule. However in the Marktoberdorf lectures and in this paper we will examine only the first of these capabilities, namely the use of the type theory terms as a metaprogramming notation. First we discuss briefly the matching problem and the value of formalized metareasoning.

A simple but important algorithm used to support automated reasoning is called *matching*: given two terms it produces a substitution, if one exists, that maps the first term to the second. In this section the matching algorithm is used to illustrate an approach to automating reasoning in which tactics for finding proofs are extracted from constructive proofs of metatheorems. Later in a subsection the algorithm is derived and verified following exactly an informal presentation of it in Section ??. The treatment of this example suggests how these theories can be soundly extended by the addition of constructive metatheorems about themselves to their libraries of results.

In this section we will describe the basic idea that automated theorem proving can be seen as the implementation of a constructive *metamathematics*. In metamathematics one proves theorems about doing mathematics. Some of these theorems have interesting computational content, such as a result claiming that a theory is decidable or a result saying that any proof of a theorem of one form can be converted to one of another form (say from prenex form into nonprenex form). We will show how some theorem proving algorithms can be understood as implementations of theorems about proofs and formulas.

One of the critical reasons for adopting this point of view is that much of the knowledge we discover about theorem proving is potentially useful to the automated system itself. In order to be used by machines, the knowledge must be formalized. Conversely, the more we can say about the algorithms we use, the more effectively we can use them. This suggests that we want to develop them in the context of a formal theory. Related reasons for doing this are discussed by Davis and Schwartz [19], Boyer and Moore [9] and Shankar [46].

One of the major advantages of formalizing knowledge about theorem proving is that we can use it to extend the stock of derived rules of inference. This in some cases allows us to avoid running theorem-proving algorithms. To the extent that results from other parts of mathematics are useful in theorem proving, as in the use of graph algorithms in congruence closure, we want to be able to prove that they are correct and preserve the correctness of the logic when they are used in derived inference rules.

We begin with an informal derivation of the match algorithm and then present a formalization of its development that Douglas Howe carried out in the Nuprl proof development system. This formalization is based directly on the informal account.

## 5.1  syntax of terms for matching problem

The matching algorithm applies to a subset of the type of terms defined above. We do not consider terms with binding operators. A simple definition of the terms we consider would be to say that a variable is a term, and if $t_1, \ldots, t_n$ are terms and $f$ is a function symbol, then $f(t_1, \ldots, t_n)$ is a term. The important point for manipulating terms is not the particular way of displaying them but rather the component parts and a means of accessing them. So for a composite term such as $f(t_1, \ldots, t_n)$, the components are $f$ and the list $(t_1, \ldots, t_n)$. The important feature of a variable is that it is the base case of the definition. So the official definition of a term specialized to this case is that it is a variable or it is a pair $\langle f, l \rangle$ where $f$ is a function symbol and $l$ is a list of terms. This is the mathematical definition, but we can agree to display a composite term as $f(l)$. If we let *Var* be the set of variables and *Fun* the set of function ymbols, then the type *Term* is defined as

$$rec(\mathit{Term}. \mathit{Var} + (\mathit{Fun} \# \mathit{Term}\ \mathit{list})).$$

The induction principle for terms is given by the principle for this recursive type. We use it to prove the following theorem.

**Theorem 1** *For all propositional functions $P$ on terms, if $\forall\, x\!: Var.\ P(x)$, and if for all Term lists $l$ and function symbols $f$, whenever $P(t)$ holds for all terms on $l$ then $P(f(l))$, then $\forall\, t\!: Term.\ P(t)$.*

We will carry out constructions based on the structure of a term. To facilitate them we introduce a case discriminator written as

$$case\ t:\ y \to a;\ f.l \to b.$$

This means that if the term $t$ is a variable then the value is $a$, where in the expression $a$, $y$ stands for the variable; if the term is composite then the value is $b$ where $f$ and $l$ are names in $b$ of the function symbol and subterm list respectively. This case analysis is used together with induction on term structure to define concepts on terms. For example, here is a definition of the notion that a variable $x$ occurs in a term $t$.

**Definition 8** *A variable $x$ occurs in a term $t$, abbreviated $x \epsilon t$, if in the case that $t$ is a variable $y$ then $y = x$, and in the case that $t$ is $f(l)$, then $x$ occurs in some term $u$ of $l$.*

The form of inductive arguments on term structure is clarified by seeing the computational meaning of induction. The induction rule defined above is not only a way to prove properties of terms, but it is also a method of computing based on the structure of a term, just as mathematical induction is a way of computing based on the structure of natural numbers. Recall that the computational form associated with induction on terms is denoted $rec\_ind(t; h, x.\, b)$. To review computing by "term induction" on a term $t$, let us carry it through for the notion of $x$ *occurs in* $t$. The expression $b$ is a case discriminator on $x$, precisely,

$$rec\_ind(t;\ occurs,\ z\ .\ case\ z{:}\ y\ \to\ x{=}y;$$
$$f,l\ \to\ for\ some\ u\ on\ l.\ occurs(u))$$

Recall that the general rule for evaluating the recursion induction form $rec\_ind(t; h, z.\, b)$ is that it reduces to

$$b[t/z, (\lambda w.\ rec\_ind(w; h, z.\, b))/h].$$

So in the case of a term such as $f(x, v)$ the computation of the recursive definition $x$ *occurs in* $f(x,v)$ is the following:

$$rec\_ind(f(x,v);\ occurs,z.\ case\ z{:}\ y\ \to\ x{=}y$$
$$f,l\ \to\ for\ some\ u\ in\ l.\ occurs(u))$$

reduces to (letting *occurs* denote $\lambda w.\ rec\_ind(w; \ldots)$)

$$
\begin{aligned}
\textit{case } f(x,v)\textit{: } y &\rightarrow\ x{=}y \\
f,l &\rightarrow\ \textit{for some u in l. occurs(u)}
\end{aligned}
$$

which reduces to

$$\textit{for some u in (x,v). occurs(u).}$$

Taking $u$ to be $x$ results in

$$
\begin{aligned}
\textit{case } x\textit{: } y &\rightarrow\ x{=}y \\
f,l &\rightarrow\ \ldots.
\end{aligned}
$$

In this case the term is a variable and the form reduces to

$$x = x$$

which is true. So $x$ does occur in $f(x, v)$.


### 5.1.1   substitutions

For the matching algorithm we need to study the substitution of terms for variables
in terms, as in substituting 2 for $x$ in $3 * x = z$ to obtain $3 * 2 = z$.

**Definition 9** *A substitution is defined as a list of pairs of a variable with a term,
such as*

$$(\langle x, 2\rangle,\ \langle y, z\rangle,\ \langle z, x + y\rangle).$$

*So a substitution is an element of the type Var $\times$ Term list. Call this type Sub.*

Given a substitution $s$ and a variable $x$ we write $s(x)$ to designate the term paired
with $x$ (the first if there is more than one). This notion can be defined precisely by
a list induction form, namely

$$s(x) = list\_ind(s; x; h, tl, v.\ \textit{if } h.1 = x \textit{ then } h.2 \textit{ else } v).$$

(In this example we let $h.1$ and $h.2$ select the first and second members of a pair
respectively.)

The application of a substitution $s$ to a term $t$ is written as $s(t)$; it is defined by
induction on the structure of $t$.

**Definition 10** *If $t$ is a variable $y$, then $s(t)$ is $s(y)$. If $t$ is $f, l$, then $s(t)$ is obtained by applying $s$ to each element of the list $l$. If we let $map(s, l)$ denote the function that applies $s$ to each term on the list $l$, then $s(t)$ is*

$$rec\_ind(t;\ h, z.\ case\ z:\ y \to s(y)\ ;\ f, l \to f(map(s, l))).$$

*Abbreviate $map(s, l)$ by $s(l)$.*

The *domain* of a substitution are those variables that appear as the first element of a pair. We write $x \in dom(s)$ to indicate that $x$ is in the domain of substitution $s$.

**Definition 11** *We say that $s_1$ is a* sub-substitution *of $s_2$, written $s_1 \subset s_2$, if for every variable $x$ such that $x \in dom(s_1)$, $x \in dom(s_2)$ and $s_1(x) = s_2(x)$. We say that a substitution is* minimal, *$min(s)$, if no variable occurs twice on the left side of a pair.*

## 5.2  matching

We are now ready to discuss the problem of (first–order) matching. This is a basic matter in automated reasoning; for instance see [11,22,8,52]. We say that term $t_1$ matches $t_2$ if there is a substitution $s$ such that $s(t_1) = t_2$. For example, $f(x, g(y, z))$ matches $f(g(y, z), g(y, z))$ using the substitution $(\langle x, g(y, z) \rangle)$. But $f(x, y)$ does not match $g(x, y)$ because the outer operators are different. We will prove that for all terms $t_1$ and $t_2$ either $t_1$ matches $t_2$, in which case we can find a minimal substitution $s$ such that $s(t_1) = s(t_2)$, or else no substitution will yield a match. More precisely, define $match?(t_1)$ if for every $t_2 \in Term$

$$\exists s: Sub.\ s(t_1) = s(t_2)\ \&\ min(s)\ \&\ \forall x: Var.\ x \in dom(s) \Leftrightarrow x \in t_1)$$
$$\lor\ \forall s: Sub.\ \neg(s(t_1) = s(t_2)).$$

The main theorem can now be stated.

**Theorem 2** *(match_thm):* $\forall t: Term.\ match?(t).$

The idea of the proof is to check the outer structure of $t_1$ and $t_2$. If $t_1$ is a variable, then the pair $\langle t_1, t_2 \rangle$ is the substitution. Otherwise $t_1$ is compound, say $\langle f_1, l_1 \rangle$. If $t_2$ is a variable then there is no match. So assume it has the form $\langle f_2, l_2 \rangle$. If $f_1 \neq f_2$, then there is no substitution. Otherwise the outcome depends on the result of trying to match $l_1$ and $l_2$.

Suppose $l_1$ is *nil*. If $l_2$ is also *nil* then the empty substitution matches $t_1$ and $t_2$; otherwise there is no match. Suppose $l_1$ is $a_1.u_1$; if $l_2$ is *nil*, then there is no match. So suppose $l_2$ is $a_2.u_2$. First match $a_1$ and $a_2$. If they do not match, then neither do $t_1$ and $t_2$. Suppose they do, and let $s$ be the substitution such that $s(a_1) = a_2$. Now try recursively to match $u_1$ and $u_2$.

If these do not match, then neither do $t_1$ and $t_2$. If $s'$ matches them, then we check whether $s$ and $s'$ are compatible substitutions, *i.e.* whether they agree on common variables. If they do, then the final substitution is the union of $s$ and $s'$. If they are incompatible, then we argue as follows that no match exists.

## 5.3  formal metamathematics

Most of the rest of this section is devoted to describing a formal proof that Douglas Howe carried out in Nuprl of the above theorem about matching. The formal proof took him about a day to construct using Nuprl. A complete listing of the proof can be obtained from Professor Howe at Cornell. The complete version is available with the distribution of the Nuprl system, and it can easily be read even by people not familiar with the details of type theory. The presentation here is based on the Marktoberdorf lecture notes which in turn came from [16].

Before we can proceed with the example, we need to give a brief description of those parts of Nuprl not covered in the type theory being used here. After the example, we discuss how programs such as matching can be incorporated in Nuprl's own inference mechanisms.

## 5.4  Nuprl

Most of the relevant type theory underlying Nuprl has been described earlier in this article. The structure of proofs was presented as an internal data type, but in order to read output from the Nuprl *system*, some notation needs to be given and related to the internal proof type. First, we know that the inference rules of Nuprl deal with *sequents*, which are displayed as

$$H_1, \ H_2, \ \ldots, \ H_n \gg P$$

where $P$ is a formula and where each $H_i$ is either a formula or a variable declaration of the form $x\colon T$ for $x$ a variable and $T$ a type. The $H_i$ are referred to as *hypotheses*, and $P$ is called the *conclusion* of the sequent. Sequents, in the context of a proof, are also called *goals*.

We have also seen in the definition of the proof type that a proof in Nuprl is a tree-structured object where each node has associated with it a sequent and a rule. We call rules in Nuprl *refinement* rules because we think of them as being applied backwards: given a goal, we *refine* it by using an inference rule whose conclusion matches the goal, obtaining *subgoals* which are the premises of the rule. The children of a node in a proof tree are the subgoals which result from the application of the refinement rule of the node. Following Bates [6] we call such logics *refinement logics*, and we consider them further in this article in our discussion of logic variables.

A key feature of Nuprl's proof structure as defined in the previous section is that a refinement rule need not be just a primitive inference rule, but can also be a *tactic* written in Nuprl itself or in the programming language ML [26]. As with a primitive refinement rules, the application of a tactic to a goal produces subgoals. Because of ML's and Nuprl's strong typing property, it is guaranteed that the subgoals imply the goal. When a tactic is used to refine a goal, the rule at the new node of the proof tree will show the text of the tactic. This gives a means for constructing higher level proofs that can serve as explanations of formal arguments.

Suppose we have proven a theorem

```
∀i,j,m,n: Int. i≤j & m≤n => m+i≤n+j
```

and named it `mono`. An example of a refinement step using a tactic which applies this theorem is the following.[2]

```
x:Int, 0≤x  >>  x+2 ≤ 2*x+2   BY (Lemma 'mono' ...)
    x:Int, 0≤x  >>  x ≤ 2*x
```

The tactic `Lemma` computes the necessary terms and applies the rules necessary to instantiate the theorem. The three dots after the tactic indicate that a general purpose tactic called the *autotactic* was applied after `Lemma`. In this example, the autotactic proved the most trivial subgoals produced by `Lemma` (that x, 2*x and 2 are integers and that 2≤2). The result of the refinement is the single subgoal shown.

## 5.5   a formal account of matching

Howe's Nuprl library containing the formalization of matching can be divided into two parts. The first part, consisting of about 150 objects, is general. It contains definitions and simple theorems about the representation of logic within Nuprl. It also contains a minimal development of the theory of lists. The second part is particular to the formalization of matching; it contains about 40 definitions and theorems, pertaining to substitution and the syntax of terms.

The following description has three parts. First is a brief discussion of the tactics that were used in building the proofs in this library. This account is a summary of a much more extensive explanation given in Howe's thesis [30]. (The entire tactic library has been documented and explained by Paul Jackson. This documentation is available with the Nuprl system.) Next is a presentation of the definitions and the simple theorems leading up to the main results of the library. Finally there is a detailed description of the proofs of the main theorem. In particular, we will examine one of the proofs of a lemma in its entirety in order to illustrate the character of reasoning about formal metatheory in Nuprl. All definitions and theorems not in the general portion of the library will be at least stated in this account.

### 5.5.1   preliminaries

Most of the work in proving the theorems in the library, in terms of number inference steps taken, was done by two general tactics. The first of these is the *autotactic*. It is usually able to completely prove subgoals involving typechecking (showing that

---

[2]We will use typewriter typeface when presenting objects constructed (or hypothetically constructed) with the Nuprl system.

a term is in a certain type or that a formula is well-formed—these properties are recursively undecidable even in the subset of Nuprl used in this article because of the generality of the type theory) and simple kinds of simple integer arithmetic and propositional reasoning. Associated with definitions in the library are theorems which give a type for the defined term. This style was adopted by Howe in his thesis in order to encorporate type checking into the autotactic; these types are used by the autotactic in proving the numerous typechecking subgoals that arise. The autotactic is usually effective in hiding the details of the type theory from the user; for example, there are no typechecking subgoals in the proofs of the main theorems described below. The autotactic is usually invoked via a Nuprl definition: when $(T \ \ldots)$ appears in a proof, it means that the autotactic was applied after the tactic $T$.

The other general tactic is called `Simp` and was used by Howe to simplify terms. This tactic can be updated from the library (through a special kind of library object that can contain an ML form) with new simplification clauses in two basic ways. First, one can specify that a theorem be used for simplification. The theorem should be a universally quantified formula of the form

$$A_1 \Rightarrow A_2 \Rightarrow \ldots \Rightarrow A_n \Rightarrow R(a, b)$$

where $R$ is an equivalence relation (such as if-and-only-if or Nuprl's built-in equality). The simplifier will then always try to rewrite terms matching $a$ to an instance of $b$. Second, one can directly specify a simplification that is computationally justified (where one term can be obtained from another by a sequence of forward or backward computation steps). For example, one can specify that the length `|h.1|` of a list with head `h` always be simplified to `|1|+1`.

The "general" portion of the match library contains numerous updates to the simplifier, mostly for propositions and some for list theory. The simplifier, like the autotactic, is invoked by a definition: the notation $(T \ \ldots \text{+s})$ indicates that the simplifier and autotactic were both applied after $T$. One of the variants of this has `sc` in place of `+s`; which means that the simplifier was applied only to the conclusions of the subgoals produced by $T$.

Several other tactics use information from the library. `Induction` is used to perform induction on a variable appearing in a goal. If there is an induction principle in the library for the type of the variable, that is used, otherwise the type should be a primitive one (not a definition instance) for which there is an induction rule. This can be seen as a simple heuristic of the kind used by Boyer and Moore and in the Clam system. `Unroll` is similar to induction except that no induction hypothesis is generated. Finally, `Decide` takes a formula $P$ and performs a case analysis on whether or not $P$ is true. Decidability ($P \vee \neg P$) is non-trivial in Nuprl, and `Decide` attempts to use theorems of the appropriate form to justify the case analysis.

There are quite a few tactics for purely logical reasoning; most of these are simple. For example, `ILeft` reduces proving a disjunction to proving the left disjunct, and `ITerm` reduces proving an existentially quantified term $\exists x{:}T.\ P(x))$ to proving $P(t)$

for some supplied term $t$. ("I" is for introduction). For analysing hypotheses there are various "elimination" tactics: E applies to most kinds of formula and performs one step of analysis; SomeE applies to "some" (or "exists") formulas; and EOn (which requires a term argument) applies to universally quantified formulas. HypCases uses a universally quantified disjunction in the hypothesis list to perform a case analysis; it takes a list of terms as arguments to instantiate the quantified variables with. The most powerful tactic for logical reasoning is Backchain which uses (universally quantified) implications in the hypothesis list in a search for a complete proof. For example, if the conclusion of a sequent is $Q$ and $P \Rightarrow Q$ is a hypothesis, then it reduces proving $Q$ to proving $P$ and then attempts to recursively prove $Q$; if this fails it tries another hypothesis of a similar form.

There are several tactics for lemma application. One of these is Lemma, an example application of which was given above. FLemma uses a forward-reasoning analogue of what Lemma does. For example, if the theorem referred to is $P \Rightarrow Q$ and $P$ is a (specified) hypothesis, then $Q$ is added as a new hypothesis. Finally, CaseLemma uses a named lemma to determine a case analysis.

Finally, there are a few miscellaneous tactics that should be mentioned. The tactic Thin and its variants are used to remove unwanted hypotheses from a goal. AndThin takes a tactic that applies to hypotheses, applies it to a hypothesis and then discards the hypothesis. The so-called *tactical* THEN is used to combine tactics; if $T_1$ and $T_2$ are tactics then the tactic $T_1$ THEN $T_2$ first applies $T_1$ then applies $T_2$ to the resulting subgoals. The last tactic we discuss is Expand. This tactic takes a list of names of definitions and expands all instances of those definitions in the goal.

### 5.5.2 definitions and simple theorems

Since the definitions given in Section 5.1 are given in type-theoretic terms, their formalizations are almost identical. In particular, the explanations given there still apply, so we will do little more here than to simply present the formal versions.

The simple theorems given below were easy to prove, and we will only present a proof of one of them. These theorems required on average about five steps each. What a "step" is is not well-defined since we can always collapse an entire proof into a single step by composing all the tactics used in the proof. We will rely on Howe's description of some representative proofs to convey what a typical step is in this setting.

The definitions for syntax are straightforward. We first define the types of representatives of functions and variables in terms of Nuprl's type of atoms (character strings).

```
Var == Atom
Fun == Atom
```

The type of term representatives is a recursive type of syntax trees:

```
Term == rec(T. Var | Fun # T list).
```

The equality relations for terms and lists of terms are frequently used, so definitions
are made for them.

```
t1=t2 == t1=t2 in Term
l1=l2 == l1=l2 in Term list
```

Note that ambiguities in notations are acceptable in Nuprl (since definition instances
do not have to be parsed). The "injections" or "term constructors" are defined by

```
x == inl(x)
f(l) == inr(<f,l>)
```

We prove an induction principle and a useful special case of it:

```
>> ∀P:Term->U1. (∀x:Var. P(x))
              => (∀l:Term list. ∀f:Fun. (∀t:l. P(t)) => P(f(l)))
              => ∀t:Term. P(t)
>> ∀P:Term->U1. ∀x:Var. P(x)
              => ∀l:Term list. ∀f:Fun. P(f(l))
              => ∀t:Term. P(t)
```

Here the proposition $\forall t{:}l.$   $P(t)$ asserts that P(t) is true for every member t of
the list l. The first of these theorems can be read as "The property $P$ of terms is
true for all terms if (1) it is true for all variables and (2) if $f$ is a member of Fun and
$l$ is a list of terms satisfying $P$ then $P$ is true of $f(l)$". For our purposes, the type
U1 can be thought of as the type of all propositions.

The form for case analysis of terms is the following.

```
case t: x→a; f,l→b == decide(t; x.a; ap. let f,l=ap in b)
```

This uses the type theory's operators for analyzing a member of a disjoint union and
for decomposing a pair. As in Section 5.1, we can now define when a variable *occurs*
in a term.

```
xϵt == rec_ind(t; P,z. case z: y → y=x; f,l→ ∃u:l. P(u))
```

We also define a version of this predicate for lists

```
xϵl == ∃t:l. xϵt
```

so that a variable occurs in a list if it occurs in some member of the list. For this definition, as with other defined recursive functions and predicates, we add appropriate clauses to the simplifier. For example, we want any term of the form `xϵf(l)` to simplify to `xϵl`.

The type of substitutions and the application of a substitution to a variable and a term are defined as before. We also define application to a list of terms.

```
Sub == Var#Term list
s(x) == list_ind(s; x; h,ll,v. if h.1=x then h.2 else v)
s(t) == rec_ind(t; h,z. case z: x→s(x); f,l→f(map(h,l)))
s(l) == map(λt.s(t), l)
```

As in Section 5.1, the notations `.1` and `.2` denote projections from a pair, and `map(h,l)` applies the function h to every element of the list l.

We will need several properties of substitutions. In particular, we define when a variable is in the domain of a substitution, when one substitution is contained in another, when a substitution is minimal (that is, no two pairs in it have the same variable component) and when two substitutions are inconsistent (that is, disagree on some variable in both their domains).

```
xϵdom(s) == ∃p:s. p.1=x
s1⊆s2 == ∀x:Var. xϵdom(s1) => xϵdom(s2) & s1(x)=s2(x)
min(s) == list_ind(s; True; h,l,v. ¬(h.1ϵdom(l)) & P)
ncst(s1,s2) == ∃x:Var. xϵdom(s1) & xϵdom(s2) & ¬(s1(x)=s2(x))
```

Finally, it is handy to have the following definition for the main theorem.

```
match?(t1) == ∀t2:Term. ∃s:Sub. s(t1)=t2 & min(s)
                             & ∀x:Var. xϵdom(s) <=> xϵt1
              ∨ ∀s:Sub. ¬(s(t1)=t2)
```

Thus `match?(t1)` asserts that for every `t2` either there is a minimal matching substitution or there is no matching substitution. The minimality and the condition that a variable be in the domain of the substitution exactly if it occurs in t are both required for the inductive proof to go through. There is also a version of the above definition for lists.

```
match?(l1) ==  ∀l2:Term list. ∃s:Sub. s(l1)=l2 & min(s)
                             & ∀x:Var. xϵdom(s) <=> xϵl1
               ∨ ∀s:Sub. ¬(s(l1)=l2)
```

This list version is defined simply for convenience, as will be made clear when we present the proof of the main theorem.

The following seven simple lemmas are incorporated into the simplifier and never have to be explicitly referenced. The fifth of these may look trivial; this is because Howe used the same display form for term equality as for equality in Var.

```
>> ∀x,y:Var. ∀s:Sub. ∀t:Term. x=y => (<x,t>.s)(y) = t
>> ∀x,y:Var. ∀s:Sub. ∀t:Term. ¬(x=y) => (<x,t>.s)(y) = s(y)
>> ∀x,y:Var. ∀s:Sub. ∀t:Term. x=y => xϵdom((<y,t>).s) <=> True
>> ∀x,y:Var. ∀s:Sub. ∀t:Term. ¬(x=y) => xϵdom((<y,t>).s) <=> xϵdom(s)
>> ∀x,y:Var. x=y <=> x=y
>> ∀f1,f2:Fun. ∀l1,l2:Term list. f1(l1)=f2(l2) <=> f1=f2 & l1=l2
>> ∀x:Var. ∀f:Fun. ∀l:Term list. x=f(l) <=> False
```

The following is the first theorem proved that has an interesting computational
interpretation. Constructively it means that we can decide whether or not two terms
are equal. From the proof of this theorem Nuprl can extract a decision procedure.

```
>> ∀t1,t2:Term. t1=t2 ∨ ¬(t1=t2)
```

The next two theorems establish that the value of a substitution on a term (list) is
characterized by its values on the variables occurring in the term (list).

```
>> ∀s1,s2:Sub. ∀t:Term.
     s1(t)=s2(t) <=> (∀x:Var. xϵt => s1(x)=s2(x))
>> ∀l:Term list. ∀s1,s2:Sub.
      s1(l)=s2(l) <=> (∀x:Var. xϵl => s1(x)=s2(x))
```

The computational content of the next simple theorem is a procedure which "looks
up" a binding for a variable $x$ in a substitution $s$, producing an indication of whether
or not the variable is in the domain of $s$ and in the former case giving the term bound
to $x$.

```
>> ∀x:Var. ∀s:Sub. ¬(xϵdom(s)) ∨ ∃t:Term. xϵdom(s) & s(x)=t
```

For future reference, the name of this theorem is sub_lookup. The proof is short and
is given in its entirety as Figures 1 to 4. The figures show the proof steps as they
would appear to a Nuprl user except that to save space we have removed from some
steps some of the hypotheses that appear in earlier steps.

The first step in the proof is in Figure 1 and is by induction on the list s. This
figure shows the goal sequent, then the rule applied, in this case a tactic, and then
the subgoal sequent with its hypothesis list numbered and displayed vertically. The
asterisk at the top left is Nuprl's indication that the proof is complete, and top is the
tree address of the proof node within the proof tree for the theorem. Note that in this
step the simplifier and autotactic together automatically proved the base case and
left the remaining subgoal in a simplified form. The step applied to this subgoal is
shown in Figure 2. Here we decompose ("eliminate") the pair p (and thin, or discard,
its hypothesis) and then perform some reduction steps to simplify the subgoal. The
next step is shown in Figure 3. In this goal we know that the property we are proving
is inductively true of s, and we have to prove it for <x1,t1>.s. We want to do a

```
* top
>> ∀x:Var. ∀s:Sub. ¬(x∈dom(s)) ∨ ∃t:Term. x∈dom(s) & s(x)=t


BY (On 's' Induction ...+s)


1* 1. x: Var
   2. s: Sub
   3. p: Var#Term
   4. ¬(x∈dom(s)) ∨ ∃t:Term. x∈dom(s) & s(x)=t
   >> ¬(p.1=x ∨ x∈dom(s))
      ∨ ∃t:Term. (p.1=x ∨ x∈dom(s)) & (p.s)(x)=t
```

Figure 1: Proof by induction on s.

```
* top 1
3. p: Var#Term
4. ¬(x∈dom(s)) ∨ ∃t:Term. x∈dom(s) & s(x)=t
>> ¬(p.1=x ∨ x∈dom(s))
   ∨ ∃t:Term. p.1=x ∨ x∈dom(s) & p.s(x)=t


BY (AndThin E 3 THEN Reduce ...)


1* 4. x1: Var
   5. t1: Term
   >> ¬(x1=x ∨ x∈dom(s))
      ∨ ∃t:Term.  (x1=x ∨ x∈dom(s)) & (<x1,t1>.s)(x)=t
```

Figure 2: Let x1 and t1 be the components of p.

```
* top 1 1
3. ¬(x∈dom(s)) ∨ ∃t:Term. x∈dom(s) & s(x)=t
4. x1: Var
5. t1: Term
>> ¬(x1=x ∨ x∈dom(s))
   ∨ ∃t:Term.  (x1=x ∨ x∈dom(s)) & (<x1,t1>.s)(x)=t


BY (Decide 'x1=x' ...+s)


1* 5. x1=x
   6. ¬(x∈dom(s)) ∨ ∃t:Term. x∈dom(s) & s(x)=t
   >> ∃t:Term. t1=t
```

Figure 3: Case analysis on whether or not x1=x.

---

```
    * top 1 1 1
    5. x1=x
    6. ¬(xεdom(s)) ∨ ∃t:Term. xεdom(s) & s(x)=t
    >> ∃t:Term. t1=t

    BY (ITerm 't1' ...)
```

Figure 4: Take t to be t1.

---

case analysis on whether x1=x, so we use the tactic Decide. The negative case was proved automatically, and we are left with a simple subgoal in the other case. The last step (Figure 4) is trivial.

The structure of the program extracted from this proof is determined by the steps we have shown. The first step (using induction) introduces list recursion. In the next step, the induction hypothesis corresponds to a recursive call of the function we are defining. To lookup the value of x in p.s we first decompose the pair p into its components x1 and t1. The use of Decide in the next step corresponds to the introduction of an "if-then-else" expression whose condition is x1=x. In the case where x1 is not equal to x, a recursive call is made (that is, the induction hypothesis is used, automatically). In the other case the value t1 is returned.

The last theorems in Howe's library before the three main theorems (which are discussed below) are added to the simplifier and forgotten.

```
    >> ∀s1,s2:Sub. ∀x:Var. ∀t:Term.
          s1⊆s2 => (<x,t>.s1 ⊆ <x,t>.s2 <=> True)
    >> ∀s1,s2:Sub. ∀x:Var. ∀t:Term.
          s1⊆s2 => ¬(xεdom(s1)) => (s1 ⊆ <x,t>.s2 <=> True)
    >> ∀s1,s2:Sub. ∀x:Var. ∀t:Term.
        s1⊆s2 => xεdom(s2) => s2(x)=t => (<x,t>.s1 ⊆ s2 <=> True)
```

### 5.5.3  matching

The match procedure is mostly the product of the last three theorems in the library.

```
    >> ∀s1,s2:Sub. min(s1) & min(s2)
        => ncst(s1,s2)
            ∨ ∃s:Sub. min(s) & s1⊆s & s2⊆s
                    & ∀x:Var. xεdom(s) => xεdom(s1) ∨ xεdom(s2)
    >> ∀l:Term list. (∀t:l. match?(t)) => match?(l)
    >> ∀t:Term. match?(t)
```

```
          * top
          >> ∀t:Term. match?(t)

          BY (On 't' Induction ...)

          1* 1. x: Var
             >> match?(x)

          2* 1. l: Term list
             2. f: Fun
             3. ∀t:l. match?(t)
             >> match?(f(l))
```

Figure 5: Proof by induction on t.

The names of these theorems in the library are sub_union, lmatch_thm and match_thm respectively. The proof of the first theorem has 23 steps, the proof of the second has 25, and the last has 13. Space limitations prevent us from presenting all of these proofs, so we will instead give the complete proof of the last theorem and just give some of the highlights of the other two. The level of inference in the three proofs is roughly the same.

The proof of match_thm is given in Figures 5 to 17. This proof is rather self explanatory, so not much additional explanation will be given. The first step (Figure 5) is by induction on t. The proof of the base case (where t is a variable) is easy and is contained in Figures 6 and 7. The first step in the proof of the induction step (Figure 8) is to apply the second of the three "main" theorems listed above, and then (Figure 9) we expand the definition of match?. In the next step (Figure 10) we "unroll" t2, doing a case analysis on whether t2 is a variable or an application. The variable case is easy (Figure 11). In the other case (Figure 12) we do a case analysis on whether the function parts of the two terms are the same.

The case where they are not the same is proved in one step (Figure 13). When they are the same we need to know if one argument list matches the other (Figure 14). When it does, we get the required matching substitution (Figure 15). Otherwise, we obtain a contradiction (Figures 16 and 17).

A discussion of the other two main theorems of the library can be found in the paper with Howe [16].


## 5.6   applying formal metamathematics

How can these results be applied to Nuprl itself? One method is to define the type of terms to match exactly the terms of Nuprl. Then the theorems such as the match

```
* top 1
1. x: Var
>> match?(x)

BY (Expand ''matchp'' ...+s)

1* 1. x: Var
   2. t2: Term
   >> ∃s:Sub. s(x)=t2 & min(s) & ∀x1:Var. x1∈dom(s) <=> x=x1
      ∨ ∀s:Sub. ¬(s(x)=t2)
```

Figure 6: Use the definition of match?.

```
* top 1 1
1. x: Var
2. t2: Term
>> ∃s:Sub. s(x)=t2 & min(s) & ∀x1:Var. x1∈dom(s) <=> x=x1
   ∨ ∀s:Sub. ¬(s(x)=t2)

BY (ILeft THENM ITerm '[<x,t2>]' ...+s)
```

Figure 7: The left disjunct is true: take s to be [<x,t2>].

```
* top 2
1. l: Term list
2. f: Fun
3. ∀t:l. match?(t)
>> match?(f(l))

BY (FLemma 'lmatch_thm' [3] ...) THEN Thin 3

1* 3. match?(l)
   >> match?(f(l))
```

Figure 8: By lmatch_thm and 3 we have match?(l).

```
* top 2 1
1. l: Term list
2. f: Fun
3. match?(l)
>> match?(f(l))
```

```
BY (Expand ''matchp'' ...+s)
```

```
1* 3. t2: Term
   4. match?(l)
   >> ∃s:Sub. f(s(l))=t2 & min(s) & ∀x:Var. x∈dom(s) <=> x∈l
      ∨ ∀s:Sub. ¬(f(s(l))=t2)
```

Figure 9: Use the definition of matchp.

```
* top 2 1 1
3. t2: Term
4. match?(l)
>> ∃s:Sub. f(s(l))=t2 & min(s) & ∀x:Var. x∈dom(s) <=> x∈l
   ∨ ∀s:Sub. ¬(f(s(l))=t2)
```

```
BY (On 't2' Unroll ...)
```

```
1* 4. x: Var
   >> ∃s:Sub. f(s(l))=x & min(s) & ∀x:Var. x∈dom(s) <=> x∈l
      ∨ ∀s:Sub. ¬(f(s(l))=x)
```

```
2* 4. 12: Term list
   5. f2: Fun
   >> ∃s:Sub. f(s(l))=f2(12) & min(s) & ∀x:Var. x∈dom(s) <=> x∈l
      ∨ ∀s:Sub. ¬(f(s(l))=f2(12))
```

Figure 10: Case analysis on the term kind of t2.

```
* top 2 1 1 1
4. x: Var
>> ∃s:Sub. f(s(l))=x & min(s) & ∀x:Var. x∈dom(s) <=> x∈l
   ∨ ∀s:Sub. ¬(f(s(l))=x)
```

```
BY (IRight ...+s)
```

Figure 11: In this case there are no matching substitutions.

```
* top 2 1 1 2
4. 12: Term list
5. f2: Fun
>> ∃s:Sub. f(s(1))=f2(12) & min(s) & ∀x:Var. x∈dom(s) <=> x∈1
   ∨ ∀s:Sub. ¬(f(s(1))=f2(12))

BY (Decide 'f=f2' ...)

1* 6. f=f2
   >> ∃s:Sub. f(s(1))=f2(12) & min(s) & ∀x:Var. x∈dom(s) <=> x∈1
      ∨ ∀s:Sub. ¬(f(s(1))=f2(12))

2* 6. ¬(f=f2)
   >> ∃s:Sub. f(s(1))=f2(12) & min(s) & ∀x:Var. x∈dom(s) <=> x∈1
      ∨ ∀s:Sub. ¬(f(s(1))=f2(12))
```

Figure 12: Either f=f2 or not.

```
* top 2 1 1 2 2
3. match?(1)
4. 12: Term list
5. f2: Fun
6. ¬(f=f2)
>> ∃s:Sub. f(s(1))=f2(12) & min(s) & ∀x:Var. x∈dom(s) <=> x∈1
   ∨ ∀s:Sub. ¬(f(s(1))=f2(12))

BY (IRight ...+s)
```

Figure 13: Clearly no match in this case.

```
* top 2 1 1 2 1
3. match?(1)
6. f=f2
>> ∃s:Sub. f(s(1))=f2(12) & min(s) & ∀x:Var. xϵdom(s) <=> xϵl
   ∨ ∀s:Sub. ¬(f(s(1))=f2(12))


BY (Expand ''lmatchp'' THEN HypCases ['12'] 3 ...)


1* 6. ∃s:Sub. s(1)=12 & min(s) & ∀x:Var. xϵdom(s) <=> xϵl
   >> ∃s:Sub. f(s(1))=f2(12) & min(s) & ∀x:Var. xϵdom(s) <=> xϵl
      ∨ ∀s:Sub. ¬(f(s(1))=f2(12))


2* 6. ∀s:Sub. ¬(s(1)=12)
   >> ∃s:Sub. f(s(1))=f2(12) & min(s) & ∀x:Var. xϵdom(s) <=> xϵl
      ∨ ∀s:Sub. ¬(f(s(1))=f2(12))
```

Figure 14: By 3 either 1 matches 12 or not.

```
* top 2 1 1 2 1 1
5. f=f2
6. ∃s:Sub. s(1)=12 & min(s) & ∀x:Var. xϵdom(s) <=> xϵl
>> ∃s:Sub. f(s(1))=f2(12) & min(s) & ∀x:Var. xϵdom(s) <=> xϵl
   ∨ ∀s:Sub. ¬(f(s(1))=f2(12))


BY ((OnLast (SomeE ''s'') THEN ILeft THENM ITerm 's' ...) ...sc)
```

Figure 15: Let s match 1 with 12. Then s matches f(1) and f2(12).

```
* top 2 1 1 2 1 2
5. f=f2
6. ∀s:Sub. ¬(s(1)=12)
>> ∃s:Sub. f(s(1))=f2(12) & min(s) & ∀x:Var. xϵdom(s) <=> xϵl
   ∨ ∀s:Sub. ¬(f(s(1))=f2(12))


BY (IRight ...+s)


1* 5. s: Sub
   6. s(1)=12
   7. ∀s1:Sub. ¬(s1(1)=12)
   >> False
```

Figure 16: 1 does not match 12, so there is no matching substitution.

```
* top 2 1 1 2 1 2 1
5. s: Sub
6. s(1)=12
7. ∀s1:Sub.  ¬(s1(1)=12)
8. f=f2
>> False

BY (EOn 's' 7 ...) THEN (Contradiction ...)
```

Figure 17: A match for `f(1)` and `f2(12)` would match 1 and 12.

theorem apply to Nuprl itself. In this form they are *enlightening* but not *useful.* To make them useful it must be possible to apply them in proving theorems. One way to do this is outlined in Howe's thesis [30]. Another method of proceeding is to formalize the metatheory of Nuprl, making the types of terms and proof two of the basic types. This is the approach we have taken here. It is this requirement that necessitated the material in the preceeding section.

# 6   Logic Programming in Type Theory

The use of constructive type theory in defining basic computing concepts is new, and the community is learning how to improve the exact definitions of the basic concepts. A similar process went on in set theory for decades before the elegant and polished accounts that we now read were developed. Some work along these lines is still being done.

In the case of type theory we can see steady development and improvement. For example the treatment of inductively defined types has improved over the past five years. The first accounts, [13,37] led to an implementation in Nuprl. Experience with these types revealed that the full notion of parameterized mutually inductive types was awkward. Subsequent work by Coquand, Paulin, and Dybjer has led to a more elegant and polished set of rules for the general case [18] [44,20], and extensive experience with the simple unparameterized case has shown just how vital the concept is in defining basic concepts [30]. Likewise new type constructors have been studied such as intersection types [Backhouse] [3] and partial types [48,14,47]. In this section we look briefly at one other possible improvement to these type theories, one that is germane to metalevel programming, namely the use of logic variables. Although these are sometimes considered to be *metavariables*, we outline an approach that treats them in the object logic.

## 6.1   refinement logics

When proving a theorem we often work backwards from the goal to axioms or assumptions. This is sometimes called *top-down development.* For instance we argue that $((A \Rightarrow B) \,\&\, (B \Rightarrow C)) \Rightarrow (A \Rightarrow C)$ by saying we need a proof of $(A \Rightarrow C)$ from a proof of $(A \Rightarrow B) \,\&\, (B \Rightarrow C)$. To prove $A \Rightarrow C$, we need a proof of $C$ from $A$ and the other assumptions. We can prove $C$ if we can prove $B$ because we are assuming $B \Rightarrow C$. But we can prove $B$ if we can prove $A$ because we assume $A \Rightarrow B$. But we assume $A$, so there is a proof.

The *tableau* proof style [49] is top-down, and Nuprl's logic is top-down. This kind of logic was also called a *refinement logic* by Bates [6]. The convenient way to present the rules of a refinement logic is to state the goal first, then the rule name and any information needed to determine the subgoals, and finally the subgoals. So for example the rule for proving $A\&B$ from assumption $H$ is written in Nuprl as

$$H >> \quad A \,\&\, B \text{ By and intro}$$
$$1. \ H >> A$$
$$2. \ H >> B.$$

A rule which requires information beyond the rule name to generate the subgoals is

$$H, \ \forall x\!:\! A.B >> G \text{ by all elimination on } a$$
$$H >> a\epsilon A$$
$$H, \forall x\!:\! A.B \ , \ B[a/x] >> G$$

We need the name of the element of $A$ to generate the subgoals.

In Nuprl the existential introduction rule also requires extra information. It is

$$H >> \quad \exists x\!:\! A.B \text{ By intro } a$$
$$H >> a\epsilon A$$
$$H >> B[a/x].$$

This rule is inconvenient because it interrupts the top-down development, forcing the user to choose $a$ early in the proof.

One of the powerful features of logic programming languages, which are also top-down, is that they employ the concept of a *logic variable* to delay existential choices. If Nurpl had logic variables, for now let's write them as capitals, $X, Y, Z, \ldots$, then the existential introduction rule could be written as

$$H >> \quad \exists x : A.B \text{ By intro } X$$
$$1. \ H >> X \epsilon A$$
$$2. \ H >> B[X/x].$$

The key assumption underlying the use of logic variables is that they are instantiated simultaneously at all occurrences. So if in the subproof of $B[X/x]$ we learn that $X$ must be a particular value $a$, and we set $X$ to $a$, then the first subgoal becomes $H >> a\epsilon A$.

We will discuss below one way to treat logic variables in refinement logics, and we apply it in a logic programming setting. But first we consider the Prolog computation mechanism as it can be explained in a refinement logic like Nuprl without logic variable.

## 6.2  Prolog proof procedure

A "program" in Prolog is a set of inductive definitions of relations. For instance here is a Prolog form of the primitive recursive function definition

$$f(0, y) = y$$
$$f(s(n), y) = h(n, f(n, y)).$$

It is a pair of relations

$$\forall y\, R(0, y, y)$$
$$\forall u, z, y(R(z, y, v) \Rightarrow R(s(z), y, h(z, v))).$$

In general a program can consist of several relations or program *clauses*, say $C_1, \ldots, C_n$. Each of them defines a relation in the form $\forall \overline{x}.(A_1 \&, \ldots, \& A_n \Rightarrow B)$ where the $A_i, B$ are atomic formulas, $n$ may be 0. We say $B$ is the "head" of its clause.

A goal clause has the form $\forall \overline{x}.(A \Rightarrow \exists \overline{y}.B)$. A constructive proof of this will find functions which compute a witness for this existential quantifier. The form of the function is simply an expression for the vector $\overline{y}$ of values in terms of the primitives. The expression is presented as a substitution for $\overline{y}$. The proof procedure can find many expressions, so it is finding more than one function.

The Prolog proof procedure can be expressed without logic variables. It is just a specific search which uses unification and backchaining. The unification can be done "under the existential quantifiers." Let Prog be the primitive recursive program above. Consider this Prolog goal

$$Prog \ >> \ \exists u.R(s(s(0)), b, u)$$

The first step is to unify $R(s(s(0)), b, u)$ against the head's of the program clauses. There are only two choices, one succeeds.

$$R(s(s(0)), b, u) \quad \text{unifies with}$$

$$R(s(z), y, h(z, v)).$$

The resulting substitution is $z := s(0), y := b$ , $u := h(s(0), v)$. We need to know whether there is a value for $v$. This generates a subgoal of the form

$$\exists v_2.R(s(0), b, v_2)$$

by backchaining on the second program clause. We can express the proof obligation by sequencing in this subgoal. The proof starts as

$$Prog >> \exists v.R(s(s(0)), b, v) \text{ By seq } \exists v_2.R(s(0), b, v_2)$$

$$1. \ Prog, \exists v_2.R(s(0), b, v_2) >> \exists u.R(s(s(0)), b, u)$$

$$2. \qquad\qquad Prog \quad >> \exists v_2.R(s(0), b, v_2).$$

Now the interesting point is that the entire first subgoal can be proved simply by elimination on $\exists v_2$ and backchaining. The first step results in the subgoal

$$Prog, \ v_2 : N, \ R(s(0), b, v_2) >> \exists u.R(s(s(0)), b, u).$$

Then we substitute properly in the second program clause (the rule is $\forall$-elimination) to obtain

$$Prog, \ v_2 : N, \ R(s(0), b, v_2), \ R(s(0), b, v_2) \Rightarrow R(s(s(0)), b, h(s(0), v_2))$$
$$>> \exists u. \ R(s(s(0)), b, u).$$

Now we know that is we take $u$ to be $h(s(0), v_2)$ the proof is finished by simple backchaining.

Notice that the definition of $u$ as $h(s(0), v_2)$ is not complete until we know $v_2$, but finding it has been relegated to the second subgoal, and substituting it into $h(s(0), v_2)$ is part of the definition of the sequent rule. So the entire problem has $h$ reduced to proving the second subgoal, and it *is of the exact same form* as the original goal. This means we can apply the same method. We call this method the *Prolog search procedure*.

In Nuprl this method can be coded as a *proof tactic*, so we can see Prolog evaluation as the result of trying a fixed tactic on goals of the form $Prog >> \exists v.B$. The tactic is guaranteed to generate only subgoals of this form; the other subgoals are automatically proved.

## 6.3 proof expressions in refinement logics

Before we can discuss the role of logic variable in refinement logics we need to consider how the justification for the goal of a sequent can be written as an algebraic expression. First consider the simple case of a step taken by *and introduction*.

$$H >> \ A\&B \text{ By intro}$$
$$1. \ H >> A$$
$$2. \ H >> B.$$

Let us write after each proved sequent an expression that codes why it is true, that is how it was proved; call these *proof expressions*. If we are looking at a complete proof of $H >> A\&B$, then $H >> A$ is proved and so is $H >> B$. Suppose the proof expression for the first is $a$ and for the second $b$. Then we can propagate up the proof tree an expression for $H >> A\&B$; we synthesize it out of expression for the subproofs. In this case we can take $pair(a; b)$ as the expression required; and we show these expressions as in this example

$$H >> \ A\&B \ : \ pair(a; b)$$
$$1. \ H >> A : a$$
$$2. \ H >> B : b.$$

We read the synthesis of proof expressions *bottom-up* and the subgoals top-down. Here are the rules for the Intuitionistic Predicate Calculus written with proof expressions indicated.

$H, x : A >> A : hyp(x)$

$H >> A\&B \ : \ pair(a; b)$       $H, s : A\&B >> G \ : \ spread(z; x, y, g)$
    $H >> A : a$                       $H, x : A, y : B >> G = g$
    $H >> B : b$

$H >> A \vee B : inl(a)$          $H, z : A \vee B >> G \ : \ decide(z; x.g_1; y.g_2)$
    $H >> A : a$                     $H, \ x : A >> G : g_1$

------------------

$or \ H >> A \vee B \ : \ inr(b)$      $H, y : B >> G : g_2$
    $H >> B : b$

$H >> A \Rightarrow B \ : \ \lambda(x.b)$       $H, f : A \Rightarrow B >> G \ \ : \ ap(f; a; y.g)$
    $H, x : A >> B : B$             $H, f : A \Rightarrow B >> A : a$
                             $H, y.B >> G : g$

$H, x : false >> G \ : \ any(x)$

$H >> G \ seq(a; x.g)$
   $H >> A : a$
   $H, \ x : A >> g$

$H >> \exists x : A.B \ : \ pair(a; p)$      $H, z : \exists x : A.B >> G \ : \ spread(z; x, y.g)$
    $H >> A : a$                      $H, xA, y : P >> G : g$
    $H >> B[a/x] : p$

$H >> \forall x : A.B \ : \ \lambda(x.b)$      $H, f : \forall x : A.B >> G \ : \ ap(f; a; y.g)$
    $H, x : A >> B$              $H, f.\forall x : A.B >> A : a$
                      $H, y.B >> G : g$

## 6.4   logic variables

One approach to introducing logic variables into a refinement logic is suggested by looking at the existential introduction rule. If we want to introduce a variable as a witness for the quantifier, then it needs to be typed and "declared." There does not seem to be a natural way to do it in the pure logic, but in the presence of proof expression we can try this idea. We allow variables to range over these proof expressions. So a possible existential introduction rule is

$$H >> \quad \exists x : A.B \text{ by intro } X$$
$$1. \ H >> A : X$$
$$2. \ H >> P[X/x]$$

The advantage of this approach is that $X$ is available in the entire proof tree, just as we want, without having to add it to the hypothesis lists. Also we need just two rules for using these variables. First we need a definition rule saying that $X$ satisfies these conditions, and it is of type $A$ and satisfies $P$. We also need a rule to refine it or specify its value further.

In order for this notion to be flexible enough, we nee to be able to indicate the dependence if $X$ on variables declared in $H$. Then we could see $X$ as a name for the function from $H$ into $A$. So the complete account of logic variables must treat them as *second-order* variables. We write $X[z_i, \ldots, z_b]$ where $z_i$ are declared in $H$. Now $X[\overline{z}]$ can be used in those contexts that provide the values needed for $\overline{z}$.

**Acknowledgements**   I would like to thank my colleagues Stuart Allen, Douglas Howe and Bill Aitken of Cornell who were coauthors on the two papers from which much of this lecture material came. I especially want to thank Doug Howe whose implementation of the matching algorithm was central to one of the sections. Indeed I have used snapshots of the proof that he prepared to explain that experience, and in this article I have borrowed from the organization of that material for our joint paper cited often in this text.

I would also like to thank Randy Pollack for many hours of stimulating discussions of type theory and especially for his help with formulating the concept of a logic variable. We may produce joint results on this topic in the near future showing how the concept defined here can be used to describe higher-order unification in a refinement logic.

# References

[1] S. Allen, R. Constable, and D. Howe. Reflecting the open-ended computation system of constructive type theory. In *Algebra, Logic and Computation*. NATO ASI Series, 1990.

[2] S. Allen, R. Constable, D. Howe, and W. Aitken. The semantics of reflected proof. *Proc. of Fifth Symp. on Logic in Comp. Sci., IEEE*, June 1990.

[3] S. F. Allen. *A non-type-theoretic semantics for type-theoretic language.* PhD thesis, Cornell University, 1987.

[4] P. Audebaud. Partial objects in the calculus of constructions. In *Sixth Symp. on Logic in Computer Science, IEEE*, Vrije University, Amsterdam, The Netherlands, 1991.

[5] D. Basin and R. Constable. Meta-logical frameworks. In *Proc. of the Second Workshop on Logical Frameworks*, Edinburgh, UK, June 1991.

[6] J. L. Bates. *A logic for correct program development.* PhD thesis, Cornell University, 1979.

[7] E. Bishop and D. Bridges. *Constructive Analysis.* NY:Springer-Verlag, 1985.

[8] W. Bledsoe and D. Loveland. *Automated Theorem Proving: After 25 Years.* American Math Soc., 1984.

[9] R. Boyer and J. Moore. Metafunctions: proving them correct and using them efficiently as new proof procedures. In *The Correctness Problem in Computer Science* , pages 103–84. NY:Academic Press, 1981.

[10] W. Buchholtz et al. Iterated inductive definitions and subsystems of analysis. *Recent Proof-Theoretical Studies, Lecture Notes in Mathematics*, 897, 1981.

[11] A. Bundy. *The Computer Modelling of Mathematical Reasoning.* NY:Academic Press, 1983.

[12] A. Bundy. A broader interpretation of logic in logic programming. In *Proc. 5th Sympo. on Logic Programming*, 1988.

[13] R. Constable and N. Mendler. Recursive Definitions in Type Theory. In *Proc. of Logics of Prog. Conf.*, pages 61–78, January 1985. Cornell TR 85-659.

[14] R. Constable and S. F. Smith. Computational foundations of basic function theory. In *Third Symp. on Logic in Somp. Sci.* IEEE, 1988. (Cornell TR 88-904).

[15] R. L. Constable et al. *Implementing Mathematics with the Nuprl Development System.* NJ:Prentice-Hall, 1986.

[16] R. L. Constable and D. J. Howe. Implementing metamathematics as an approach to automatic theore m proving. In R. Banerji, editor, *Formal Techniques in Artificial Intelligence: A Source Boo k*, pages 45–76. Elsevier Science Publishers (North-Holland), 1990.

[17] T. Coquand and G. Huet. The calculus of construction. *Information and Computation*, 76:95–120 , 1988.

[18] T. Coquand and C. Paulin-Mohring. Inductively defined types. preprint.

[19] M. Davis and J. T. Schwartz. Metamathematical extensibility for theorem verifiers and proof checkers. *Comp. Math. with Applications*, 5:217–230, 1979.

[20] P. Dybjer. Inductive sets and families in Martin-Löf's type theory and their set-theoretic semantics. In *Proceedings of the B.R.A. Workshop on Logical Frameworks*, Sophia-Antipolis, France, June 1990. (To appear.).

[21] S. Feferman. Formal theories for transfinite iterations of generalized inductive definitions and some subsystems of analysis. In *Proc. Conf. Intuitionism and Proof Theory*, pages 303–326, Buffalo, NY, 1970. North-Holland.

[22] J. Gallier. Logic for computer science. In *Foundations of Automatic Theorem Proving*. Harper and Row, 1986.

[23] J.-Y. Girard. Une extension de l'interpretation de godel a l'analyse, et son application a l'elimination des coupures dans l'analyse et la theorie des types. In *2nd Scandinavian Logic Symp.*, pages 63–69. NY:Springer-Verlag, 1971.

[24] J.-Y. Girard, Y. Lafont, and P. Taylor. *Proofs and Types*. Cambridge University Press, 1988.

[25] M. Gordon. Hol: A machine oriented formalization of higher order logic. Technical report, Cambridge University, 1985. TR 68.

[26] M. Gordon, R. Milner, and C. Wadsworth. Edinburgh LCF: a mechanized logic of computation. *Lecture Notes in Computer Science*, 78, 1979.

[27] T. Griffin. A formulas-as-types notion of control. In *POPL*, 1990.

[28] S. Hayashi and H. Nakano. *PX: A Computational Logic*. Foundations of Computing. MIT Press, Cambridge, MA, 1988.

[29] D. Howe and J. Chirimar. Nuprl mathematics library. unpublished, 1989.

[30] D. J. Howe. *Automating Reasoning in an Implementation of Constructive Type Theory*. PhD thesis, Cornell University, 1988.

[31] D. J. Howe. Equality in lazy computation systems. *Proc. Fourth Symp. Logic in Computer Science, IEEE*, 1989.

[32] T. B. Knoblock and R. L. Constable. Formalized metareasoning in type theory. In *Proc. of the First Annual Symp. on Logic in Computer Science*. IEEE , 1986.

[33] J. Lipton. Logic programming in the nuprl type theory environment. Technical report, Cornell University, Ithaca, NY, 1991. To appear as a technical report, summer 1991.

[34] Z. Luo. Ecc, an extended calculus of construction. In *Proc. Fourth Symp. on Logics in Computer Science, IEEE*, Washington, DC, June 1989.

[35] P. Martin-Lof. An intuitionistic theory of types: predicative part. In *Logic Colloquium '73.*, pages 73–118. Amsterdam:North-Holland, 1973.

[36] P. Martin-Lof. Constructive mathematics and computer programming. In *Sixth International Congress for Logic, Methodology, and Philosophy of Science*, pages 153–75. Amsterdam:North Holland, 1982.

[37] P. Mendler. Recursive Types and Type Constraints in Second-Order Lambda Calculus. *Proc. in Second Symp. on Logic in Comp. Sci., IEEE*, pages 30–36, June 1987.

[38] P. Mendler. *Inductive Definition in Type Theory*. PhD thesis, Cornell University, Ithaca, NY, 1988.

[39] D. Miller and G. Nadathur. A logic programming approach to manipulating formulas and programs. In *IEEE Symp. on Logic Programming*. ACM, 1987.

[40] C. Murthy. *Extracting Constructive Content for Classical Proofs*. PhD thesis, Cornell University, Dept. of Computer Science, 1990. TR 89-1151.

[41] C. Murthy. An evaluation semantics for classical proofs. In *LICS, '91*, Amsterdam, The Netherlands, July 1991.

[42] C. Murthy and J. Russell. A constructive proof of higman's lemma. Technical Report TR 89-1049, Cornell University, Ithaca, NY 14853, January 1989.

[43] B. Nordstrom, K. Peterson, and J. Smith. *Programming in Martin-Lof's Type Theory*. Oxford Sciences Publication, Oxford, 1990.

[44] C. Paulin-Mohring. *Extraction in the calculus of constructions*. PhD thesis, University of Paris VII, 1989.

[45] F. Pfenning. Elf: a language for logic definition and verfied metaprogramming. *LICS*, pages 313–321, 1989.

[46] N. Shanker. Towards mechanical metamathematics. *J. Automated Reasoning*, 1(4):407–434, 1985.

[47] S. Smith. *Partial Objects in Type Theory*. PhD thesis, Cornell University, Ithaca, NY, 1989.

[48] S. Smith and R. L. Constable. Partial objects in constructive type theory. In *Symposium on Logic in Computer Science.*, pages 183–93. Washington, D.C.:IEEE., 1987.

[49] R. M. Smullyan. *First–Order Logic.* Springer–Verlag, New York, 1968.

[50] A. Troelstra. Metamathematical investigation of intuitionistic mathematics. *Lecture Notes in Mathematics*, 344, 1973.

[51] R. Weyrauch. Prolegomena to a theory of formal reasoning. *Artificial Intelligence*, 13:133–170., 1980.

[52] L. Wos, R. Overbeek, L. Ewing, and J. Boyle. *Automated Reasoning.* Prentice–Hall, Englewood Cliffs, NJ, 1984.

# Laws of Programming

C.A.R. Hoare[1], I.J. Hayes[3], He Jifeng[1], C.C. Morgan[1], A.W. Roscoe[1], J.W. Sanders[1], I.H. Sorenson[2], J.M. Spivey[1], and B.A. Sufrin[1]

[1]Programming Research Group, Oxford University Computing Laboratory, 11 Keble Road, Oxford, OX1 3QD, England.

[2]BP International Limited, Sunbury Research Centre, Chertsey Road, Sunbury on Thames, Middlesex, TW16 7LN, England.

[3]Department of Computer Science, University of Queensland, St. Lucia, Queensland 4067, Australia.

**Abstract:** A complete set of algebraic laws is given for Dijkstra's nondeterministic sequential programming language. Iteration and recursion are explained in terms of Scott's domain theory as fixed points of continuous functionals. A calculus analogous to weakest preconditions is suggested as an aid to deriving programs from their specifications.

**Keywords:** Algebra, imperative programming

## Introduction

Some computer scientists have abandoned the search for rational laws to govern conventional procedural programming. They tend to recommend the use of functional programming [2] or logic programming [10] as alternatives. Here we shall substantiate a claim that conventional procedural programs are mathematical expressions, and that they are subject to a set of laws as rich and elegant as those of any other branch of mathematics, engineering, or natural science.

Furthermore, we suggest that a comprehensive set of algebraic laws serves as a useful formal definition (axiomatization) of a set of related mathematical notations, and specifically of a programming language — a suggestion due originally to Igarishi [8]. The algebraic laws provide an interface between the user of the language and the mathematical engineer who designs it. Of course, the mathematician should also design a model of the language, to check completeness and consistency of the laws, to provide a framework for the specifications of programs, and for proofs of correctness [1, 3, 5].

Here are some of the familiar laws of arithmetic, which apply to multiplication of real numbers:

(1) Multiplication is *symmetric*, or in symbols,

$$x \times y = y \times x, \quad \text{for all numbers } x \text{ and } y.$$

It is conventional in quoting laws to omit the phrase "for all $x$ and $y$ in the relevant set".

(2) Multiplication is *associative*, or in symbols,

$$x \times (y \times z) = (x \times y) \times z.$$

It is conventional to omit brackets for associative operators and write simply $x \times y \times z$

(3) Multiplication by zero always gives zero:

$$0 \times x = 0$$

Zero is said to be a *fixed point* or *zero* of multiplication.

(4) Multiplication by one leaves a number unchanged:

$$1 \times x = x.$$

The number one is said to be an *identity* or a *unit* for multiplication.

(5) Division is the *inverse* of multiplication:

$$y \times (x/y) = x, \quad \text{provided } y \neq 0.$$

Another law relating multiplication and division is $\quad z/(x \times y) = (z/x)/y,$ provided $y \neq 0$ and $x \neq 0$.

(6) Multiplication *distributes* through addition:

$$(x + y) \times z = (x \times z) + (y \times z).$$

It is usual for brackets to be omitted on the right-hand side of this equation, on the convention that a distributive operator binds tighter than the operator through which it distributes.

(7) Multiplication by a nonnegative number is monotonic, in the sense that it preserves ordering in its other operand, or in symbols,

$$x \leq y \Rightarrow x \times z \leq y \times z, \quad \text{provided } z \geq 0.$$

If either factor is reduced, the value of the product does not increase.

(8) Multiplication is *continuous* in the sense that it preserves the limit of any convergent sequence of numbers:

$$\left( \lim_{n \to \infty} x_n \right) \times y = \lim_{n \to \infty} (x_n \times y), \qquad \text{provided } x_n \text{ converges.}$$

(9) If we define

$$x \cap y \;=\; \text{the lesser of } x \text{ and } y$$
$$x \cup y \;=\; \text{the greater of } x \text{ and } y,$$

then we have the following laws:

$$
\begin{array}{lll}
x \cap y & = \; y \cap x & \text{idempotence} \\
(x \cap y) \geq z & \equiv \; x \geq z \wedge y \geq z & \text{greatest lower bound} \\
(x \cup y) \leq z & \equiv \; x \leq z \wedge y \leq z & \text{least upper bound} \\
x \cap (y \cup z) & = \; (x \cap y) \cup (x \cap z) & \text{distribution} \\
x \cup (y \cap z) & = \; (x \cup y) \cap (x \cup z) & \text{distribution.}
\end{array}
$$

The mathematician or engineer will be intimately familiar with all these laws, having used them frequently and intuitively. The applied mathematician, scientist, or engineer will also be familiar with many relevant laws of nature and will use them explicitly to find solutions for otherwise intractable problems. Ignorance of such laws would be regarded as a disqualification from professional practice. What then are the laws of programming, which provide the formal basis for the profession of software engineering?

Many programmers may be unable to quote even a single law. Many who have suffered the consequences of unreliable programs may claim that programmers do not observe any laws. This accusation is both unfair and inappropriate. The laws of programming are like the laws of arithmetic. They describe the properties of programs expressible in a suitable notation, just as the laws of arithmetic describe the properties of numbers, for example, in decimal notation. It is the responsibility of programming language designers and implementors to ensure that programs obey the appropriate collection of useful, elegant, and clearly stated laws.

The designers of computer hardware have a similar responsibility to ensure that their arithmetic units obey laws such as the monotonicity of multiplication (7). Regrettably, several computer designs have failed to do so. Similarly, many current programming languages fail to obey even the most obvious laws such as those expounded in this paper. Occam [11] is one of the first languages to be deliberately designed to obey such mathematical laws. The language used in this paper is simpler than Occam, in that it omits communication and concurrency. Laws that are not valid in the more complex language will be noted.

## The Language

In order to formulate mathematical laws, it is necessary to introduce some notation for describing programs. We shall use a notation (programming language) that is especially concise and suitable for its purpose, based on the language introduced in [4]. It has three kinds of primitive command and five methods of composing larger commands (programs). The *SKIP* command is denoted II. Execution of this command terminates successfully, leaving everything unchanged.

The *ABORT* command is denoted $\perp$. It places no constraint on the behavior or misbehavior of the executing machine, which may do anything, or fail to do anything; in particular, it may fail to terminate. Thus $\perp$ represents the behavior of a broken machine, or a program that has run wild. The most important property of *ABORT* is that it is a program that no one would ever want to use or write. It is like a singularity in a mathematical system that must, at all costs, be avoided by the competent engineer. In order to prove the absence of such an error, one must use a mathematical theory that admits its existence.

In the *Assignment* command, let $x$ be a list of distinct variables, and let E be a list of the same number of expressions. The assignment $x := $ E is executed by evaluating all the expressions of E (with all variables taking their most recently assigned values) and then assigning the value of each expression to the variable at the same position in the list $x$. This is known as multiple or simultaneous assignment. We assume that expressions are evaluated without side effects and stipulate that the values of the variables in the list $x$ do not change until all the evaluations are complete. For simplicity, we shall also assume that all operators in all expressions are defined for all values of their arguments, so that the evaluation of an expression always terminates successfully. This assumption will be relaxed in the section on undefined expressions.

*Sequential composition.* If P and Q are programs, (P; Q) is a program that is executed by first executing P. If P does not terminate, neither does (P; Q). If and when P terminates, Q is started; and then (P; Q) terminates when Q does.

*Conditional.* If P and Q are programs and b is a Boolean expression, then (P◁b▷Q) is a program. It is executed by first evaluating b. If b is true, then P is executed, but if b is false, then Q is executed instead. The more usual notation for a conditional is **if** b **then** P **else** Q. We have chosen an infix notation ◁b▷ because it simplifies expression of the relevant algebraic laws.

*Nondeterminism.* If P and Q are programs, then $(P \cup Q)$ is a program that is executed by executing either P or Q. The choice between them is arbitrary. The programmer has deliberately postponed the decision, possibly to a later stage in the development of the program, or possibly has even delegated the decision to the machine that executes the program.

*Iteration.* If P is a program and b is a Boolean expression, then $(b * P)$ is a program. It is executed by first evaluating b; if b is false, execution terminates successfully, and nothing is changed. But, if b is true, the machine proceeds to execute P; $(b * P)$. A more conventional notation for iteration is **while** b **do** P.

*Recursion.* Let X be the name of a program that we will define by recursion, and let $F(X)$ (containing occurrences of the name X) be a program text defining its behavior. Then $\mu X.F(X)$ is a program that behaves like $F(\mu X.F(X))$; that is, all recursive occurrences of the program name have been replaced by the whole recursive program. This fact is formalized in the following law:

$$\mu X.F(X) = F(\mu X.F(X)).$$

In mathematics, substitution of equals is always allowed and may be repeated indefinitely:

$$\mu X.F(X) = F(\mu X.F(X)) = F(F(\mu X.F(X))) = \dots$$

This is essentially the way that recursively defined programs are executed by computer. Of course, iteration is only a special case of recursion:

$$b * P = \mu X.(P;\ X) \triangleleft b \triangleright \text{II}.$$

Iteration is simpler and more familiar than general recursion, and so is worth treating separately. An understanding of (or liking for) recursion is not needed for appreciation of this article.

As an example of the use of these notations, here is a program to compute the quotient $q$ and remainder $r$ of division of nonnegative $x$ by positive $y$. It offers a choice of methods, one of which terminates when $y = 0$.

$$(q, r := 0, x; (r \geq y * (q, r := q + 1, r - y))) \cup ((q, r := x \div y, x \text{ rem } y) \triangleleft y \neq 0 \triangleright q := 0).$$

The free use of brackets around the subexpressions of a computer program may seem unfamiliar, but follows naturally from our decision to treat programs as mathematical formulas. Sometimes it is convenient to omit brackets, provided they can be reinserted by precedence rules such as the following:

$$
\begin{array}{ll}
, & \text{binds tightest} \\
:= & \\
* & \\
; & \\
\triangleleft \text{b} \triangleright & \\
\cup & \text{binds loosest.}
\end{array}
$$

Normal arithmetic operators bind tightest of all. Thus the example quoted above could have been written without any brackets.

The notations of our language can be defined in terms of Dijkstra's language of guarded commands:

$$
\begin{aligned}
\text{P} \cup \text{Q} &= \quad \textbf{if } \text{true} \rightarrow \text{P} \,\square\, \text{true} \rightarrow \text{Q } \textbf{fi} \\
\text{P} \triangleleft b \triangleright \text{Q} &= \quad \textbf{if } b \rightarrow \text{P} \,\square\, \neg b \rightarrow \text{Q } \textbf{fi} \,, \text{ where } \neg \text{b is the negation of } b \\
\text{b} * \text{P} &= \quad \textbf{do } b \rightarrow \text{P } \textbf{od} \,.
\end{aligned}
$$

Conversely, guarded commands can be defined in terms of the notations given above, for example:

$$
\begin{aligned}
\textbf{if } \text{b} \rightarrow \text{P} \,\square\, \text{c} \rightarrow \text{Q } \textbf{fi} &= ((\text{P} \cup \text{Q}) \triangleleft \text{c} \triangleright \text{P}) \triangleleft \text{b} \triangleright (\text{Q} \triangleleft \text{c} \triangleright \bot) \\
\textbf{do } \text{b} \rightarrow \text{P} \,\square\, \text{c} \rightarrow \text{Q } \textbf{od} &= (\text{b} \vee \text{c}) * (\textbf{ if } \text{b} \rightarrow \text{P} \,\square\, \text{c} \rightarrow \text{Q } \textbf{fi }).
\end{aligned}
$$

Thus any program expressed in Dijkstra's language can be translated into our language, but this may cause an explosion in the length of the code for guarded command sets. Any program expressed in our notation, but restricted to iteration as the only form of recursion, can be translated into Dijkstra's language. We make no claim of notational superiority, except that our notation permits more succinct expression of some of the laws.

This description of the commands of our programming language is quite informal and deliberately fails to give a mathematical definition of the concept of a program. Experienced programmers will understand our intention. It would be inappropriate to postpone a study of the laws of arithmetic until after giving the traditional foundationary definitions of the concept of a number (e.g. that a cardinal is a class of sets related by one-one functions). Indeed, even a mathematician gains a clearer, deeper, and more useful understanding of a concept by the study of its properties rather than its formal definition. One of the objectives of this article is to propose that the details of the design of a programming language can also be neatly explained by algebraic laws.

## Technical Notes

It is usual in theoretical texts to make a sharp distinction between concrete notations like numerals (expressed in a variety of bases) and the abstract objects for which they stand (e.g. the natural numbers). In this article the word *program* will stand for an abstract concept, roughly equated with *the range of possible observable effects of executing its text in a variety of initial states*. When we wish to emphasize the specific concrete text being manipulated by program transformation, we will call it a *program text*. But, in general, we shall take a relaxed attitude toward these formal distinctions, as do most applied mathematicians and engineers.

The laws given in this article can be regarded as a completely formal algebraic specification of our programming language. The laws are strong enough to permit each program, not involving recursion, to be reduced to a normal form. A natural partial ordering is defined for normal forms, and programs can be identified with ideals in this partial ordering. The ideal construction is somewhat reminiscent of the Dedekind definition of a real number as a certain set of rationals.

Among theoreticians, it is common to identify a non-deterministic program with a relation, namely, the set of all pairs of states of a machine such that if the program starts in the first state of the pair then it can end in the second state. For example, II would be the identity relation, and *ABORT* would be the universal relation. However, this definition would not be correct for our language, since it would invalidate several of the laws (e.g. see (3) under "Sequential Composition" and (4) under "Limits"). In order to make these laws true, it is necessary to define programs as a particular *subset* of relations that have special properties. For example,

- a program is a *total* relation;

- the image of each state is finite or universal.

Nontermination has to be represented by a fictitious "state at infinity" that can be "reached" only by a non-terminating program. Also, if the fictitious state is in the image of a state, then that image is universal. These properties are preserved by all the operators of our language. For further details see [7].

The fictitious state involves both controversy and complexity, both of which are irrelevant to the needs of practising programmers. For practising engineers, mathematical laws are also more relevant than the elaborate models constructed in a study of foundations.

Another problem with models is that they change quite radically when the programming language is extended, for example, by introduction of input and output commands. Such extensions can be (and perhaps should be) designed to preserve the validity of most of the laws obeyed by the simpler language, just as the arithmetic of real numbers shares many of the properties of integer arithmetic. This has been done for Occam in [11].

## Summary

The laws presented here not only apply to concrete programs, as expressed in the notations of the programming language described in the first section; most of them also apply to program specifications, expressible in a wider range of more powerful notations. Additional laws are given to assist in the stepwise development of designs from specifications and programs from designs. In fact, we shall study a series of four classes of object, where each class includes its predecessor in the series and obeys all or almost all of the same laws.

(1) Finite programs are expressible in the notations of the programming language, excluding iteration and recursion. Laws for finite programs are given in the next section. They are sufficiently powerful to permit every finite program text to be reduced to a simple normal form.

(2) Concrete programs are expressible in the full programming language, including recursion.

(3) Abstract programs are expressible by means of programming notations plus an additional operator for denoting a limit of a convergent sequence of consistent programs. The relevant concepts and laws are those of domain theory and are explained under "Domain Properties".

(4) The remaining class is that of specifications. This is the most general class because there is no restriction on the notations in which they may be expressed. Any well-defined operator of mathematics or logic may be freely used, even including negation. The laws that apply to specifications are useful in the stepwise development of designs and programs to meet their specifications. The price of the greatest notational freedom of expression of specifications is that it is possible (and easy) to write specifications that cannot be satisfied by any program. Another source of potential difficulty is that specifications fail to obey some of the laws that are valid for programs.

The distinction between these classes may seem complicated, but is actually as simple as the familiar distinctions made between different classes of numbers.

(1) Finite programs can be compared to rational numbers. Algebraic laws permit all arithmetic expressions to be reduced to a ratio of coprime integers, whose equality can be easily established.

(2) Concrete programs are like algebraic real numbers, which are definable within a restricted notational framework (as solutions of polynomial equations). They constitute a denumerable set.

(3) Abstract programs are like real numbers; they enjoy the property that convergent sequences have a limit. For many purposes (e.g., calculus) real numbers are far more convenient to reason with than algebraic numbers. They form a nondenumerable set.

(4) Specifications may be compared to complex numbers, where more operators (e.g., square root) are total functions. The acceptance of imaginary numbers may be difficult at first, because they cannot be represented in the one-dimensional real continuum. Furthermore, they fail to obey such familiar laws as $x^2 \geq 0$. Nevertheless, it pays to use them in definition, calculation, and proof, even for problems where the eventual answers must be real. In the same way, specifications are useful

(even necessary) in requirements analysis and program development, even though they will never be executed by computer.

It might seem preferable to report a case study in which the laws had been used to assist in the development of a correct program of substantial size. Unfortunately, this is not possible. The task of writing a substantial program requires more application-orientated mathematics than the elementary algebra presented in this article. One would not expect to illustrate the laws of arithmetic by a case study in the design of a bridge. The laws of programming are broad and shallow, like the laws of arithmetic. They should be learned and used intuitively, like the grammatical rules of a foreign language.

# Algebraic Laws

About 30 algebraic laws for finite programs (programs expressible without iteration or recursion) will be presented in this section. The laws are sufficiently powerful to permit every finite program to be reduced to a simple normal form, which can be used to test whether any two such texts denote the same program.

We shall adopt the following conventions for the range of variables:

| | |
|---|---|
| P, Q, R | stand for programs. |
| b, c, d | stand for Boolean expressions. |
| e, f, g | stand for single expressions. |
| E, F, G | stand for lists of expressions. |
| $x, y, z$ | stand for lists of assignable program variables, where no variable appears more than once in the combined list $x, y, z$. |

Furthermore, $x$ is the same length as E, $y$ the same length as F, and $z$ the same length as G.

## Technical Notes

(1) The phrase "P stands for a program" may be ambiguous. Does P stand for some mathematical object that can be expressed in several different ways in a certain programming notation? Or does P stand for the *text* of a program, which must replace P before any law containing P is used? A reasonable answer to these questions is that it does not matter. Consider the analogy of the mathematical equation

$$n + m = m + n.$$

The variables $n$ and $m$ are normally considered to represent actual numbers, independent of the way in which they are expressed. But the law is equally valid if $n$ and $m$ are replaced by numerals, that is, sequences of digits in ternary notation, for example. Furthermore, it is equally valid if $n$ is consistently replaced by any other arithmetic expression, such as $2 \times p^2 + q$ in $2 \times p^2 + q + m = m + 2 \times p^2 + q$.

(2) Similar questions may be asked about the meaning of the equations that embody the algebraic laws of programming and that assert the identity of two programs written in different ways. Clearly it cannot be the *texts* of the programs that are

stated to be equal, but rather the *meanings*. The meaning of a program can be roughly understood as the possible effects of its execution in each initial state of the computer. For example, the following equations are true:

$$
\begin{aligned}
(x := 037) &= (x := 37) \\
(x := y \times y + 2 \times y \times z + z \times z) &= (x := (y + z) \times (y + z)) \\
(x := 0; x := 1) &= (x := 1).
\end{aligned}
$$

In each case, the two programs are equal, even though one of them may be slower in execution than the other. This reflects a deliberate decision to abstract from questions of execution speed, with the explicit objective of allowing inefficient program texts to be replaced by more efficient representations of the same program. Thus the laws of programming can be used as correctness-preserving transformations, or as justification for automatic code optimization. That is the practical reward for designing, implementing, and using languages with rigorously mathematical properties.

(3) On several occasions we shall introduce new notations into the laws; these are not included in the programming language, but describe programs that could be so expressed. This is a fairly familiar mathematical practice. For example, a polynomial in $x$ can be defined syntactically as a text that is either a constant or a polynomial multiplied by $x$ and added to a constant. Thus the following are polynomials:

$$0, \ 7, \ 7 \times x + 4, \ (7 \times x + 4) \times x + 17, \ \ldots$$

However, a polynomial is often more conveniently described using summation ($\Sigma$) and exponentiation,

$$\sum_{i \leq n} a_i \times x^i,$$

even though these notations are not included in the formal language of polynomials.

## Nondeterminism

The laws governing nondeterministic choice apply to all kinds of choice. The equations given below assert the identity of the whole range of choices described by their left- and right-hand sides. Of course, a particular selection made from the left-hand side may differ from a particular selection made from the right-hand side. This is also true of two selections made from the *same* text. So, in the presence of nondeterminism, two executions of the same program text do not necessarily give the same result.

(1) Clearly, it does not make any difference in what order a choice is offered. "Tea or coffee?" is the same as "coffee or tea?"

$$P \cup Q = Q \cup P \qquad \text{(symmetry)}$$

(2) A choice between three alternatives (tea, coffee, or cocoa) can be offered as first a choice between one alternative and the other two, followed (if necessary) by a choice between the other two; and it does not matter in which way the choices are grouped.

$$P \cup (Q \cup R) = (P \cup Q) \cup R \qquad \text{(associativity)}$$

(3) A choice between one thing and itself offers no choice at all (Hobson's choice).

$$P \cup P = P \qquad \text{(idempotence)}$$

(4) The *ABORT* command already allows completely arbitrary behavior, so an offer of further choice makes no difference to it.

$$\perp \cup P = \perp \qquad \text{(zero } \perp\text{)}$$

This law is sometimes known as Murphy's Law, which states, "If it can go wrong, it will"; the left-hand side describes a machine that *can* go wrong (or can behave like P), whereas the right-hand side might be taken to describe a machine that *will* go wrong. But the true meaning of the law is actually worse than this: The program $\perp$ will not always go wrong — only when it is most disastrous for it to do so! The abundance of empirical evidence for law (4) suggests that it should be taken as the first law of computer programming.

A choice between $n + 1$ alternatives can be expressed more briefly by the indexed notation

$$\cup_{i \leq n} P_i = P_0 \cup P_1 \cup \cdots \cup P_n.$$

The indexed notation need not be included in the programming language. Nevertheless, it is useful in formulating the laws, since it enables a single law to be applied to an arbitrary number of alternatives. In each application of the laws to an actual program text, the list of alternatives would be written in full.


## The Conditional

For each given Boolean expression b, the choice operator $\triangleleft b \triangleright$ specifies a choice between two alternatives with one written on each side. The first two laws clearly express the criteria for making this choice, that is, the truth or falsity of b:

(1) $P \triangleleft \text{true} \triangleright Q = P.$

(2) $P \triangleleft \text{false} \triangleright Q = Q.$

Like $\cup$, the conditional is idempotent and associative:

(3) $P \triangleleft b \triangleright P = P.$

(4) $P \triangleleft b \triangleright (Q \triangleleft b \triangleright R) = (P \triangleleft b \triangleright Q) \triangleleft b \triangleright R.$

Furthermore, it satisfies the less familiar laws

(5) $P \triangleleft b \triangleright Q = Q \triangleleft \neg b \triangleright P,$ where $\neg b$ is the negation of b;

(6) $P \triangleleft c \triangleleft b \triangleright d \triangleright Q = (P \triangleleft c \triangleright Q) \triangleleft b \triangleright (P \triangleleft d \triangleright Q),$

where $c \triangleleft b \triangleright d$ is a conditional expression, giving value c if b is true and d if b is false; and

(7) $P \triangleleft b \triangleright (Q \triangleleft b \triangleright R) = P \triangleleft b \triangleright R.$

These laws may be checked by considering the two cases when b is true and when it is false. For example, law (7) states that the middle operand Q is not selected in either case.

Suppose one of the operands of a conditional offers a nondeterministic choice between P and Q. Then it does not matter whether this choice is made before evaluation of the condition or afterward, since the value of the condition is not affected by the choice:

(8) $(P \cup Q) \triangleleft b \triangleright R = (P \triangleleft b \triangleright R) \cup (Q \triangleleft b \triangleright R)$.

From this can be deduced a similar law for the right operand of $\triangleleft b \triangleright$:

(9) $R \triangleleft b \triangleright (P \cup Q) = (R \triangleleft b \triangleright P) \cup (R \triangleleft b \triangleright Q)$.

$$\text{PROOF. LHS} = (P \cup Q) \triangleleft \neg b \triangleright R \qquad \text{(by (5))}$$
$$= (P \triangleleft \neg b \triangleright R) \cup (Q \triangleleft \neg b \triangleright R) = \text{RHS} \quad \text{(by (5) and (8))}. \qquad \square$$

An operator that distributes like this through $\cup$ is said to be *disjunctive*.

Any operation that does not change the value of the Boolean expression b will distribute through $\triangleleft b \triangleright$. An example is nondeterministic choice. It does not matter whether the choice is exercised before or after evaluation of b:

(10) $(P \triangleleft b \triangleright Q) \cup R = (P \cup R) \triangleleft b \triangleright (Q \cup R)$.

For the same reason, a conditional $\triangleleft c \triangleright$ distributes through another conditional with a possibly different condition $\triangleleft b \triangleright$:

(11) $(P \triangleleft b \triangleright Q) \triangleleft c \triangleright R = (P \triangleleft c \triangleright R) \triangleleft b \triangleright (Q \triangleleft c \triangleright R)$.

Using these laws we can prove the theorem

(12) $(P \triangleleft c \triangleright R) \triangleleft b \triangleright (Q \triangleleft d \triangleright R) = (P \triangleleft b \triangleright Q) \triangleleft c \triangleleft b \triangleright d \triangleright R$.

$$\text{PROOF. RHS} = ((P \triangleleft b \triangleright Q) \triangleleft c \triangleright R) \triangleleft b \triangleright ((P \triangleleft b \triangleright Q) \triangleleft d \triangleright R) \quad \text{(by (6))}$$
$$= ((P \triangleleft c \triangleright R) \triangleleft b \triangleright (Q \triangleleft c \triangleright R))$$
$$\triangleleft b \triangleright ((P \triangleleft d \triangleright R) \triangleleft b \triangleright (Q \triangleleft d \triangleright R)) \qquad \text{(by (11))}$$
$$= \text{LHS} \qquad \text{(by (7) and (4))}. \qquad \square$$

# Sequential Composition

Sequential composition is associative; to perform three actions in order, one can either perform the first action followed by the other two or the first two actions followed by the third.

(1) $P; (Q; R) = (P; Q); R$ \qquad (associativity)

To precede or follow a program P by the command II (which changes nothing) does not change the effect of program P.

(2) $(II; P) = (P; II) = P$ \qquad (unit II)

To precede or follow a program P by the command ⊥ (which may do anything whatsoever) results in a program that may do anything whatsoever — it may even behave like
P!

$$(3) \quad (\bot; P) = (P; \bot) = \bot \qquad\qquad (\text{zero } \bot)$$

The law $P; \bot = \bot$ states that we are not able to observe anything that P does before
$P; \bot$ reaches ⊥. This law will not be true for a language in which P can interact with
its environment, for example, by input and output.

The informal explanation of this law is weak. Perhaps it is better explained as a
moral law. The program ⊥ is one that the programmer has a duty to avoid. Equally,
the sequential programmer has the duty to avoid both $(\bot; P)$ and $(P; \bot)$. It is pointless
to draw distinctions between programs that must be avoided anyway.

A machine that selects between P and Q, and then performs R when the selected
alternative terminates, cannot be distinguished from one that initially selects whether
to perform P followed by R or Q followed by R:

$$(4) \quad (P \cup Q); R = (P; R) \cup (Q; R).$$

For the same reason, composition distributes rightward through ∪:

$$R; (P \cup Q) = (R; P) \cup (R; Q).$$

In summary, sequential composition is a disjunctive operator.

Evaluation of a condition is not affected by what happens afterwards, and therefore
composition distributes leftward through a conditional:

$$(5) \quad (P \triangleleft b \triangleright Q); R = (P; R) \triangleleft b \triangleright (Q; R).$$

However, it does not distribute rightward through a conditional, so in general it is not
true that $R; (P \triangleleft b \triangleright Q) = (R; P) \triangleleft b \triangleright (R; Q)$. On the left-hand side, b is evaluated after
executing R, whereas on the right-hand side it is evaluated before R; and in general,
prior execution of R can change the value of b.

## Assignment

It is a law of mathematics that the value of an expression is unchanged when the
variables it contains are replaced by expressions or constants denoting the values of
each variable. If $F(x)$ is a list of expressions and E is a list of the values of the variables
$x$, then $F(E)$ is a copy of F in which every occurrence of each variable of $x$ is replaced by
a copy of the expression occupying the same position in the list E. This convention is
used in the first law of assignment, which permits merging of two successive assignments
to the same variables:

$$(1) \quad (x := E; x := F(x)) = (x := F(E)).$$

For example, $\quad (x := x - 1 ; x := 2 \times x + 1) = (x := 2 \times (x - 1) + 1).$

The second law states that the assignment of the value of a variable back to itself does not change anything:

(2) $(x := x) = \text{II}$.

> This law is false for a language in which access to an uninitialized variable leads to a different effect, for example, abortion.

In fact, such a vacuous assignment can be added to any other assignment without changing its effect (recall that $x$ and $y$ are disjoint):

(3) $(x, y := E, y) = (x := E)$.

Finally, the list of variables and expressions may be subjected to the same permutation without changing the effect of the assignment:

(4) $(x, y, z := E, F, G) = (y, x, z := F, E, G)$.

> *Corollary.* $(x, y := E, F) = (y, x := F, E)$ (when $z$ is the empty list).

These four laws together are sufficient to reduce any sequence of assignments to a single assignment. For example,

$$
\begin{aligned}
(x, y &:= F, G \,;\, y, z &&:= H(x, y), J(x, y)) \\
&= (x, y, z := F, G, z \,;\, x, y, z := && x, H(x, y), J(x, y)) && \text{(by (3) and (4))} \\
&= (x, y, z := F, H(F, G), J(F, G)) && && \text{(by (1))}.
\end{aligned}
$$

Assignment distributes rightward through a conditional, changing occurrences of the assigned variables in the condition

(5) $(x := E \,;\, (P \triangleleft b(x) \triangleright Q)) = ((x := E \,;\, P) \triangleleft b(E) \triangleright (x := E \,;\, Q))$.

A conditional joining two assignments (to the same variables) may be replaced by a single assignment of a conditional expression to the same variables:

(6) $((x := E) \triangleleft b \triangleright (x := F)) = (x := (E \triangleleft b \triangleright F))$.

The conditional distributes down to the individual components of a list of expressions:

(7) $(e, E) \triangleleft b \triangleright (f, F) = (e \triangleleft b \triangleright f), (E \triangleleft b \triangleright F)$.

Using these laws, we can eliminate conditionals from sequences of assignments by driving them into the expressions. For example,

(8) $(x := E \,;\, (x := F(x) \triangleleft b(x) \triangleright x := G(x))) = (x := (F(E) \triangleleft b(E) \triangleright G(E)))$.

The following theorem will also be useful in reduction to normal forms:

(9) $((x := E \triangleleft b \triangleright \bot); (x := F(x) \triangleleft c(x) \triangleright \bot)) = (x := F(E) \triangleleft c(E) \triangleleft b \triangleright \text{false} \triangleright \bot)$.


PROOF.

$$
\begin{aligned}
\text{LHS} &= (x := E \,;\, ((x := F(x) \triangleleft c(x) \triangleright \bot) \triangleleft b \triangleright (\bot; (x := F(x) \triangleleft c(x) \triangleright \bot)))) \\
& \qquad\qquad\qquad \text{(by (5) under ``Sequential Composition'')} \\
&= (x := F(E) \triangleleft c(E) \triangleright \bot) \triangleleft b \triangleright \bot \\
& \qquad\qquad \text{(by (1) and (5) and (3) under ``Sequential Composition'')} \\
&= \mathit{RHS} \qquad \text{(by (2) and (6) under ``The Conditional'')}. \qquad \square
\end{aligned}
$$

# Undefined Expressions

If the notations of the programming language include expressions that may be undefined for some values of their operands, then some of the laws quoted above need to be slightly weakened. We assume that the range of expressions allowed in the programming language is sufficiently narrow that for each expression (or list of expressions) E there is a Boolean expression (which we will denote $\mathcal{D}$E) that is true in all circumstances in which evaluation of E would succeed and false in all circumstances in which evaluation of E would fail. Thus evaluation of $\mathcal{D}$E itself will always succeed (this would not be possible in a language with arbitrary programmer-defined functions). Note that $\mathcal{D}$ is not assumed to be a notation of the programming language. Here are some examples:

$$
\begin{aligned}
\mathcal{D}\ \text{true} \;=\; \mathcal{D}\ \text{false} \;&=\; \text{true} \\
\mathcal{D}(\text{E} + \text{F}) \;&=\; \mathcal{D}\text{E} \wedge \mathcal{D}\text{F} \\
\mathcal{D}(\text{E}/\text{F}) \;&=\; \mathcal{D}\text{E} \wedge \mathcal{D}\text{F} \wedge \text{F} \neq 0 \\
\mathcal{D}(\text{E} \triangleleft \text{b} \triangleright \text{F}) \;&=\; \mathcal{D}\text{b} \wedge (\mathcal{D}\text{E} \triangleleft \text{b} \triangleright \mathcal{D}\text{F}) \\
\mathcal{D}\mathcal{D}\text{E} \;&=\; \text{true}.
\end{aligned}
$$

Now we stipulate that the effect of attempting to evaluate an expression outside its domain is wholly arbitrary, so

(1) $x := \text{E} = (x := \text{E} \triangleleft \mathcal{D}\text{E} \triangleright \perp)$,

(2) $\text{P} \triangleleft \text{b} \triangleright \text{Q} = (\text{P} \triangleleft \text{b} \triangleright \text{Q}) \triangleleft \mathcal{D}\text{b} \triangleright \perp$, and

(3) $\text{P} \triangleleft \text{b} \triangleright \perp = \text{P} \triangleleft \text{b} \triangleleft \mathcal{D}\text{b} \triangleright \text{false} \triangleright \perp$ .

In view of this, some of the preceding laws need alteration, as follows:

(4) $\text{P} \triangleleft \text{b} \triangleright \text{P} = \text{P} \triangleleft \mathcal{D}\text{b} \triangleright \perp$                 (see (3) under "The Conditional").

(5) $(\text{P} \triangleleft \text{b} \triangleright \text{Q}) \triangleleft \text{c} \triangleright \text{R} = ((\text{P} \triangleleft \text{c} \triangleright \text{R}) \triangleleft \text{b} \triangleright (\text{Q} \triangleleft \text{c} \triangleright \text{R})) \triangleleft \mathcal{D}\text{b} \triangleright (\perp \triangleleft \text{c} \triangleright \text{R})$
                                              (see (11) under "The Conditional").

(6) $(x := \text{E}; x := \text{F}(x)) = (x := \text{F}(\text{E}) \triangleleft \mathcal{D}\text{E} \triangleright \perp)$      (see (1) under "Assignment").

(7) $x := \text{E}; (x := \text{F}(x) \triangleleft \text{b}(x) \triangleright x := \text{G}(x)) = x := (\text{F}(\text{E}) \triangleleft \text{b}(\text{E}) \triangleright \text{G}(\text{E})) \triangleleft \mathcal{D}\text{E} \triangleright \perp$
                                              (see (8) under "Assignment").

Theorem (9) under "Assignment" also requires modification; also the proof (but not the statement) of (12) under "The Conditional."

Reasoning with undefined expressions can be complicated and needs some care. But there are also some rewards. For example, the fact that the minimum of an empty set is undefined permits a simple formulation of Dijkstra's linear search theorem [4, pp. 105–106]:

(8) $(i := 0\,; \neg\text{b}(i) * (i := i + 1)) = (i := \min\{i \mid \text{b}(i) \wedge i \geq 0\})$.

Note that, in general, the minimum function cannot be implemented or included in any programming language. So the right-hand side of (8) should be regarded rather as an exact specification of the program on the left.

# Normal Form

To illustrate the power of the laws given so far, we can use them to reduce every finite program text of our language to a simple normal form. In normal form a program looks like

$$(\cup_{i \le n}\, x := E_i) \lhd b \rhd \bot,$$

where $b \Rightarrow \mathcal{D}E_i$ for all $i \le n$, and $\Rightarrow$ denotes logical implication. Without loss of generality, we can ensure that in this context $\mathcal{D}b =$ true by replacing b if necessary by $(b \lhd \mathcal{D}b \rhd$ false$)$ (by (3) under "Undefined Expressions"). A notable feature of the normal form is that the sequential composition operator does not appear in it.

To explain how to reduce a program text to normal form, it is sufficient to show how each primitive command can be written in normal form and how each operator, when applied to operands in normal form, yields a result expressible in normal form. The section on assignment explained how all assignments of a program can be adapted so that they all have the same list of variables on the left; so we can assume this has already been done.

(1) *SKIP.*    $\text{II} = ((x := x) \lhd \text{true} \rhd \bot)$
$$\text{(by (2) under ``Assignment'' and (1) under ``The Conditional'').}$$

(2) *ABORT.*    $\bot = (x := x \lhd \text{false} \rhd \bot)$    (by (2) under "The Conditional").

(3) *Assignment.*    $(x := E) = (x := E \lhd \mathcal{D}E \rhd \bot)$
$$\text{(by (1) under ``Undefined Expressions'').}$$

(4) *Nondeterminism.*
$$(\text{P} \lhd b \rhd \bot) \cup (\text{Q} \lhd c \rhd \bot)$$

$$= (\text{P} \cup (\text{Q} \lhd c \rhd \bot)) \lhd b \rhd (\bot \cup (\text{Q} \lhd c \rhd \bot))$$
$$\text{(by (10) under ``The Conditional'')}$$

$$= ((\text{P} \cup \text{Q}) \lhd c \rhd (\text{P} \cup \bot)) \lhd b \rhd \bot$$
(by (10) under "The Conditional," and (1) and (4) under "Nondeterminism")

$$= ((\text{P} \cup \text{Q}) \lhd c \rhd \bot) \lhd b \rhd \bot \qquad \text{(by (1) and (4) under ``Nondeterminism'')}$$

$$= (\text{P} \cup \text{Q}) \lhd c \lhd b \rhd \text{false} \rhd \bot \qquad \text{(by (2) and (6) under ``The Conditional'').}$$

Here, P and Q stand for lists of assignments separated by $\cup$, so $\text{P} \cup \text{Q}$ is just the union of these two lists. The condition $(c \lhd b \rhd \text{false})$ is equivalent to $(c \wedge b)$. Since the operands are normal forms, this is defined everywhere and implies that all expressions in $\text{P} \cup \text{Q}$ are also defined.

(5) *Conditional.* $(\text{P} \lhd c \rhd \bot) \lhd b \rhd (\text{Q} \lhd d \rhd \bot) = (\text{P} \lhd b \rhd \text{Q}) \lhd c \lhd b \rhd d \rhd \bot$
$$\text{(by (12) under ``The Conditional'').}$$
If $\text{P} = \cup_{i \le n}\, x := E_i$ and $\text{Q} = \cup_{j \le m}\, x := F_j$, then
$\text{P} \lhd b \rhd \text{Q} = \cup_{i \le n} \cup_{j \le m} (x := E_i \lhd b \rhd x := F_j)$
$$\text{(by (8) and (9) under ``The Conditional'')}$$

$$= \cup_{i \le n} \cup_{j \le m}\, x := (E_i \lhd b \rhd F_j) \qquad \text{(by (6) under ``Assignment'').}$$
Since $c \Rightarrow \mathcal{D}E_i$ and $d \Rightarrow \mathcal{D}F_j$, it follows that $c \lhd b \rhd d \Rightarrow \mathcal{D}(E_i \lhd b \rhd F_j)$,   for all $i$ and $j$. Thus the LHS of (5) is reducible to normal form.

(6) *Sequential composition.* $((\bigcup_{i \leq n} x := E_i) \triangleleft b \triangleright \bot); ((\bigcup_{j \leq m} x := F_j(x)) \triangleleft c(x) \triangleright \bot)$
can be reduced (by distribution through $\bigcup$) to

$\bigcup_{i \leq n} \bigcup_{j \leq m} ((x := E_i \triangleleft b \triangleright \bot); (x := F_j(x) \triangleleft c(x) \triangleright \bot))$

$\quad = \bigcup_{i \leq n} \bigcup_{j \leq m} (x := F_j(E_i) \triangleleft c(E_i) \triangleleft b \triangleright \text{false} \triangleright \bot)$ (by (9) under "Assignment").

The method described in (4) can be used to distribute the unions into the conditional, obtaining

$$(\bigcup_{i \leq n} \bigcup_{j \leq m} x := F_j(E_i)) \triangleleft (\wedge_{i \leq n} c(E_i)) \triangleleft b \triangleright \text{false} \triangleright \bot,$$

where the conjunction notation $\wedge$ can be defined by induction:

$$\begin{aligned}
\wedge_{i \leq 0}\, c_i &= c_0 \\
\wedge_{i \leq n+1}\, c_i &= (\wedge_{i \leq n} c_i) \triangleleft c_{n+1} \triangleright \text{false}.
\end{aligned}$$

This completes the proof that all finite programs are reducible.

The importance of normal forms is that they provide a complete test as to whether two finite program texts denote the same program. The two programs are first reduced to normal form. If the normal forms are equal, so are the programs; otherwise they are unequal.

Two normal forms

$$(\bigcup_{i \leq n} x := E_i) \triangleleft b \triangleright \bot \quad \text{and} \quad (\bigcup_{j \leq m} x := F_j) \triangleleft c \triangleright \bot$$

are equal if and only if

$$b = c \text{ and } \{v \mid \exists i \leq n.\, v = E_i\} = \{w \mid \exists j \leq m.\, w = F_j\},$$

where these equations must hold for all values of the variables contained in the expressions $b, c, E_i$, and $F_j$. The truth of these equations may not be decidable, as in integer arithmetic, for example. The results shown here establish only relative completeness.

# Domain Properties

In this section we introduce iteration and recursion, using the methods of [12].

## The Ordering Relation

As a preliminary we shall explore the properties of an ordering relation $\supseteq$ between programs.

    *Definition.* $\qquad\qquad P \supseteq Q \triangleq P \cup Q = P.$

This means that Q is a more deterministic program than P. Everything that Q can do, P may also do, and everything that Q cannot do, P may also fail to do. So Q is in all respects a more predictable program and more controllable than P. In any circumstance where P reliably serves some useful purpose, P may be replaced by Q, in the certainty that it will serve the same purpose. But not vice versa. There may be some purposes for which Q is adequate, but P, given its greater non-determinism, cannot be relied on. Thus $P \supseteq Q$ means that, for any purpose, Q is better than P, or at least as good. From this point on, we will use the comparative "better" by itself, with the understanding that it means "better or at least as good." The relation $\supseteq$ is not a total ordering on programs, because it is not true for all P and Q that $P \supseteq Q$ or $Q \supseteq P$; P may be better than Q for some purposes, and Q may be better than P for others. However, $\supseteq$ is a *partial* order, in that it satisfies the following laws:

$$(1) \qquad P \supseteq P \qquad\qquad\qquad\qquad\qquad (reflexivity).$$
$$(2) \qquad P \supseteq Q \wedge Q \supseteq P \Rightarrow P = Q \quad (antisymmetry).$$
$$(3) \qquad P \supseteq Q \wedge Q \supseteq R \Rightarrow P \supseteq R \quad (transitivity).$$

These laws can be proved directly from the definition, together with the laws for $\cup$.
PROOF.

$$(1) \qquad P \cup P \qquad = \quad P \qquad\qquad\qquad (\text{idempotence of } \cup).$$
$$(2) \qquad (P \cup Q = P) \quad \wedge \quad (Q \cup P = Q)$$
$$\Rightarrow \quad P = P \cup Q = Q \cup P = Q \quad (\text{symmetry of } \cup).$$
$$(3) \qquad (P \cup Q = P) \quad \wedge \quad (Q \cup R = Q) \qquad (\text{antecedent})$$
$$\Rightarrow \quad P \cup R = (P \cup Q) \cup R \quad (\text{first antecedent})$$
$$= P \cup (Q \cup R) \quad (\text{associativity of } \cup)$$
$$= P \cup Q \qquad\quad (\text{second antecedent})$$
$$= P \qquad\qquad\quad (\text{first antecedent}) \qquad \square$$

The *ABORT* command $\perp$ is the most nondeterministic of all programs. It is the least predictable, the least controllable, and in short, for all purposes, the worst:

$$(4) \quad \perp \supseteq P.$$

$$\text{PROOF.} \qquad\qquad\qquad \perp \cup P = \perp. \qquad\qquad\qquad \square$$

The machine that behaves either like P or like Q is, in general, worse than both of them:

$$(5) \quad (P \cup Q) \supseteq P \text{ and } (P \cup Q) \supseteq Q.$$

$$\text{PROOF.} \qquad (P \cup Q) \cup P \quad = \quad P \cup (Q \cup P) \quad (\text{associativity})$$
$$= \quad P \cup (P \cup Q) \quad (\text{symmetry})$$
$$= \quad (P \cup P) \cup Q \quad (\text{associativity})$$
$$= \quad P \cup Q \qquad\quad (\text{idempotence}). \qquad \square$$

In fact, $P \cup Q$ is the best program that has property (5). Any program R that is worse than both P and Q is also worse than $P \cup Q$, and vice versa:

$$(6) \quad R \supseteq (P \cup Q) \equiv (R \supseteq P \wedge R \supseteq Q).$$

$$\text{PROOF.} \quad \text{LHS} \Rightarrow R \supseteq P \qquad\qquad\qquad (\text{by transitivity from (5)})$$
$$\text{LHS} \Rightarrow R \supseteq Q \qquad\qquad\qquad (\text{similarly})$$
$$\text{RHS} \Rightarrow (R \cup P = R) \wedge (R \cup Q = R) \quad (\text{by definition of } \supseteq)$$
$$\Rightarrow (R \cup P) \cup (R \cup Q) = R \cup R \quad (\text{by adding the equations})$$
$$\Rightarrow R \cup (P \cup Q) = R \qquad (\text{by properties of } \cup)$$
$$\Rightarrow \textit{LHS} \qquad\qquad\qquad (\text{by definition of } \supseteq). \qquad \square$$

If $P \supseteq Q$ this means that Q is, in all circumstances, better than P. It follows that wherever P appears within a larger program, it can be replaced by Q, and the only consequence will be to improve the larger program (or at least to leave it unchanged). For example,

$$
\begin{aligned}
(7) \quad \text{If } P \supseteq Q \quad \text{then} \qquad P \cup R &\supseteq Q \cup R \\
\wedge (P; R) &\supseteq (Q; R) \\
\wedge (R; P) &\supseteq (R; Q) \\
\wedge (P \triangleleft b \triangleright R) &\supseteq (Q \triangleleft b \triangleright R) \\
\wedge (R \triangleleft b \triangleright P) &\supseteq (R \triangleleft b \triangleright Q) \\
\wedge (b * P) &\supseteq (b * Q).
\end{aligned}
$$

In summary, the law quoted above states that all the operators of our small programming language are *monotonic*, in the sense that they preserve the $\supseteq$ ordering of their operands. In fact, every operation that distributes through $\cup$ is also monotonic.

THEOREM. *If F is any function from programs to programs, and for all programs P and Q, F(P $\cup$ Q) = F(P) $\cup$ F(Q), then F is monotonic.*

PROOF.
$$
\begin{aligned}
P \supseteq Q \;\Rightarrow\; & P \cup Q = P && \text{(by definition of } \supseteq) \\
\Rightarrow\; & F(P) \cup F(Q) \;=\; F(P \cup Q) && \text{(by distribution of } F) \\
& \qquad\qquad\;=\; F(P) && \text{(by property of } =) \\
\Rightarrow\; & F(P) \supseteq F(Q) && \text{(by definition of } \supseteq). \qquad \square
\end{aligned}
$$

One important fact about monotonicity is that every function defined from composition of monotonic functions is also monotonic. Since all the operators of our programming language are monotonic, every program composed by means of these operators is monotonic in each of its components. Thus, if any component is replaced by a possibly better one, the effect can only be to improve the program as a whole. If the new program is also more efficient than the old, the benefits are increased.

## Least Upper Bounds

We have seen that $P \cup Q$ is the best program *worse* than both P and Q. Suppose we want a program *better* than both P and Q. In general, there will be no such program. Consider the two assignments $x := 1$ and $x := 2$. These programs are incompatible, and there is no program better for all purposes than both. If the final value of $x$ should be 1, the first program is suitable, but the second cannot possibly give an acceptable result. On the other hand, if the final value should be 2, the first program is totally unsuitable. Consider two nondeterministic programs:

$$
\begin{aligned}
P &= (x := 1 \cup x := 2 \cup x := 3), \\
Q &= (x := 2 \cup x := 3 \cup x := 4).
\end{aligned}
$$

In this case there exists a program better than both, namely, $x := 2$. In fact there exists a worst program that is better than both, and we will denote this by $P \cap Q$:

$$
P \cap Q = (x := 2 \cup x := 3).
$$

Two programs P and Q are said to be *compatible* if they have a common improvement; and then their worst common improvement is denoted $P \cap Q$. This fact is summarized in the law

(1)  $(P \supseteq R) \wedge (Q \supseteq R) \equiv (P \cap Q) \supseteq R$.

*Corollary.*  $P \supseteq (P \cap Q) \wedge Q \supseteq (P \cap Q)$   (by (1) under "The Ordering Relation").

The operator $\cap$, wherever it is defined, is idempotent, symmetric, and associative, and has identity $\perp$ .

(2)  $P \cap P = P$.

(3)  $P \cap Q = Q \cap P$.

(4)  $P \cap (Q \cap R) = (P \cap Q) \cap R$.

(5)  $\perp \cap P = P$.

The $\cap$ operator generalizes to any finite set of compatible programs

$$S = \{P, Q, \ldots, T\}.$$

Provided that there exists a program better than all of them, the worst such program is denoted $\cap S$ :

$$\cap S = P \cap Q \cap \cdots \cap T,$$

provided $P \supseteq R \wedge Q \supseteq R \wedge \cdots \wedge T \supseteq R$, for some R.

It follows that

(6)  $(\forall P \in S . P \supseteq R) \equiv (\cap S \supseteq R)$, provided $\cap S$ is defined.

It is important to recognize that $\cap$ is not a combinator of our programming language, and that $P \cap Q$ is not strictly a program, even if P and Q are compatible programs. For example, let P be a program that assigns an arbitrary ascending sequence of numbers to an array, and let Q be a program that subjects the array to an arbitrary permutation. Then $P \cap Q$ would be a program that satisfies both these specifications and consequently would sort the array into ascending order. Unfortunately, programming is not that easy. As in other branches of engineering, it is not generally possible to make many different designs, each satisfying one requirement of a specification, and then merge them into a single product satisfying all requirements. On the contrary, the engineer has to satisfy all requirements in a single design. This is the main reason why designs and programs get complicated.

These problems do not arise with pure specifications, which may be freely connected by the conjunction *and*. $P \cap Q$ may be regarded as an *abstract* program, or a specification of a program that accomplishes whatever P accomplishes *and* whatever Q accomplishes, and fails only when *both* P and Q fail. $P \cap Q$ specifies a program that (if it exists) is, for all purposes, better than both P and Q.

In fact, conjunction is the most useful operator for structuring large specifications. It should be included in any language designed for that purpose. Clarity of specification, achieved by using the conjunction operator, is much more important than using a specification language that can be implemented. It is unfortunate that conjunction is so difficult to implement. As a consequence, the formulation of appropriate specifications, and the design of programs to meet them (i.e., software engineering), will always be a serious intellectual challenge.

## Limits

Suppose S is a nonempty (possibly infinite) set, and for every pair of its members, S actually contains a member better than both. Such a set is said to be *directed*.

*Definition.*    S is directed means that

$$(S \neq \{\ \}) \wedge \forall P, Q \in S.\ \exists R \in S.\ P \supseteq R \wedge Q \supseteq R.$$

Examples of directed sets are

$$
\begin{array}{ll}
\{P\} & \text{a set with only one member,} \\
\{P, P \cup Q\} & \text{since } P \cup Q \supseteq P \text{ and } P \supseteq P, \text{ and} \\
\{P, Q, R\} & \text{where } P \supseteq R \wedge Q \supseteq R.
\end{array}
$$

If S is finite and directed, then clearly it contains a member that is better than all the other members, so $\cap S$ is defined and $\cap S \in S$.

If S is directed but infinite, then it does not necessarily contain a best member. Nevertheless, the set has a limit $\cap S$: it is the worst program better than all members of S. The set S is like a convergent sequence of numbers, tending toward a limit that is not a member of the sequence. By selecting members of the set, it is possible to approximate its limit as closely as we please. Since no metric has been introduced, "closeness" has to be interpreted indirectly; that is, for any *finite* program P worse than $\cap S$ there exists a member of S better than P.

One interesting property of the limit of a directed set of programs is that it is preserved by all the operators of our programming language; such operators are therefore said to be *continuous*.

(1) $(\cap S) \cup Q = \cap\{P \cup Q \mid P \in S\}.$

(2) $(\cap S) \triangleleft b \triangleright Q = \cap\{P \triangleleft b \triangleright Q \mid P \in S\}.$

(3) $(\cap S); Q = \cap\{P; Q \mid P \in S\}.$

(4) $Q; (\cap S) = \cap\{Q; P \mid P \in S\}.$

(5) $b * (\cap S) = \cap\{b * P \mid P \in S\}.$

A fact about continuity is that any composition of continuous functions is also continuous. Let X stand for a program, and let $F(X)$ be a program constructed solely by means of continuous operators, and possibly containing occurrences of X. If S is directed, it follows that

(6) $F(\cap S) = \cap\{F(X) \mid X \in S\}.$

# Iteration and Recursion

Given a program P and a Boolean expression b, we can define by induction an infinite set

$$\{Q_n \mid n \geq 0\},$$

where

$$Q_0 = \perp,$$
$$Q_{n+1} = (P; Q_n) \triangleleft b \triangleright \text{II}, \quad \text{for all } n \geq 0.$$

From these definitions it is clear that $Q_n$ is a program that behaves like $(b * P)$ up to $n$ iterations of the body P, but breaks on the $n$th iteration, and can do anything $(\perp)$.

Clearly, therefore

$$Q_n \supseteq Q_{n+1}, \quad \text{for all } n,$$

(which can be proved formally by induction). Consequently, the set $\{Q_n \mid n \geq 0\}$ is directed, and by taking $n$ large enough, we can approximate as closely as we please to the behavior of the loop $(b * P)$. The loop itself can be defined as the limit of all its approximations:

(1) $b * P = \cap \{Q_n \mid n \geq 0\}$.

The same technique can be used to define a more general form of recursion. Let X stand for the name of the recursive program that we wish to construct, and let $F(X)$ define the intended behavior of the program. Within $F(X)$, each occurrence of X stands for a call on the whole recursive program again. As before, we can construct a series of approximations to the behavior of the recursive program:

$$F^0(Q) = Q,$$

$$F^{n+1}(Q) = F(F^n(Q)), \quad \text{for all } n \geq 0.$$

$F^n(\perp)$ behaves correctly provided that the recursion depth does not equal or exceed $n$. Because $F$ is monotonic and $F^0(\perp) \supseteq F^1(\perp)$, it follows that

$$F^n(\perp) \supseteq F^{n+1}(\perp), \quad \text{for all } n.$$

Consequently, $\{F^n(\perp) \mid n \geq 0\}$ is a directed set, and we define the recursive program (denoted by $\mu X.F(X)$) as its limit:

(2) $\mu X.F(X) = \cap \{F^n(\perp) \mid n \geq 0\}$.

In accordance with the explanation given above, iteration is a special case of recursion:

(3) $b * P = \mu X.(P; X) \triangleleft b \triangleright \text{II}$.

The most important fact about a recursively defined program is that each of the recursive calls is equal to the whole progam again, or more formally that $\mu X.F(X)$ is a solution of the equation $X = F(X)$. This is stated in the following law:

(4) $\mu X.F(X) = F(\mu X.F(X))$.

$$
\begin{aligned}
\text{PROOF. RHS} \quad &= \quad F(\cap\{F^n(\bot) \mid n \geq 0\}) &&\text{(by definition of } \mu) \\
&= \quad \cap\{F(F^n(\bot)) \mid n \geq 0\} &&\text{(by continuity of } F) \\
&= \quad \cap(\{F^{n+1}(\bot) \mid n \geq 0\} \cup \{\bot\}) &&\text{(by definition of } F^{n+1} \text{ and} \\
& &&\text{(5) under ``Least Upper Bounds'')} \\
&= \quad \cap\{F^n(\bot) \mid n \geq 0\} &&\text{(since } F^0(\bot) = \bot) \\
&= \quad \text{LHS} &&\text{(by definition).} \qquad \square
\end{aligned}
$$

*Corollary.* $\qquad \mathrm{b} * \mathrm{P} = (\mathrm{P}; (\mathrm{b} * \mathrm{P})) \triangleleft \mathrm{b} \triangleright \mathrm{II}$ .

In general, there will be more than one solution of the equation $X = F(X)$. Indeed, for the equation $X = X$, absolutely every program is a solution. But, of all the solutions, $\mu X.F(X)$ is the worst:

(5) $Y = F(Y) \Rightarrow \mu X.F(X) \supseteq Y$.

$$
\begin{aligned}
\text{PROOF.}(Y = F(Y)) \quad &\Rightarrow \quad (\bot \supseteq Y) \wedge (Y = F(Y)) \\
&\Rightarrow \quad (F(\bot) \supseteq F(Y)) \wedge (Y = F(Y)), \quad \text{(by } F \text{ monotonic)} \\
&\Rightarrow \quad (F(\bot) \supseteq Y).
\end{aligned}
$$

By induction it follows that, for any $n \geq 0$,

$$
\begin{aligned}
Y = F(Y) \quad &\Rightarrow \quad F^n(\bot) \supseteq F^n(Y) \wedge Y = F^n(Y) \\
&\Rightarrow \quad F^n(\bot) \supseteq Y \\
&\Rightarrow \quad \cap\{F^n(\bot) \mid n \geq 0\} \supseteq Y \, \text{(by (6) under ``Least Upper Bounds.'')} \; \square
\end{aligned}
$$

## Specifications

Two important concepts have been introduced thus far. The first is that a specification describes the intended behavior of a program, without giving any guidance as to how the program might be executed. Secondly, a concrete program P may be *better* than a specification S; so whenever a program that behaves like S is required, the concrete program P will serve the purpose. In this case, we can say that P *satisfies* the specification S, or in symbols $S \supseteq P$. It is the responsibility of the programmer, when given a specification S, to find a program P that satisfies S. The practical purpose of a calculus of specifications is to aid in the development of programs. Specifications do not have to be executed by machine, so there is no reason to confine ourselves to the notations of a particular programming language. Also, it is not necessary to confine ourselves to specifications that can be satisfied. As an extreme example, we introduce the specification $\top$, which cannot be satisfied by any program whatsoever.

To accept the risk of asking the impossible has as its reward that the $\cap$ operator is defined on all specifications: Wherever R and S are inconsistent, the result of $(R \cap S)$ is $\top$. Furthermore, if S is *any* set of specifications, then

$\cap S$ is the specification that requires *all* R in S to be satisfied, and

$\cup S$ is the specification that requires *some* R in S to be satisfied.

These specifications are limits of S:

(1) $Q \supseteq \cup S \equiv \forall R \in S.\, Q \supseteq R.$

(2) $\cap S \supseteq Q \equiv \forall R \in S.\, R \supseteq Q.$

The $\supseteq$ ordering applies to specifications, just as it does to programs, but it can be interpreted in a new sense. If $S \supseteq R$, it means that S is a weaker specification and easier to meet than R. Any program that satisfies R will serve for S, but it may be that more programs will satisfy S. Thus $\bot$ is the easiest specification, satisfied by any program, and $\top$ is the most difficult, because it is impossible.

We do not introduce any specific notation or language for specifications; we permit any language that describes a relationship between the values of variables before and after execution of a program. Thus we may use the notations of a programming language, possibly extended even by noncomputable operators, or we may use predicates as in predicative programming [6] or predicate pairs as in VDM [9]. We assume that the specification language includes at least all the notations of our programming language, so that each program is its own strongest specification.

Thus all programs are specifications, but not necessarily vice versa. As a consequence, there are certain laws that are true for programs, but *not* for specifications. In particular, law (3) in " Sequential Composition" and law (4) in "Limits" are *not* valid for specifications. However, we believe it is reasonable to insist that specifications obey all the laws of the calculus of relations [13].

## Weakest Prespecification

Specifications may be constructed in terms of all the operators available for concrete programs. For example, if R and S are specifications, them (R; S) is a specification satisfied by any program (P; Q), where P satisfies R and Q satisfies S (it can also be satisfied by other programs). This fact is extremely useful in the top-down development of programs (also known as stepwise refinement). Suppose, for example, that the original task is to construct a program that meets the specification W. Perhaps we can think of a way to decompose this task into two simpler subtasks specified by R and S. The correctness of the decomposition can be proved by showing that $W \supseteq R; S$. This proof should be completed before embarking on design for the subtasks R and S. Then similar methods can be used to find programs P and Q that solve these subtasks, such that $R \supseteq P$ and $S \supseteq Q$. It follows immediately from monotonicity of sequential composition that P; Q is a program that will solve the original task W, that is, $W \supseteq (P; Q)$.

In approaching the task W, suppose we think of a reasonable specification for the second of the two subtasks S, but we do not know the exact specification of the first subtask. It would be useful to calculate R from S and W. Therefore we define the *weakest prespecification* $S \backslash W$ to be the weakest specification that must be met by the first subprogram R in order that the composition (R; S) will accomplish the original task W. This fact is expressed in symbols:

(1) $W \supseteq (S \backslash W); S$

$(S \backslash W)$ is a sort of left quotient of W by S; the divisor S can be cancelled by postmultiplication, and the result will be the same as W or better.

Here are some examples of weakest prespecifications, where $x$ is an integer variable:

$(x := 3 \times x)\backslash(x := 6 \times y) = (x := 2 \times y)$, because $(x := 2 \times y; x := 3 \times x) = (x := 6 \times y)$.

$(x := 2 \times x)\backslash(x := 3) = \top$, since 3 is odd and cannot be the result of doubling an integer.

$(x := 2 \times x)\backslash(x := 3 \cup x := 4) = (x := 2)$, because $(x := 3 \cup x := 4) \supseteq x := 4 = (x := 2; x := 2 \times x)$.

The law given above does not uniquely define S\W. But, of all the solutions for X in the inequality $W \supseteq (X; S)$, the solution S\W is defined as the weakest one, because this is the easiest to achieve. Thus, to find such a solution, a necessary and sufficient condition is that the solution should satisfy S\W:

(2) $\quad W \supseteq (X; S) \equiv (S\backslash W) \supseteq X.$

Thus in developing a sequential program to meet specification W, there is no loss of generality in taking S\W as the specification of the left operand of sequential composition, given that S is the specification of the right operand. That is why it is called the *weakest* prespecification. For the remainder of this article, for convenience we will omit the word *weakest*.

The prespecification P\R, where P is a program, plays a role very similar to Dijkstra's weakest precondition. It satisfies the analogue of several of his conditions. In the following three laws, P must be a *program*.

The accomplishment of an impossible task is still impossible, even with the help of P:

(3) $\quad P\backslash\top = \top.$

To accomplish two tasks with the help of P, one must write a program that accomplishes both of them simultaneously:

(4) $\quad P\backslash(R1 \cap R2) = (P\backslash R1) \cap (P\backslash R2).$

This distributive law extends to limits of arbitrary sets:

$$P\backslash(\cap S) = \cap\{P\backslash R \mid R \in S\}.$$

Finally, consider a set of specifications $S = \{R_i \mid i \geq 0\}$ such that $R_{i+1} \supseteq R_i$.

(5) $\quad P\backslash(\cup S) = \cup\{P\backslash R_i \mid i \geq 0\}.$

The following laws are very similar to the corresponding laws for weakest preconditions. The program II changes nothing. Anything one wants to achieve after II must be achieved before:

(6) $\quad II\backslash R = R.$

To achieve R with the aid of $P \cup Q$, both of them must achieve it.

(7) $\quad (P \cup Q)\backslash R = (P\backslash R) \cap (Q\backslash R).$

To achieve R with the aid of $(P; Q)$, one must achieve $(Q\backslash R)$ with the aid of P:

(8) $\quad (P; Q)\backslash R = P\backslash(Q\backslash R).$

The corresponding law for the conditional requires a new operator on specifications:

(9) $(P \triangleleft b \triangleright Q) \backslash R = (P \backslash R) \triangleleft \breve{b} \triangleright (Q \backslash R)$,

where $S \triangleleft \breve{b} \triangleright T$ specifies a program as follows: If b is true *after* execution, it has behaved in accordance with specification S. If b is false afterwards, it has behaved in accordance with specification T. $P \triangleleft \breve{b} \triangleright Q$ is not a program, even if P and Q are; in fact it may not even be implementable. Consider the example

$$x := \text{ false } \triangleleft \breve{x} \triangleright x := \text{ true.}$$

The \ operator has a dual, /, called postspecification: $(R/S)$ is the weakest specification of a program X such that $R \supseteq (S; X)$. Its properties are very similar to those of \; for example,

(1) $R \supseteq S; (R/S)$,

(2) $R \supseteq (S; X) \equiv (R/S) \supseteq X$,

(3) $\top/P = \top$,    if P is a program,

(4) $(R1 \cap R2)/P = (R1/P) \cap (R2/P)$,

(5) $R/II = R$,

(6) $R/(P \cup Q) = (R/P) \cap (R/Q)$, and

(7) $R/(P; Q) = (R/P)/Q$.

## General Inverse

The prespecification and postspecification are, in a sense, the right and left inverses of sequential composition. This type of inverse can be given for any operator $F$ that distributes through arbitrary unions. It is defined as follows:

(1) $F^{-1}(R) = \cup\{P \mid R \supseteq F(P)\}$.

This is not an exact inverse of $F$, but it satisfies the law

(2) $R \supseteq F(F^{-1}(R))$.

PROOF.   $\begin{aligned} \text{RHS} &= F(\cup\{P \mid R \supseteq F(P)\}) &&\text{(by definition of } F^{-1}) \\ &= \cup\{F(P) \mid R \supseteq F(P)\} &&\text{(by distribution of } F) \\ &\subseteq R &&\text{(by set theory).} \end{aligned}$ □

Since $F^{-1}(R)$ is the union of all solutions for X in the inequation $R \supseteq F(X)$, it must be the weakest (most general) solution:

(3) $R \supseteq F(X) \equiv F^{-1}(R) \supseteq X$.

The condition that $F$ must distribute through $\cup$ is essential to the existence of the inverse $F^{-1}$. To show this, consider the counterexample:

$$F(X) \quad = \quad X; X \tag{1}$$
$$P \quad = \quad x := x \tag{2}$$
$$Q \quad = \quad x := -x. \tag{3}$$

$F$ is a function that may require more than one execution of its operand. When applied to the nondeterministic choice of two programs P or Q, each execution may involve a different choice. Consequently, $F$ does not distribute, as shown by the following example:

$$
\begin{aligned}
F(\mathrm{P} \cup \mathrm{Q}) \quad &= \quad (\mathrm{P} \cup \mathrm{Q}); (\mathrm{P} \cup \mathrm{Q}) && \text{(by definition of } F) \\
&= \quad (\mathrm{P}; \mathrm{P}) \cup (\mathrm{P}; \mathrm{Q}) \cup (\mathrm{Q}; \mathrm{P}) \cup (\mathrm{Q}; \mathrm{Q}) && \text{(by ; disjunctive)} \\
&= \quad (x := x; x := x) \cup (x := x; x := -x) \\
&\qquad \cup (x := -x; x := x) \cup (x := -x; x := -x) \\
&= \quad (x := x) \cup (x := -x.) \\
\text{But } F(\mathrm{P}) \cup F(\mathrm{Q}) \quad &= \quad (x := x; x := x) \cup (x := -x; x := -x) \\
&= \quad x := x.
\end{aligned}
$$

Since $\mathrm{P} \supseteq F(\mathrm{P})$ and $\mathrm{P} \supseteq F(\mathrm{Q})$, it follows that

$$
\cup \{ \mathrm{X} \mid \mathrm{P} \supseteq F(\mathrm{X}) \} \supseteq \mathrm{P} \cup \mathrm{Q} \qquad \text{(by set theory).}
$$

By law (3) and the definition of $F^{-1}(\mathrm{P})$, we could conclude that $\mathrm{P} \supseteq F(\mathrm{P} \cup \mathrm{Q})$, which is false. The contradiction shows that $F$ does not have an inverse, even in the weak sense described by law (3).

The inverse $F^{-1}(\mathrm{R})$ (when it exists) could be of assistance in the top-down development of a program to meet the specification R. Suppose it is decided that the top-level structure of the program is defined by $F$. Then it will be necessary to calculate $F^{-1}(\mathrm{R})$ and use it as the specification of the component program X, secure in the knowledge that the final program $F(\mathrm{X})$ will meet the original specification R.

Unfortunately, the method does not generalize to a structure $F$ with two or more components. It is necessary to fix all but one of the components before calculating the inverse.

## Conclusion

The laws presented here should assist programmers in the reasoning necessary to develop programs that meet their specifications. They should also help with optimization by algebraic transformation. The basic insight is that programs themselves, as well as their specifications, are mathematical expressions. Therefore they can be used directly in mathematical reasoning in the same way as expressions that denote familiar mathematical concepts, such as numbers, sets, functions, groups, categories, among others. It is also very convenient that programs and specifications are treated together in a homogeneous framework; the main distinction between them is that programs are a subclass of specification expressed in such severely restricted notations that they can be input, translated, and executed by a general-purpose digital computer.

However, we admit the exposition of this article does have deficiencies. One theoretical weakness is that the laws are presented as self-evident axioms or postulates, intended to command assent from those who already understand the properties of programs they express. For sequential programs and their specifications, the relevant mathematical definitions would be formulated within the classical theory of relations [7,13]. The use of such definitions to prove the laws enumerated in this article yields a valuable reassurance that the laws are consistent. Furthermore, the definitions give additional insight into the mathematics of programming and how it may be applied in practice. Specifically they suggest additional useful laws, and establish that a given set of laws is *complete* in the sense that some clearly defined subset of all truths about programming can be deduced directly from the laws, without appeal to the possibly greater complexity of the definitions. This could be extremely useful to the practising programmer, who does not have to know the foundations of the subject any more than the scientist has to know about the definition of real numbers in terms of Dedekind cuts.

Although nearly one hundred laws are given in this article, we are still a long way from knowing how to apply them directly to the design of correct and efficient programs on the scale required by modern technology. Gaining practical experience in the application of these mathematical laws to programming is the way to go. The search for deeper and more specific theorems that can be used more simply on limited but not too narrow ranges of problems should continue. That is the way that applied mathematics, as well as pure mathematics, has made such great progress in the last two thousand years. If we follow this example, perhaps we may make more rapid progress in Computing Science, both in theoretical research and in its practical application.

# References


1. Backhouse, R.C. Program Construction and Verification. Prentice-Hall International, London, 1986.

2. Backus, J. Can programming be liberated from the von Neumann style? Commun. ACM 21, 8(Aug. 1978), 613-641.

3. de Bakker, J.W. Mathematical Theory of Program Correctness. Prentice-Hall International, London, 1980.

4. Dijkstra, E.W. A Discipline of Programming. Prentice-Hall, Englewood Cliffs, N.J., 1976.

5. Gries, D. The Science of Programming. Springer-Verlag, New York, 1981.

6. Hehner, E.C.R. Predicative programming parts I and II. Commun. ACM 27, 2 (Feb 1984), 134-151.

7. Hoare, C.A.R., and He, J. Weakest prespecification. Tech. Monogr. PRG-44, Programming Research Group, Oxford Univ., 1985.

8. Igarishi, S. An axiomatic approach to equivalence problems of algorithms with applications. Rep., Computer Centre, Univ. of Tokyo, 1968.

9. Jones, C.B. Software Development: A Rigorous Approach. Prentice–Hall International, London, 1980.

10. Kowalski, R.A. The relation between logic programming and logic specification. In Mathematical Logic and Programming Languages, C.A.R. Hoare and J.C. Shepherdson, Eds. Prentice–Hall International, London, 1985, pp. 11–27.

11. Roscoe, A.W. Laws of Occam programming. Tech. Monogr. PRG–53, Programming Research Group, Oxford Univ., 1986.

12. Scott, D.S. Outline of a mathematical theory of computation. Tech. Monogr. PRG–2, Programming Research Group, Oxford Univ., 1970.

13. Tarski, A. On the calculus of relations. J.Symbolic Logic 6 (1941), 73–89.

With acknowledgements to the co-authors: I.J. Hayes, He Jifeng, C.C. Morgan, A.W. Roscoe, J.W. Sanders, I.H. Sorensen, J.M. Spivey and B.A. Sufrin.

Published in the August 1987 issue of Communications of the ACM.

# SOME APPLICATIONS OF POINTER ALGEBRA

Bernhard Möller

Institut für Mathematik der Universität Augsburg
Universitätsstr.2, W-8900 Augsburg, Germany


## 1 Introduction

It is well-known that algorithms involving pointers are both difficult to write and to verify. One reason is that, due to the implict connections through paths within a pointer structure, the side effects of a pointer assignment are usually much harder to survey than those of an ordinary assignment. Second, a careless assignment may destroy the last link to a substructure which thus is lost forever. Now, not only is it easy to make such errors; it is also very hard to find them. With this paper we want to show that these difficulties can be greatly reduced by making the store, which is an implicit global parameter in procedural languages, into an explicit parameter and by passing to an applicative treatment using a suitable algebra of operations on the store.

The storage state of a von Neumann machine can be viewed as a total mapping from addresses to certain values. A part of such a state that forms a logical unit may then be represented by a partial submapping of that mapping. This gives the possibility of describing the state in a modularized way as the union of the submappings for its logical subunits. In the case of pointer structures this means that the usual "spaghetti" structure of the complete state can be (at least partly) disentangled. Therefore we use the algebra of partial maps as our tool for specifying and developing pointer algorithms in a formal and yet convenient way.

In the first part, we restrict ourselves to the case of singly linked lists. However, the approach is not limited to such simple structures: In [Berger et al. 89] we have derived an efficient and intricate garbage collection algorithm for a storage structure that allows the representation of arbitrary graphs. The second part presents some essential steps of this development.


## 2 Preliminaries

Notationally and conceptually, we closely follow the (ALGOL variant of) the language CIP-L (cf. [Bauer et al. 85, 89]). CIP-L has a mathematical semantics that associates with each expression $E$ the set $\mathcal{B}[\![E]\!]$ of possible values; $\mathcal{B}[\![E]\!]$ is called the **breadth** of $E$. The special value $\perp$ models the possibility of an erroneous or nonterminating computation. We set

$$\mathrm{DEFINED}[\![E]\!] \overset{\text{def}}{\Leftrightarrow} \perp \notin \mathcal{B}[\![E]\!] \,,$$
$$\mathrm{DETERMINATE}[\![E]\!] \overset{\text{def}}{\Leftrightarrow} |\mathcal{B}[\![E]\!]| = 1 \,,$$

and call $E$ **defined** resp. **determinate** if $\mathrm{DEFINED}[\![E]\!]$ resp. $\mathrm{DETERMINATE}[\![E]\!]$ holds. For convenience, the semantical values are also considered as expressions. The details of the semantic description can be found in [Bauer et al. 85].

Based on the breadth, two fundamental relations between expressions are defined: $E_1$ and $E_2$ are called **equivalent** if $\mathcal{B}[\![E_1]\!] = \mathcal{B}[\![E_2]\!]$; we denote this by $E_1 \equiv E_2$. This "strong" or meta-equality is not to be confused with "weak" (i.e. strict) equality tests $=$ in the language itself: The formula

$$\bot \equiv \bot$$

is valid. However,

$$\bot = \bot \;\not\equiv\; \textbf{true} \;;$$

rather we have

$$\bot = \bot \;\equiv\; \bot \;.$$

Equivalences are also denoted in the form of transformation rules, viz. as

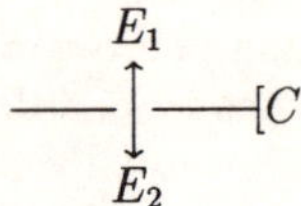

where $C$ is a (possibly empty) list of applicability conditions, i.e., of conditions sufficient for the validity of the equivalence. More generally, $E_2$ is called a **descendant** of $E_1$ if $\mathcal{B}[\![E_2]\!] \subseteq \mathcal{B}[\![E_1]\!]$; we denote this by $E_1 \sqsupseteq_D E_2$ and, in the form of a transformation rule, as

$$
\begin{array}{c}
E_1 \\
\rule{2cm}{0.4pt}\!\!\downarrow\!\!\rule{1cm}{0.4pt}\!\!-\!\!\{C \\
E_2
\end{array}
$$

where $C$ again is a list of applicability conditions.

As an important aid in specifying and developing recursive routines we use assertions about the objects involved. They are formulated as Boolean expressions of the language. Given such an expression $P$, we use the notation

$$P \;\triangleright\; E$$

as an abbreviation for the expression

$$\textbf{if } P \textbf{ then } E \textbf{ else } \bot \textbf{ fi} \;.$$

A collection of useful algebraic properties of this construct can be found in [Möller 89]. Our principal use of it is within parameter restrictions for functions (cf. [Bauer, Wössner 82, Bauer et al. 85]): Let $R$ be a Boolean expression possibly involving the identifier $x$. Then the declaration

$$\textbf{funct } f \;\equiv\; (\textbf{m } x : R)\textbf{n} : E$$

of function $f$ with parameter $x$ restricted by $R$ and with body $E$ is by definition equivalent to

$$\textbf{funct } f \;\equiv\; (\textbf{m } x)\textbf{n} : R \;\triangleright\; E \;.$$

This means that $f$ is undefined for all arguments $x$ that violate the restriction $R$; i.e., $R$ acts as a precondition for $f$. If $f$ is recursive, $R$ has to hold also for the parameters of the recursive calls to ensure definedness; hence in this case $R$ corresponds to invariants as known from imperative programming. Analogous constructions apply to statements and procedures.

As our next piece of notation we introduce the conditional conjunction **cand** and the conditional disjunction **cor** defined by

$$x \text{ \textbf{cand} } y \quad \overset{\text{def}}{\equiv} \quad \text{\textbf{if} } x \text{ \textbf{then} } y \text{ \textbf{else} false fi}$$
$$x \text{ \textbf{cor} } y \quad \overset{\text{def}}{\equiv} \quad \text{\textbf{if} } x \text{ \textbf{then} true \textbf{else} } y \text{ \textbf{fi}} .$$

So they are asymmetric, sequentially evaluated variants of the usual boolean operations $\wedge$ and $\vee$. We mainly use them to shield partialities. E.g., the expression

$$x \neq 0 \text{ \textbf{cand} } 10/x = 2$$

is defined for all $x$, whereas

$$x \neq 0 \wedge 10/x = 2$$

is undefined for $x \equiv 0$, since then $10/x$ is undefined and $=$ and $\wedge$ are strict. Of course, in addition **cand** and **cor** are usually more efficient than $\wedge$ and $\vee$, since, as the above definition shows, they may save evaluation of the second operand.


# 3  The Algebra of Partial Maps

The use of algebraic operations on maps for describing the effect of a program dates back at least to [Reynolds 79]. The most useful operation in our setting, viz. map union, however, seems to have been neglected until recently [Berger et al. 89, Pepper, Möller 89].

A (**partial**) **map** $m$ from a set $M$ to a set $N$ is a subset of $M \times N$ such that $(x,y) \in m \wedge (x,z) \in m \Rightarrow y \equiv z$. Some of our notation derives from this set view of maps. E.g., by $\emptyset$ we denote the empty partial map from $M$ to $N$. For finite maps we assume a strict boolean-valued equality test $=$.

Let $m : P \longrightarrow Q$ be a partial map. We write $\downarrow m, \uparrow m$ for **domain** and **range** of $m$, resp. For $s \subseteq P$, $[s \mapsto y]$ is the constant map $\{(x,y) \mid x \in s\}$. In using this notation we omit singleton set braces, i.e., we write $[x \mapsto y]$ instead of $[\{x\} \mapsto y]$. Note that $[x \mapsto y] \equiv \{(x,y)\}$. To cope with partialities in an algebraically convenient way, we define, for maps $m, n$ and elements $x, y$,

$$[m(x) \mapsto n(y)] \overset{\text{def}}{\equiv} \emptyset$$

if $x \notin \downarrow m$ or $y \notin \downarrow n$.

The **restriction** of a map $m : M \longrightarrow N$ to a set $s \subseteq M$ is

$$m|s \overset{\text{def}}{\equiv} m \cap (s \times N) .$$

Moreover,

$$m \ominus s \overset{\text{def}}{\equiv} m|\overline{s} .$$

Here again we omit singleton set braces, i.e., we write $m \ominus x$ instead of $m \ominus \{x\}$. Note that both $m|s \subseteq m$ and $m \ominus s \subseteq m$.

The intersection $m \cap n$ of two maps is again a map. Restriction interacts with intersection as follows:

$$(m \cap n)|s \equiv m|s \cap n|s$$
$$(m \cap n) \ominus s \equiv m \ominus s \cap n \ominus s$$
$$m|(s \cap t) \equiv m|s \cap m|t$$
$$m \ominus (s \cap t) \equiv m \ominus s \cup m \ominus t .$$

Two maps $m, n : M \longrightarrow N$ are **compatible** if $m|(\downarrow m \cap \downarrow n) \equiv n|(\downarrow m \cap \downarrow n)$. This holds in particular if $\downarrow m \cap \downarrow n \equiv \emptyset$. We can give a technically simpler characterization of compatibility:

**Lemma 3.1**

$m$ and $n$ are compatible iff $m|\downarrow n \equiv n|\downarrow m$.

**Proof:** We observe that

$$
\begin{aligned}
& m|(\downarrow m \cap \downarrow n) \\
\equiv\; & m|\downarrow m \cap m|\downarrow n \\
\equiv\; & m \cap m|\downarrow n \\
\equiv\; & \quad (\text{since } m|\downarrow n \subseteq m) \\
& m|\downarrow n \;.
\end{aligned}
$$

Now the claim is immediate. $\blacksquare$

The union $m \cup n$ of two maps $m, n$ is again a map iff $m$ and $n$ are compatible. More generally, for a family $(m_i)_{i \in I}$ of maps ($I$ may even be infinite) the union $\bigcup_{i \in I} m_i$ is a map iff the maps $m_i$ are pairwise compatible. If $I \equiv \emptyset$, we set $\bigcup_{i \in I} m_i \equiv \emptyset$ as well. It should be clear that $\emptyset$, $[. \mapsto .]$, and $\bigcup$ form a complete set of constructors for the set of partial maps, since we have

$$
m \equiv \bigcup_{x \in \downarrow m} [x \mapsto m(x)] \;.
$$

The operation of map union is the key tool in obtaining a modular description of pointer structures, since it allows viewing a (total) storage state as the union of those of its (partial) substates that form logical units. This aspect of modularization is reflected by a large number of distributive laws that allow propagation of operations to substates of a state. For the operations introduced so far we have:

$$
\begin{aligned}
\downarrow(m \cup n) &\equiv \downarrow m \cup \downarrow n & \uparrow(m \cup n) &\equiv \uparrow m \cup \uparrow n \\
(m \cup n)|s &\equiv m|s \cup n|s & (m \cup n) \ominus s &\equiv m \ominus s \cup m \ominus t \;.
\end{aligned}
$$

The following decomposition property is the key to recursions over maps:

$$
m \equiv m|s \cup m \ominus s \;.
$$

Another important operation is **map overwriting** (see e.g. [Jones 80]): Given maps $m, n : M \longrightarrow N$ we define

$$
m \triangleleft n \stackrel{\text{def}}{\equiv} (m \ominus \downarrow n) \cup n \;.
$$

Hence,

$$
(m \triangleleft n)(x) \equiv \text{if } x \in \downarrow n \text{ then } n(x) \text{ else } m(x) \text{ fi} \;.
$$

In other words, $m \triangleleft n$ results from $m$ by changing the values according to the prescription of $n$ (if any). For example, forming $m \triangleleft [x \mapsto y]$ makes $y$ the value corresponding to $x$. This operation will be our main tool for describing selective updating. We now prove a number of useful properties of this operation.

**Lemma 3.2**

$\downarrow(m \triangleleft n) \equiv \downarrow m \cup \downarrow n$

**Proof:**
$$\downarrow(m \twoheadleftarrow n)$$
$$\equiv \quad \downarrow(m \ominus \downarrow n \cup n)$$
$$\equiv \quad \downarrow(m \ominus \downarrow n) \cup \downarrow n$$
$$\equiv \quad \downarrow m \backslash \downarrow n \cup \downarrow n$$
$$\equiv \quad \downarrow m \cup \downarrow n$$

$\blacksquare$

Maps form a monoid under overwriting:

## Lemma 3.3 (Monoid)

(1) $\emptyset \twoheadleftarrow m \equiv m \twoheadleftarrow \emptyset \equiv m$

(2) $(l \twoheadleftarrow m) \twoheadleftarrow n \equiv l \twoheadleftarrow (m \twoheadleftarrow n)$

**Proof:** (1)
$$\emptyset \twoheadleftarrow m$$
$$\equiv \quad \emptyset \ominus \downarrow m \cup m$$
$$\equiv \quad \emptyset \cup m$$
$$\equiv \quad m$$

$$m \twoheadleftarrow \emptyset$$
$$\equiv \quad m \ominus \downarrow \emptyset \cup \emptyset$$
$$\equiv \quad m \ominus \emptyset$$
$$\equiv \quad m$$

(2)
$$(l \twoheadleftarrow m) \twoheadleftarrow n$$
$$\equiv \quad (l \ominus \downarrow m \cup m) \ominus \downarrow n \cup n$$
$$\equiv \quad (l \ominus \downarrow m) \ominus \downarrow n \cup m \ominus \downarrow n \cup n$$
$$\equiv \quad l \ominus (\downarrow m \cup \downarrow n) \cup m \ominus \downarrow n \cup n$$
$$\equiv \quad \text{(by Lemma 3.2)}$$
$$l \ominus \downarrow(m \twoheadleftarrow n) \cup (m \twoheadleftarrow n)$$
$$\equiv \quad l \twoheadleftarrow (m \twoheadleftarrow n)$$

$\blacksquare$

## Lemma 3.4 (Overwriting and union)

$m \twoheadleftarrow n \equiv n \twoheadleftarrow m$ iff $m$ and $n$ are compatible.

In this case, $m \twoheadleftarrow n \equiv m \cup n$.

**Proof:** ($\Rightarrow$) Assume that $m \ominus \downarrow n \cup n \equiv n \ominus \downarrow m \cup m$. Then
$$m \ominus \downarrow n \equiv (n \ominus \downarrow m \cup m) \backslash n \equiv m \backslash n \,.$$

Now
$$m | \downarrow n$$
$$\equiv \quad \text{(decomposition)}$$
$$m \backslash (m \ominus \downarrow n)$$
$$\equiv \quad m \backslash (m \backslash n)$$
$$\equiv \quad m \cap n \,.$$

Since the assumption is symmetric in $m$ and $n$, we may substitute $n, m$ for $m, n$ and hence obtain also $n|{\downarrow}m \equiv n \cap m$. Now $m, n$ are compatible by commutativity of $\cap$ and Lemma 3.1.

($\Leftarrow$) By the assumption and Lemma 3.1, $m|{\downarrow}n \equiv n|{\downarrow}m$. Hence

$$
\begin{aligned}
& m \twoheadleftarrow n \\
\equiv\ & m \ominus {\downarrow}n \cup n \\
\equiv\ & \text{(decomposition)} \\
& m \backslash (m|{\downarrow}n) \cup n \\
\equiv\ & m \backslash (n|{\downarrow}m) \cup n \\
\equiv\ & \text{(since } n|{\downarrow}m \subseteq n) \\
& m \cup n \,.
\end{aligned}
$$

Now use symmetry.

$\blacksquare$

To proceed further, we now state some general facts about binary operations. Let $\oplus$ be a binary operation. Define

$$
\begin{aligned}
x \le_\oplus y & \stackrel{\text{def}}{\Leftrightarrow} x \oplus y \equiv y \\
x\ _\oplus{\le}\ y & \stackrel{\text{def}}{\Leftrightarrow} x \oplus y \equiv x
\end{aligned}
$$

**Lemma 3.5**

(1) If $\oplus$ is idempotent then $_\oplus{\le}, \le_\oplus$ are reflexive.

(2) If $\oplus$ is commutative then $_\oplus{\le}, \le_\oplus$ are antisymmetric and converses of each other.

(3) If $\oplus$ is associative then $_\oplus{\le}, \le_\oplus$ are transitive.

(4) If $\oplus$ has a neutral element 0 then 0 is least w.r.t. $\le_\oplus$ and greatest w.r.t. $_\oplus{\le}$.

Now, since $\twoheadleftarrow$ is idempotent and associative with neutral element $\emptyset$, the relations $_\twoheadleftarrow{\le}, \le_\twoheadleftarrow$ are preorders. It turns out that they have interesting equivalent characterizations:

**Lemma 3.6**

(1) $m \twoheadleftarrow n \equiv n \Leftrightarrow {\downarrow}m \subseteq {\downarrow}n$

(2) $m \twoheadleftarrow n \equiv m \Leftrightarrow n \subseteq m$

**Proof:** (1)

$$
\begin{aligned}
& m \twoheadleftarrow n \equiv n \\
\Rightarrow\ & {\downarrow}(m \twoheadleftarrow n) \equiv {\downarrow}n \\
\Rightarrow\ & {\downarrow}m \cup {\downarrow}n \equiv {\downarrow}n \\
\Leftrightarrow\ & {\downarrow}m \subseteq {\downarrow}n \,.
\end{aligned}
$$

On the other hand, ${\downarrow}m \subseteq {\downarrow}n$ implies $m \ominus {\downarrow}n \equiv \emptyset$ and hence $m \twoheadleftarrow n \equiv m \ominus {\downarrow}n \cup n \equiv n$.

(2) From $m \twoheadleftarrow n \equiv m \ominus {\downarrow}n \cup n \equiv m$ we conclude $n \subseteq m$. Conversely, $n \subseteq m$ implies $m|{\downarrow}n \equiv n$ and hence

$$
\begin{aligned}
& m \\
\equiv\ & \text{(decomposition)}
\end{aligned}
$$

$$m \ominus {\downarrow}n \cup m|{\downarrow}n$$
$$\equiv\ m \ominus {\downarrow}n \cup n$$
$$\equiv\ m \dashleftarrow n \ .$$

∎

## Lemma 3.7 (Sequentialization)

$l \dashleftarrow (m \cup n) \equiv (l \dashleftarrow m) \dashleftarrow n$ provided $m$ and $n$ are compatible.

**Proof:** Immediate from Lemma 3.4 and associativity of $\dashleftarrow$.  ∎

## Lemma 3.8 (Annihilation)

$m \subseteq l \ \Rightarrow\ l \dashleftarrow (m \cup n) \equiv l \dashleftarrow n$ provided $m$ and $n$ are compatible.

**Proof:** Immediate from sequentialization and Lemma 3.6.  ∎

## Lemma 3.9 (Distributivity)

$(l \cup m) \dashleftarrow n \equiv (l \dashleftarrow n) \cup (m \dashleftarrow n)$ provided $l$ and $m$ are compatible.

**Proof:**
$$(l \cup m) \dashleftarrow n$$
$$\equiv\ (l \cup m) \ominus {\downarrow}n \cup n$$
$$\equiv\ l \ominus {\downarrow}n \cup m \ominus {\downarrow}n \cup n$$
$$\equiv\ l \ominus {\downarrow}n \cup n \cup m \ominus {\downarrow}n \cup n$$
$$\equiv\ (l \dashleftarrow n) \cup (m \dashleftarrow n) \ .$$

∎

The following property allows localizing side effects to that part of a store they really affect:

## Lemma 3.10 (Localization)

${\downarrow}l \ \cap\ {\downarrow}n \ \equiv\ \emptyset$ implies $(l \cup m) \dashleftarrow n \equiv l \cup (m \dashleftarrow n)$
provided $l$ and $m$ are compatible.

**Proof:**
$$(l \cup m) \dashleftarrow n$$
$$\equiv\ (l \cup m) \ominus {\downarrow}n \cup n$$
$$\equiv\ l \ominus {\downarrow}n \cup m \ominus {\downarrow}n \cup n$$
$$\equiv\ l \cup m \ominus {\downarrow}n \cup n$$
$$\equiv\ l \cup (m \dashleftarrow n) \ .$$

∎

The map operations introduced enjoy a vast number of further useful algebraic laws. Some of them can be found in [Berger et al. 89].

# 4  Chains

As an example of how to describe pointer structures within the algebra of maps we now study singly linked lists. We abstract from the concrete contents of the records in such a list and consider only their interrelationship through the pointers, since this is the only source of problems in pointer algorithms. Then a **state** simply is a finite partial map $m : \mathbf{cell} \longrightarrow \mathbf{cell}$ where **cell** is the set of storage cells; the set of states is denoted by **state**. A single cell $x$ together with its contents $y$ is modeled by the map $[x \mapsto y]$.

By a **chain** we mean a (finite) cycle-free singly linked list. Such a chain contains a number of cells in a certain order prescribed by the links in the list. This induces a sequence structure on these cells: The first element in the sequence is the head cell, followed by the others in the order of traversal. Since there is no cycle, the sequence is repetition-free.

In treating chains one frequently uses a special chain terminator common to all chains considered (e.g., nil in Pascal). Let therefore $\square \in$ **cell** be a distinguished element, called the **anchor**. The elements of $\mathbf{cell}\backslash\{\square\}$ are called **proper cells**. In the sequel we require $\square \notin \downarrow m$ for all states $m$ considered. This means that $\square$ may never be assigned a contents and hence never be dereferenced; it will always be an empty cell, whence our notation. Moreover, this implies that there can be no $\square$ cell properly within a chain; if present, $\square$ terminates the respective list. A chain is called **anchored** if it is empty or ends with $\square$, i.e., its last proper cell contains $\square$.

By the above considerations, anchored chains are in exact correspondence with non-empty repetition-free sequences of proper cells. Given such a sequence, we can construct an anchored chain using

$$\mathbf{funct}\ chain\ \equiv\ (\mathbf{cellsequ}\ s : ischainable(s))\,\mathbf{state} : \bigcup_{i=1}^{|s|} [s[i] \mapsto s[i+1]]\ .$$

By $|s|$ we denote the length of $s$ and by $s[i]$ the $i$-th element of $s$; if $i > |s|$ or $i = 0$, we set $s[i] \stackrel{\mathrm{def}}{\equiv} \square$. The predicate $ischainable$ is given by

$$\begin{aligned}
&\mathbf{funct}\ ischainable\ \equiv\ (\mathbf{cellsequ}\ s)\,\mathbf{bool} : \\
&\quad \mathbf{if}\ s = \diamond\ \mathbf{then}\ \mathbf{true} \\
&\quad\quad\quad\quad \mathbf{else}\ first(s) \neq \square\ \wedge\ first(s) \notin rest(s)\ \wedge\ ischainable(rest(s))\ \mathbf{fi}\ ,
\end{aligned}$$

where $\diamond$ denotes the empty sequence, $first(s) \stackrel{\mathrm{def}}{\equiv} s[1]$, and $rest(s) \stackrel{\mathrm{def}}{\equiv} s[2 : |s|]$. Note that $first(\diamond) \equiv \square$. We have

**Lemma 4.1**
    (1) $chain(<x> + s)\ \equiv\ [x \mapsto first(s)] \cup chain(s)$.
    (2) $chain(t + <x> + s)\ \supseteq\ [x \mapsto first(s)]$

Here, $<x>$ is the singleton sequence consisting just of $x$, and $+$ denotes concatenation.

Conversely, given a cell $x$ and a state $m$, we can retrieve the sequence of cells in the sublist starting from $x$ (if any) using

$$\begin{aligned}
&\mathbf{funct}\ sequ\ \equiv\ (\mathbf{cell}\ x, \mathbf{state}\ m)\,\mathbf{cellsequ} : \\
&\quad \mathbf{if}\ x \notin \downarrow m\ \mathbf{then}\ \diamond\ \mathbf{else}\ <x> + sequ(m(x), m)\ \mathbf{fi}\ .
\end{aligned}$$

Note that this function will not terminate if the sublist within $m$ starting from $x$ contains a cycle. In our applications this will not occur. For more general use, however, one should base this on a non-strict functional language in which the algorithm then would return a periodically infinite sequence of cells. Then a cell $y$ can be reached from $x$ following the links of $m$ (zero or more times) iff $y \in sequ(x, m) \equiv \mathbf{true}$, where

$$\begin{aligned}
&\mathbf{funct}\ .\in.\ \equiv\ (\mathbf{cell}\ y, \mathbf{cellsequ}\ s)\,\mathbf{bool} : \\
&\quad \mathbf{if}\ s = \diamond\ \mathbf{then}\ \mathbf{false}\ \mathbf{else}\ y = first(s)\ \mathbf{cor}\ y \in rest(s)\ \mathbf{fi}\ .
\end{aligned}$$

To characterize the case where the sublist starting from $x$ in $m$ is an anchored chain, we use

> **funct** *isanchored* $\equiv$ (**cell** $x$, **state** $m$) **bool** :
>   **if** $x \notin\ \downarrow m$ **then** $x = \square$ **else** *isanchored*$(m(x), m)$ **fi** .

This function again doesn't terminate if there is a cycle, and it yields **false** if it runs into a "dangling reference", i.e., a cell different from $\square$ not having any contents. Note that *isanchored*$(\square, m)$ $\equiv$ **true**. This is reasonable, since $sequ(\square, m) \equiv \diamond$. Moreover, we get

**Lemma 4.2**
  (1) $isanchored(x, m)\ \wedge\ x \notin\ \downarrow m\ \equiv\ x = \square$ .
  (2) $first(sequ(x, m))\ \equiv\ x$ provided $isanchored(x, m)\ \equiv$ **true** .

We say that a pair $(x, m)$ **represents** a chainable sequence $s$ if $sequ(x, m) \equiv s$. The corresponding representation function is

> **funct** *chainrep* $\equiv$ (**cellsequ** $s$ : *ischainable*$(s)$)(**cell**, **state**) :
>   $(first(s), chain(s))$

Note that $chainrep(\diamond) \equiv (\square, \emptyset)$. Moreover,

**Lemma 4.3**
  (1) $chainrep(s)$ represents $s$ provided $ischainable(s) \equiv$ **true**, i.e.,
      $sequ(chainrep(s))\ \equiv\ s$ provided $ischainable(s) \equiv$ **true** .
  (2) $chain(sequ(x, m))\ \subseteq\ m$ provided $isanchored(x, m)\ \equiv$ **true** .

We now investigate how union and overwriting affect chains. To this end we need some more notation. For cell $x$ and state $m$ we define the set $m^*(x)$ of all cells reachable from $x$ via links in $m$ by

$$m^*(x) \stackrel{\text{def}}{\equiv} \bigcup_{i \in \mathbb{N}} m^i(x)$$

where

$$m^0(x) \stackrel{\text{def}}{\equiv} \{x\}$$
$$m^{i+1} \stackrel{\text{def}}{\equiv} (m^i(x))\!\uparrow m ,$$

and $s\!\uparrow m$ denotes the image of a set $s$ under $m$. Note that

$$m^*(x)\ \equiv\ set(sequ(x, m)) \cup \{\square\}$$

provided $isanchored(x, m) \equiv$ **true**. Moreover,

$$m^*(x)\ \subseteq\ \{x\} \cup \uparrow m .$$

**Lemma 4.4**
  $sequ(x, m \cup n)\ \equiv\ sequ(x, m)$ provided $m^*(x) \cap \downarrow n\ \equiv\ \emptyset$.

**Proof:** We use the well-known technique of computational induction (see e.g. [Manna 74]) for proving properties of recursively defined functions: Consider a continuous (or admissible) predicate $P$ and a recursive definition

$$\textbf{funct } g\ \equiv\ \tau[g]$$

with least fixpoint $\mu_\tau$ of the associated functional $\tau$. If we can show $P[\Omega]$ (where $\Omega$ is the totally undefined function) and $\forall f : P[f] \Rightarrow P[\tau[f]]$, then we may infer $P[\mu_\tau]$ as well. Here, we choose the predicate

$$P[f] \stackrel{\text{def}}{\Leftrightarrow} \forall x, m, n :$$
$$m^*(x) \cap \downarrow n \equiv \emptyset \Rightarrow f(x, m \cup n) \equiv f(x, m) .$$

The induction base $P[\Omega]$ is trivial. For the induction step assume $P[f]$ and $m^*(x) \cap \downarrow n \equiv \emptyset$. This implies, in particular, $x \notin \downarrow n$. The functional belonging to the definition of $sequ$ is

$$\tau \; : \; f \; \mapsto \; (\text{cell } x, \text{state } m) \, \text{cellsequ} :$$
$$\text{if } x \notin \downarrow m \text{ then } \diamondsuit \text{ else } \mathord{<}x\mathord{>} + f(m(x), m) \text{ fi} .$$

We calculate

$$\begin{aligned}
& \tau[f](x, m \cup n) \\
\equiv \; & \text{if } x \notin \downarrow(m \cup n) \text{ then } \diamondsuit \\
& \qquad\qquad \text{else } \mathord{<}x\mathord{>} + f((m \cup n)(x), m \cup n) \text{ fi} \\
\equiv \; & \text{if } x \notin \downarrow m \cup \downarrow n \text{ then } \diamondsuit \\
& \qquad\qquad \text{else } \mathord{<}x\mathord{>} + f((m \cup n)(x), m \cup n) \text{ fi} \\
\equiv \; & \quad (\text{since } x \notin \downarrow n) \\
& \text{if } x \notin \downarrow m \text{ then } \diamondsuit \text{ else } \mathord{<}x\mathord{>} + f(m(x), m \cup n) \text{ fi} \\
\equiv \; & \quad (\text{by the induction hypothesis } P[f], \text{ since } m^*(m(x)) \subseteq m^*(x)) \\
& \text{if } x \notin \downarrow m \text{ then } \diamondsuit \text{ else } \mathord{<}x\mathord{>} + f(m(x), m) \text{ fi} \\
\equiv \; & \tau[f](x, m) .
\end{aligned}$$

$\blacksquare$

**Corollary 4.5**

$$sequ(x, m) \equiv sequ(x, m \leftarrow [u \mapsto v]) \text{ provided } u \notin m^*(x) .$$

**Proof:** First we note that from the definition it is immediate that $.^*$ is monotonic:

$$l \subseteq n \Rightarrow \forall x : l^*(x) \subseteq n^*(x) .$$

Hence $u \notin m^*(x)$ implies $u \notin (m \ominus u)^*(x)$, so that

$$\begin{aligned}
& sequ(x, m) \\
\equiv \; & \quad (\text{decomposition}) \\
& sequ(x, m \ominus u \cup [u \mapsto m(u)]) \\
\equiv \; & \quad (\text{by the above lemma}) \\
& sequ(x, m \ominus u) \\
\equiv \; & \quad (\text{by the above lemma}) \\
& sequ(x, m \ominus u \cup [u \mapsto v]) \\
\equiv \; & sequ(x, m \leftarrow [u \mapsto v]) .
\end{aligned}$$

$\blacksquare$

# 5  Concatenation of Chains "in Situ"

## 5.1  Specification and First Explicit Solution

We now want to specify and develop an algorithm for concatenating two non-overlapping anchored chains "in situ". First we give the precondition for our desired function:

> funct $disjoint$ $\equiv$ (cell $x$, cell $y$, state $m$) bool :
> $(isanchored(x,m)$ $\wedge$ $isanchored(y,m))$ cand
> $set(sequ(x,m)) \cap set(sequ(y,m)) = \emptyset$ .

Here, $set(s)$ gives the set of all cells occurring in sequence $s$. So we consider a state $m$ in which the sublists starting from $x$ and $y$ are anchored chains the sets of proper cells of which are disjoint. We want to form a new state in which the concatenation of these two sublists is overwritten onto *the same* set of proper cells; moreover, the order of traversal within the sublists should be preserved, and all cells from the sublist of $x$ should precede all cells in the sublist of $y$. This can be specified by

> funct $conc$ $\equiv$ (cell $x$, cell $y$, state $m$ : $disjoint(x,y,m)$) state :
> $m \leftarrow chain(sequ(x,m) + sequ(y,m))$ .

So the proper cells of the subchains are collected in the right order, the resulting sequence is chained, and this chain is overwritten onto $m$ *re-using the same cells*. Hence, no copying is involved and we really are specifying concatenation "in situ".

We now want to develop an algorithm from this specification. Our first goal is a recursion without the "detour" through sequences. We try to obtain it by the familiar unfold/fold technique. We consider the following cases:

**Case 1:** $x \notin \downarrow m$, i.e., $x = \square$ by $isanchored(x,m)$ and Lemma 4.2(1). Then $sequ(x,m) \equiv \diamond$, and hence

$$\begin{aligned}
& m \leftarrow chain(sequ(x,m) + sequ(y,m)) \\
\equiv\ & m \leftarrow chain(sequ(y,m)) \\
\equiv\ & \quad (\text{since } chain(sequ(y,m)) \subseteq m) \\
& m \ .
\end{aligned}$$

**Case 2:** $x \in \downarrow m$ cand $m(x) \notin \downarrow m$, i.e., $m(x) = \square$. Then $sequ(x,m) \equiv {<}x{>} +sequ(m(x),m)$ $\equiv {<}x{>}$, and hence

$$\begin{aligned}
& m \leftarrow chain(sequ(x,m) + sequ(y,m)) \\
\equiv\ & m \leftarrow chain({<}x{>} +sequ(y,m)) \\
\equiv\ & \quad (\text{by Lemma 4.1(1)}) \\
& m \leftarrow ([x \mapsto y] \cup chain(sequ(y,m))) \\
\equiv\ & \quad (\text{annihilation, since } chain(sequ(y,m)) \subseteq m \text{ by Lemma 4.3(2)}) \\
& m \leftarrow [x \mapsto y] \ .
\end{aligned}$$

**Case 3:** $x \in \downarrow m$ cand $m(x) \in \downarrow m$. Then $sequ(x,m) \equiv {<}x{>} +sequ(m(x),m)$, and hence

$$\begin{aligned}
& m \leftarrow chain(sequ(x,m) + sequ(y,m)) \\
\equiv\ & m \leftarrow chain({<}x{>} +sequ(m(x),m) + sequ(y,m))
\end{aligned}$$

134

$$\equiv \quad \text{(by Lemma 4.1(1))}$$
$$m \mathbin{+\mkern-10mu\leftarrow} ([x \mapsto m(x)] \cup chain(sequ(m(x), m) + sequ(y, m)))$$
$$\equiv \quad \text{(annihilation, since } [x \mapsto m(x)] \subseteq m)$$
$$m \mathbin{+\mkern-10mu\leftarrow} chain(sequ(m(x), m) + sequ(y, m))$$
$$\equiv \quad \text{(fold } conc)$$
$$conc(m(x), y, m) \ .$$

Altogether we have

> **funct** $conc \equiv$ (**cell** $x$, **cell** $y$, **state** $m : disjoint(x, y, m)$) **state** :
>    **if** $x = \square$ **then** $m$
>         **else if** $m(x) = \square$ **then** $m \mathbin{+\mkern-10mu\leftarrow} [x \mapsto y]$
>                 **else** $conc(m(x), y, m)$ **fi fi** .

Termination of this recursion follows from $isanchored(x, m)$. It is quite reassuring that the fundamental unfold/fold technique for deriving recursions also applies to pointer algorithms in this setting.

## 5.2 Introducing Selective Updating

Since we have even obtained a tail-recursive version, we are already very close to an imperative program. To get there, we introduce a procedure specified by

> **proc** $powconc \equiv$ (**var state** $m$, **cell** $x, y : disjoint(x, y, m)$)
>    $m := conc(m, x, y)$ .

Note that this clearly specifies $m$ as a transient parameter, whereas $x$ and $y$ are passed by value. Therefore the imperative version of $powconc$ needs local variables for $x$ and $y$, whereas it may operate on $m$ directly. This is described by the following schematic rule for passing from a procedure that calls a tail-recursive function to a procedure with a loop in its body:

> **proc** $p \equiv$ (**var m** $a$, **n** $b : P$) :   $a := f(a, b)$
> **where**
> **funct** $f \equiv$ (**m** $a$, **n** $b$) **m** :
>    **if** $C$ **then** $T$ **else** $f(K, L)$ **fi**
>
> ───────────────────── ↕ ─────────────────────┤ NEW$[\![B]\!]$
>
> **proc** $p \equiv$ (**var m** $a$, **n** $B : P[\![B$ **for** $b]\!]$) :
>    $\lceil$ **var n** $b := B$ ;
>     **while**$\neg$ $C$ **do** $(a, b) := (K, L)$ **od** ;
>     $a := T$                    $\rfloor$ .

Note that $a, b$, and $B$ may stand for tuples of identifiers. $P, C, T, K$, and $L$ stand for expressions of appropriate types, possibly involving $a$ and $b$. The condition NEW$[\![B]\!]$ states that $B$ has to be a (tuple of) fresh identifier(s). The rule does not match our current form of $conc$; however, this is easily transformed into

$$\textbf{funct } conc \;\equiv\; (\textbf{cell } x, \textbf{cell } y, \textbf{state } m : disjoint(x,y,m))\,\textbf{state}:$$

$$\textbf{if } x = \square \textbf{ cor } m(x) = \square$$
$$\quad \textbf{then if } x = \square \textbf{ then } m \textbf{ else } m \nleftarrow [x \mapsto y] \textbf{ fi}$$
$$\quad \textbf{else } \; conc(m(x), y, m) \qquad\qquad\qquad\qquad \textbf{fi }.$$

Now the rule applies leading to

$$\textbf{proc } powconc \;\equiv\; (\textbf{var state } m, \textbf{cell } X, Y : disjoint(X,Y,m)):$$
$$\lceil\; (\textbf{var cell } x, y) := (X, Y)\;;$$
$$\quad \textbf{while } x \neq \square \textbf{ cand } m(x) \neq \square \textbf{ do } (m, x, y) := (m, m(x), y) \textbf{ od }\;;$$
$$\quad m := \textbf{if } x = \square \textbf{ then } m \textbf{ else } m \nleftarrow [x \mapsto y] \textbf{ fi} \qquad\qquad \rfloor\;.$$

Our final version results from eliminating useless assignments of the form $z := z$ as well as the variable $y$ which never is changed:

$$\textbf{proc } powconc \;\equiv\; (\textbf{var state } m, \textbf{cell } X, Y : disjoint(X,Y,m)):$$
$$\lceil\; \textbf{var cell } x := X\;;$$
$$\quad \textbf{while } x \neq \square \textbf{ cand } m(x) \neq \square \textbf{ do } x := m(x) \textbf{ od }\;;$$
$$\quad \textbf{if } x = \square \textbf{ then skip}$$
$$\qquad\qquad \textbf{else } m := m \nleftarrow [x \mapsto Y] \textbf{ fi} \qquad\qquad \rfloor\;.$$

If we write the assignment

$$m := m \nleftarrow [x \mapsto Y]$$

in a Pascal-like way as

$$x \uparrow := Y\;,$$

(where $m$ now is an implicit parameter), we see that we actually have derived a version with selective updating. By standard transformation techniques one can derive a version with a simpler loop test that avoids double inspection of cells.

In the derivation we have not made use of any assumptions about absence of sharing. Indeed, if in $m$ there are pointers from other data structures to (parts of) the lists headed by $x$ and $y$, there will be indirect side effects on these pointers. However, since by the specification we know the value of the complete store after execution of our procedure, we can *calculate* these effects using our algebraic laws. Also, one can easily write stronger preconditions that exclude sharing if this is desired.

# 6 Chain Reversal

## 6.1 Specification and First Explicit Solution

Next we want to derive a procedure for reversing a non-empty chain "in situ". Again we first specify a purely applicative version. The reverse of a chain should contain exactly the same proper cells as the original chain, however, in reverse order of traversal. We can express this as follows:

$$\textbf{funct } reverse \;\equiv\; (\textbf{cell } x, \textbf{state } m : isanchored(x,m))\,\textbf{state}:$$
$$\quad m \nleftarrow chain(rev(sequ(x,m)))$$

where $rev$ is the reversal function on sequences:

$$\textbf{funct } rev \equiv (\textbf{cellsequ } s)\textbf{cellsequ} :$$
$$\textbf{if } s = \Diamond \textbf{ then } \Diamond \textbf{ else } rev(rest(s)) + \,<\!first(s)\!>\, \textbf{ fi} \,.$$

Let us now derive a recursion for *reverse*. The basic idea for the development is to adapt the standard technique for making *rev* tail-recursive. There one defines a generalized function *rrev* with an additional parameter that accumulates the intermediate results:

$$rrev(s,t) \stackrel{\text{def}}{\equiv} rev(s) + t \,.$$

*rev* is embedded into *rrev* by

$$rev(s) \equiv rrev(s, \Diamond) \,.$$

A straightforward unfold/fold derivation using associativity of $+$ leads to the tail-recursion

$$\begin{aligned}
& rrev(s,t) \\
\equiv \quad & \textbf{if } s = \Diamond \textbf{ then } t \\
& \qquad \textbf{else } rrev(rest(s), <\!first(s)\!> + t) \ \textbf{fi} \,.
\end{aligned}$$

In the case of *reverse* we now proceed similarly; however, we do not carry the accumulating submap itself as a parameter, but just its head cell. Hence we define

$$\textbf{funct } rreverse \equiv (\textbf{cell } x, y, \textbf{state } m : disjoint(x,y,m))\,\textbf{state} :$$
$$m \twoheadleftarrow chain(rev(sequ(x,m)) + sequ(y,m)) \,.$$

An appropriate embedding is

$$reverse(x,m) \equiv rreverse(x, \square, m) \,,$$

since $sequ(\square, m) \equiv \Diamond$.

As before, we now perform a case analysis.

**Case 1:** $x = \square$. Then $sequ(x,m) \equiv \Diamond$, and hence $rev(sequ(x,m)) \equiv \Diamond$. Thus

$$\begin{aligned}
& m \twoheadleftarrow chain(rev(sequ(x,m)) + sequ(y,m)) \\
\equiv \quad & m \twoheadleftarrow chain(sequ(y,m)) \\
\equiv \quad & \qquad (\text{since } chain(sequ(y,m)) \subseteq m \text{ by Lemma 4.3(2)}) \\
& m \,.
\end{aligned}$$

**Case 2:** $x \neq \square$. Then $sequ(x,m) \equiv <\!x\!> + sequ(m(x), m)$, and hence $rev(sequ(x,m)) \equiv rev(sequ(m(x),m)) + <\!x\!>$. Thus

$$\begin{aligned}
& chain(rev(sequ(x,m)) + sequ(y,m)) \\
\equiv \quad & chain(rev(sequ(m(x),m)) + <\!x\!> + sequ(y,m)) \\
\equiv \quad & \qquad (\text{by Corollary 4.5, since } disjoint(x,y,m) \text{ implies } x \notin sequ(y,m)) \\
& chain(rev(sequ(m(x),m)) + <\!x\!> + sequ(y, m \twoheadleftarrow [x \mapsto y])) \\
\equiv \quad & \qquad (\text{definition of } sequ) \\
& chain(rev(sequ(m(x),m)) + sequ(x, m \twoheadleftarrow [x \mapsto y])) \\
\equiv \quad & \qquad (\text{by Corollary 4.5, since by cyclefreeness } x \notin sequ(m(x),m)) \\
& chain(rev(sequ(m(x), m \twoheadleftarrow [x \mapsto y])) + sequ(x, m \twoheadleftarrow [x \mapsto y])) \,.
\end{aligned}$$

Now we obtain

$$m \leftarrow chain(rev(sequ(x, m)) + sequ(y, m))$$
$$\equiv \quad m \leftarrow chain(rev(sequ(m(x), m)) + <x> + sequ(y, m))$$
$$\equiv \qquad \text{(by Lemma 4.1(2) and Lemma 4.2(2))}$$
$$m \leftarrow ([x \mapsto y] \cup chain(rev(sequ(m(x), m)) + <x> + sequ(y, m)))$$
$$\equiv \qquad \text{(sequentialisation)}$$
$$m \leftarrow [x \mapsto y] \leftarrow chain(rev(sequ(m(x), m)) + <x> + sequ(y, m))$$
$$\equiv \qquad \text{(by the above subderivation)}$$
$$m \leftarrow [x \mapsto y] \leftarrow chain(rev(sequ(m(x), m \leftarrow [x \mapsto y])) + sequ(x, m \leftarrow [x \mapsto y]))$$
$$\equiv \qquad \text{(fold } rreverse)$$
$$rreverse(m(x), x, m \leftarrow [x \mapsto y]) \ .$$

Altogether, we have

$$\textbf{funct } rreverse \ \equiv \ (\textbf{cell } x, y, \textbf{state } m : disjoint(x, y, m)) \, \textbf{state} :$$
$$\textbf{if } x = \Box \textbf{ then } m$$
$$\textbf{else } \ rreverse(m(x), x, m \leftarrow [x \mapsto y]) \ \textbf{fi} \ .$$

Again we have arrived at an (obviously terminating) tail recursion.

## 6.2   A Version With Selective Updating

Specifying a procedure

$$\textbf{proc } powrev \ \equiv \ (\textbf{var state } m, \textbf{cell } x : isanchored(x, m)) :$$
$$m := reverse(x, m) \ ,$$

we obtain, as in the previous section, the final version

$$\textbf{proc } powrev \ \equiv \ (\textbf{var state } m, \textbf{cell } X : isanchored(X, m)) :$$
$$\lceil \ (\textbf{var cell } x, y) := (X, \Box) \ ;$$
$$\textbf{while } x \neq \Box$$
$$\textbf{do } (x, y, m) := (m(x), x, m \leftarrow [x \mapsto y]) \ \textbf{od} \ \rfloor \ .$$

Note that sequentialization of the collective assignment would require an auxiliary variable. This is a spot of frequent error in attempts to write down this algorithm straightforwardly without deriving it. The systematic derivation allows us to avoid such errors by using the standard knowledge about the treatment of collective assignments.

This program describes a well-known algorithm for reversing a list "in situ". Whereas verification purely at the procedural level is by no means easy (see e.g. [Burstall 72, Levy 78]), in particular if all the details were to be filled in, we have derived and thereby verified the program by a fairly short and simple formal calculation using standard transformation techniques.

## 7   A Garbage Collection Problem

We now want to present the specification and parts of the derivation of a garbage collection algorithm; the full details are given in [Berger et al. 89]. For earlier attempts at similar problems cf. [Broy, Pepper 82], [Dewar et al. 82], [van Diepen, de Roever 86]. Differing from all of these,

[Berger et al. 89] develops the algorithm to a level which can actually be transcribed directly into machine code allowing the use of overwriting, address arithmetic, and the like. Again, the algebra of partial maps is the most important tool.

The situation in which garbage collection becomes necessary is the following: The store, which accomodates a large number of records referencing one another through pointers, is exhausted, i.e., there is (almost) no more free storage left for the allocation of new records. Usually there is a distinguished set of **entry pointers** to the pointer structure which is given by the values of the currently active variables of the program that operates on the store. Only those records reachable through chains of references from the entry pointers actually need to be saved; all other records are inaccessible and thus the corresponding storage can be reclaimed.

What does garbage collection mean in a more abstract sense? To explain this, we liberate ourselves from the concrete contents of the records in the store and consider only their interrelationship through the pointers. This leads to a graph-like structure $G$ in which the nodes correspond to the records, and the arcs correspond to the pointers. If, by some process, we can distinguish a proper subgraph $G'$ of $G$ such that $G'$ contains all the nodes accessible from the entry nodes, then $G$ can be said to contain garbage about which we could as well forget. In this case, garbage collection means to compute $G'$ from $G$ and to operate on $G'$ successively.

Getting more concrete again, we work with representations of such graphs. The task then consists in computing a representation of $G'$ from one of $G$. In this paper we treat representations, called **states**, of graphs in a linear storage. For a state, the restriction to a substate usually leads to gaps in the storage; i.e., there are cells the contents of which have no meaning for the represented graph. Now, one possibility of garbage collection consists in detecting these gaps and compactifying the meaningful part by copying it to an initial interval of the storage; then a contiguous rest of the storage becomes free for further use.

We give a formal specification of this problem at the level of graphs together with a notion of their representation in a linear storage. Then we treat in detail the copying algorithm involved which is the center of the whole algorithm derived in [Berger et al. 89].


## 7.1   Storage Graphs

Storage graphs are intended to model the accessibility relations between records as given by the pointers in the records. Since the fields of a record are ordered and may contain repetitions of pointers, the usual notion of a directed graph where each node is connected to a *set* of successor nodes is not adequate for our purposes. Rather we consider *sequences* of successor nodes. Also, since we shall have to deal with arbitrary parts of such storage graphs, we generalize in another direction by allowing the successor map to "leave" the part under consideration.

Let **node** be a set of "nodes". Given a set $M$, we denote by $M^*$ the set of all finite sequences of $M$-elements. Then **pseudo-graph** is a partial map $G : \textbf{node} \longrightarrow \textbf{node}^*$ such that $\downarrow G$ is finite. For a node $x \in \downarrow G$ the nodes in $G(x)$ are called the **immediate successors** of $x$. We set

$$nodes(G) \stackrel{\text{def}}{=} \downarrow G \cup \bigcup_{x \in \downarrow G} set(G(x)) \, .$$

A **storage graph** then is a pseudo-graph $G$ such that $nodes(G) \subseteq \downarrow G$. The more general notion of pseudo-graphs also allows "dangling references" which will occur e.g. during the copying phase of our garbage collection algorithm when only part of the accessible cells have been copied to their new locations.

## 7.2 Allocations

We also want to talk about the representation of such pseudo-graphs in a linear memory. Hence we now assume that **cell** is denumerable and linearly ordered by some ordering $\leq$ in which $\square$ is the least element. Without loss of generality we assume **cell** to be the set $\mathbb{N} \cup \{\infty\}$ of natural numbers under the usual ordering, enlarged by a greatest element $\infty$; then $\square \equiv 0$.

Let now $G$ be a pseudo-graph. We want to represent $G$ by a state. The idea is to store each node of $G$ together with its successors in a block of contiguous cells; the node itself is marked by cell contents $\square$. In practice, this leading cell frequently is used for storing information about a record, such as its length, type information, and the like. However, as stated in the introduction, we abstract from such details; $\square$ seems an adequate substitute here.

An **allocation** of $G$ is an injective partial map $g : \mathbf{node} \longrightarrow \mathbf{cell}$ such that $nodes(G) \subseteq {\downarrow}g$ and $\square \notin {\uparrow}g$. It is supposed to assign to each node in ${\downarrow}G$ the starting cell of its block; the additional condition ensures that $\square$ can in fact be used to characterize block beginnings. Given an allocation $g$ of $G$ we define for $x \in {\downarrow}G$

$$block(x,g) \stackrel{\mathrm{def}}{\equiv} [g(x) \mapsto \square] \cup \bigcup_{i \in [1:|G(x)|]} [(g(x)+i) \mapsto g(G(x)[i])] .$$

Define for a subset $s \subseteq \mathbf{cell}$

$$s^{\vee} \stackrel{\mathrm{def}}{\equiv} \{y \mid \exists\, x \in s : x \leq y\}$$
$$s^{\wedge} \stackrel{\mathrm{def}}{\equiv} \{y \mid \exists\, x \in s : y \leq x\} .$$

We call $s$ an **interval** if $s \equiv s^{\vee} \cap s^{\wedge}$. For abbreviation we write $\check{x}, \hat{x}$ instead of $\{x\}^{\vee}, \{x\}^{\wedge}$.

**Corollary 7.1**
   ${\downarrow}block(x,g)$ is the interval $[g(x) : g(x) + |G(x)|]$.

An allocation $g$ of $G$ is **overlap-free** if

$$x \not\equiv y \ \Rightarrow\ {\downarrow}block(x,g) \cap {\downarrow}block(y,g) \equiv \emptyset .$$

For an overlap-free allocation we can extend the function $block$ to sets $s \subseteq {\downarrow}G$ by setting

$$block(s,g) \stackrel{\mathrm{def}}{\equiv} \bigcup_{x \in s} block(x,g) .$$

Then the following state is a **representation** of $G$:

$$blockrep(G,g) \stackrel{\mathrm{def}}{\equiv} block({\downarrow}G,g) .$$

We call a state $m$ a **pseudo-graph state** if $m \equiv blockrep(G,g)$ for some pseudograph $G$ and some overlap-free allocation $g$ of $G$. Then

$$keys(m) \stackrel{\mathrm{def}}{\equiv} \square{\downarrow}m ,$$

the inverse image of $\square$ under $m$, denotes the set of **keys** of $m$, i.e., the set of cells that are the beginnings of blocks. Moreover we define

$$followers(m,y) \equiv \mathbf{if}\ y+1 \in keys(m)\ \mathbf{then}\ \diamondsuit\ \mathbf{else}\ <y+1> +followers(m,y+1)\ \mathbf{fi}.$$

This is the sequence of cells following $y$ in the block to which $y$ belongs.

The following lemma shows how a graph can be reconstructed from its block representation:

**Lemma 7.2**

Let $g$ be an overlap-free allocation of $G$ and let $m \equiv blockrep(G, g)$. Then

$$\downarrow G \equiv keys(m) \downarrow g \qquad \text{and}$$
$$G \equiv \bigcup_{x \in keys(m)} [x \downarrow g \mapsto g^{-1} * (m * followers(m, x))]$$

where $keys(m) \downarrow g$ is the inverse image of $keys(m)$ under $g$ and for a function $f : M \longrightarrow N$ we denote by $f*$ its unique homomorphic extension mapping $M^*$ to $N^*$, i.e.,

$$f * (<x_1, \ldots, x_n>) \equiv <f(x_1), \ldots, f(x_n)> \ .$$

Assume now that $nodes(G)$ is linearly ordered by some order $\leq$. Then an allocation $g$ is called **order-preserving** if

$$\forall \ x, y \in nodes(G) : x \leq y \ \Rightarrow \ g(x) \leq g(y) \ .$$

**Lemma 7.3**

Let $g$ be an order-preserving allocation. Then for $x, y \in nodes(G)$ we have

1. $x < y \ \Rightarrow \ g(x) < g(y)$ .

2. $x \leq y$ iff $g(x) \leq g(y)$, i.e., $nodes(G)$ and $\uparrow g$ are order-isomorphic.

**Proof:**   1. is immediate from the injectivity of $g$.

2. We only need to show ($\Leftarrow$). Assume $g(x) \leq g(y)$ but $x \not\leq y$. By linearity of $\leq$ then $y < x$ and hence also $g(y) < g(x)$ by 1. Contradiction!

∎

This lemma holds for arbitrary order-preserving injections between linear orders. For the special case of graph linearization we get

**Lemma 7.4**

Let $g$ be an overlap-free and order-preserving allocation. Then

$$x < y \ \Rightarrow \ \downarrow block(x, g) < \downarrow block(y, g) \ ,$$

where for subsets $s, t$ of an ordered set

$$s < t \ \stackrel{\text{def}}{\Leftrightarrow} \ \forall \ x \in s : \forall \ y \in t : x < y \ .$$

**Proof:** Since $g$ is order-preserving, $g(x) < g(y)$. Let now $u \in \downarrow block(x, g)$ and $v \in \downarrow block(y, g)$ and assume $v \leq u$. Since $g(y) \leq v$, we have then $g(x) \leq g(y) \leq u$ and hence $g(y) \in \downarrow block(x, g)$, since $\downarrow block(x, g)$ is an interval. But then $\downarrow block(x, g) \cap \downarrow block(y, g) \neq \emptyset$, a contradiction. ∎

We now want to characterize contiguous block representations. Call an overlap-free allocation $g$ of $G$ **gap-free** if $\downarrow blockrep(G, g)$ is an interval. $g$ is called **perfect** if it is overlap-free, order-preserving, and gap-free.

For a finite linearly ordered set $M$ with greatest element $\infty$ we define for $s \subseteq M$ and $x \in M \backslash \{\infty\}$

$$succ_s(x) \ \stackrel{\text{def}}{\equiv} \ min((s \backslash \hat{x}) \cup \{\infty\}) \ .$$

**Lemma 7.5**

Let $g : \textsf{node} \longrightarrow \textsf{cell}$ be a perfect allocation of $G$ and $x \in {\downarrow}G$ such that $x$ is not the maximum of ${\downarrow}G$. Then

$$g(succ_{{\downarrow}G}(x)) \;\equiv\; g(x) + |G(x)| + 1 \;.$$

**Proof:** Since $g$ is order-preserving and injective, we have $g(x) < g(succ_{{\downarrow}G}(x))$. The previous lemma now implies ${\downarrow}block(x,g) < {\downarrow}block(succ_{{\downarrow}G}(x),g)$ and, in particular, ${\downarrow}block(x,g) < \{succ_{{\downarrow}G}(x)\}$. Let

$$z \;\overset{\text{def}}{\equiv}\; max({\downarrow}block(x,g)) \;\equiv\; g(x) + |G(x)| \;.$$

Assume $z < u < g(succ_{{\downarrow}G}(x))$ for some $u$. Since ${\downarrow}blockrep(G,g))$ is an interval, we get $u \in {\downarrow}blockrep(G,g)$, say $u \in {\downarrow}block(w,g)$ for some $w \in {\downarrow}G$. Then $g(w) \leq u < g(succ_{{\downarrow}G}(x))$ and hence $w < succ_{{\downarrow}G}(x)$, since $g$ is order-preserving and injective. This is equivalent to $w \leq x$. But then ${\downarrow}block(w,g) \leq {\downarrow}block(x,g)$ which, by Lemma 7.4, contradicts $g(x) \leq z < u \in {\downarrow}block(w,g)$. Therefore

$$g(succ_{{\downarrow}G}(x)) \;\equiv\; z+1 \;\equiv\; g(x) + |G(x)| + 1 \;.$$

$\blacksquare$

Set, for $x \in {\downarrow}G$,

$$\|x\| \;\overset{\text{def}}{\equiv}\; |G(x)| + 1 \;.$$

**Corollary 7.6**

Let $g : \textsf{node} \longrightarrow \textsf{cell}$ be a perfect allocation of $G$ and $x \in {\downarrow}G$ such that $|\; \check{x} \cap {\downarrow}G| > i$. Then

$$g(succ^{i}_{{\downarrow}G}(x)) \;\equiv\; g(x) + \sum_{j<i} \|succ^{j}_{{\downarrow}G}(x)\| \;.$$

**Proof:** Induction on $i$ using the above lemma.

$\blacksquare$

A pseudo-graph state $m$ is called **compressed** if

$$\textstyle{\downarrow}m \;\equiv\; ({\downarrow}m)^{\wedge}\backslash\{\square\} \;,$$

i.e., if its domain is an initial interval of $\textsf{cell}\backslash\{\square\}$, which implies that there are no gaps in ${\downarrow}m$. By the above corollary, given a pseudo-graph $G$ there is exactly one perfect allocation $g$ of $G$ such that $blockrep(G,g)$ is compressed; $g$ is called the **compressing allocation** of $G$.


## 7.3 Formal Specification of the Garbage Collection Problem

When garbage collection becomes necessary, there is a set of immediate entries into the store. All blocks reachable from these entries need to be saved whereas everything else is garbage to be removed. We first treat the reachability problem at the level of storage graphs.

Let $G$ be a pseudo-graph and let $x, y \in \textsf{node}$. A sequence $p \in \textsf{node}^*$ is called a **path in $G$ from** $x$ **to** $y$ iff the predicate $ispath_G(p,x,y)$ holds, where

$$\begin{aligned}
ispath_G(p,x,y) \;\overset{\text{def}}{\Leftrightarrow}\; & |p| > 0 \;\wedge\; set(p)\backslash\{last(p)\} \subseteq {\downarrow}G \;\wedge\; \\
& x = first(p) \;\wedge\; y = last(p) \;\wedge\; \\
& \forall\, i \in [1 : |p| - 1] : p[i+1] \in G(p[i]) \;.
\end{aligned}$$

We define

$$\text{nodeset} \stackrel{\text{def}}{=} \{\, s \mid s \subseteq \text{node} \wedge |s| < \infty \,\} \,.$$

Given a pseudo-graph $G$, a node $x \in nodes(G)$ is **reachable** from some set $s \in \text{nodeset}$ iff the predicate $isreachable_G(x, s)$ holds where

$$isreachable_G(x, s) \stackrel{\text{def}}{\Leftrightarrow} \exists\, z \in s, p \in \text{node}^* : ispath_G(p, z, x) \,.$$

The function $rnset_G : \text{nodeset} \longrightarrow \text{nodeset}$, defined by

$$rnset_G(s) \stackrel{\text{def}}{=} \{\, x \in \text{node} \mid isreachable_G(x, s) \,\}$$

computes the set of nodes reachable from a given set.

Now, for a storage graph $G$ and a set $s \subseteq \downarrow G$, the **subgraph of $G$ reachable from** $s$ is $G_s \stackrel{\text{def}}{=} G|rnset_G(s)$. It is easily verified that the pseudo-graph $G_s$ indeed is a storage graph.

Consider now a storage graph $G$ with a perfect allocation $g$ and a set $s \subseteq \downarrow G$. Moreover, set $n \stackrel{\text{def}}{=} blockrep(G, g)$. Then the **garbage collection problem** consists in computing the reachable subgraph $G_s$ together with the compressing allocation $g_s$ of $G_s$ as well as the corresponding state $n_s \stackrel{\text{def}}{=} blockrep(G_s, g_s)$. In fact, ultimately we are interested in an algorithm that computes $n_s$ directly from $n$.


## 7.4   A First Analysis of the Problem

Assume $G, g, n$ and $G_s, g_s, n_s$ as in Section 7.3. Define

$$n_1 \stackrel{\text{def}}{=} blockrep(G_s, g) \,.$$

Thus, $n_1$ is the accessible but not yet compressed part of the storage. To compute $n_s$ from $n_1$, define a collapsing map $k : \downarrow n_1 \longrightarrow \text{cell}$ by

$$k(g(x) + i) \stackrel{\text{def}}{=} g_s(x) + i$$

for $x \in \downarrow G_s$ and $0 \leq i \leq |G_s(x)|$.

**Lemma 7.7**

    (1) $k$ is well-defined. Moreover, $k$ is an order-embedding and $\uparrow k \equiv \downarrow n_s$ (which is an initial interval of $\text{cell}\backslash\{\square\}$).

    (2) $n_s \circ k \equiv k \circ n_1$.

**Proof:**  (1) is obvious.

$$
\begin{aligned}
(2) \qquad & n_s(k(g(x) + i)) \\
\equiv\ & n_s(g_s(x) + i) \\
\equiv\ & g_s(G(x)[i]) \\
\equiv\ & k(g(G(x)[i])) \\
\equiv\ & k(n_1(g(x) + i)) \,.
\end{aligned}
$$

$\blacksquare$

The preceding considerations suggest a decomposition of the problem into the following parts:

1. Compute $G_s$ from $G$ and $s$ (reachability).

2. Compute $n_1 \equiv blockrep(G_s, g)$ from $G_s$ and $g$.

3. Compute $k$ from $n_1$.

4. Compute $n_s$ from $n_1$ and $k$ (copying).

Diagrammatically the situation can be described as follows:

$$
\begin{array}{ccccc}
G & \stackrel{reach}{\Longrightarrow} & G_s & & \\
\Big\| g & & \Big\| g & \searrow g_s & \\
blockrep(G,g) & \supseteq & blockrep(G_s,g) & \stackrel{k}{\Longrightarrow} & blockrep(G_s,g_s)
\end{array}
$$

Here the double arrows indicate that the respective functions are of second order, since their arguments, viz. pseudo-graphs and states, are mappings themselves.

# 8 Copying Pointer Structures

## 8.1 Statement of the Problem

We now treat the task of copying a state to another part of a memory. For a map $p$ we define

$$
set(p) \stackrel{\mathrm{def}}{\equiv} \downarrow p \cup \uparrow p .
$$

Let now $m, n$ be states. We call $n$ a **copy** of $m$ if there is a total bijection

$$
k : set(m) \cup \{\Box\} \longrightarrow set(n) \cup \{\Box\}
$$

such that $k(\Box) \equiv \Box$ and the following diagram commutes:

$$
\begin{array}{ccc}
\downarrow m & \stackrel{m}{\longrightarrow} & \uparrow m \\
\Big\downarrow k & & \Big\downarrow k \\
\downarrow n & \stackrel{n}{\longrightarrow} & \uparrow n
\end{array}
$$

This means that $k \circ m \equiv n \circ k$ and, since $k$ is bijective, that $n \equiv k \circ m \circ k^{-1}$. Hence, given $k$, we can compute $n$ from $m$ in two passes: First we form $k \circ m$ which means that the cell contents as given by $m$ are updated to contain the corresponding cells of the copy ("pointer relocation pass"); then we compose with $k^{-1}$ which means the actual transport of the new contents to the new locations ("copying pass").

For the treatment of these two passes the following properties of general maps are useful:

**Lemma 8.1**

Consider a map $l : M \longrightarrow N$.

$$(1) \quad l \equiv \bigcup_{x \in \downarrow l} [x \mapsto l(x)] \qquad \text{(domain-oriented representation)}$$

$$(2) \quad l \equiv \bigcup_{z \in \uparrow l} [z {\downarrow} l \mapsto z] \qquad \text{(range-oriented representation)}$$

$$(3) \quad q \circ \bigcup_{i \in I} l_i \equiv \bigcup_{i \in I} (q \circ l_i)$$

$$(4) \quad (\bigcup_{i \in I} l_i) \circ r \equiv \bigcup_{i \in I} (l_i \circ r)$$

$$(5) \quad l^{-1} \equiv \bigcup_{x \in \downarrow l} [l(x) \mapsto x] \qquad \text{provided } l \text{ is injective.}$$

## 8.2 Copying Pass

Given $p \equiv k \circ m$, the copying pass is easily performed. First, by totality of $k$, we have $\downarrow p \equiv \downarrow m \; (\subseteq \downarrow k)$. Now

$$p \circ k^{-1}$$

$$\equiv \qquad \text{(by Lemma 8.1(5))}$$

$$p \circ \bigcup_{x \in \downarrow k} [k(x) \mapsto x]$$

$$\equiv \qquad \text{(by Lemma 8.1(3))}$$

$$\bigcup_{x \in \downarrow k} p \circ [k(x) \mapsto x]$$

$$\equiv \bigcup_{x \in \downarrow k} [k(x) \mapsto p(x)]$$

$$\equiv \bigcup_{x \in \downarrow p} [k(x) \mapsto p(x)] \, ,$$

since $[k(x) \mapsto p(x)] \equiv \emptyset$ for $x \in \downarrow k \backslash \downarrow p \equiv \downarrow k \backslash \downarrow m$. We set

$$copass(p, k) \stackrel{\text{def}}{\equiv} \downarrow p \subseteq \downarrow k \;\; \triangleright \bigcup_{x \in \downarrow p} [k(x) \mapsto p(x)] \, .$$

## 8.3 Pointer Relocation

The more difficult subtask consists in computing the composition $k \circ m$ efficiently. According to Lemma 8.1 there are essentially two ways of forming $k \circ m$:

1. domain-oriented:

$$k \circ m \equiv \bigcup_{x \in \downarrow m} [x \mapsto k(m(x))]$$

   If we look at the union as a loop, this way of forming $k \circ m$ needs an explicit representation of $k$, since the same value of $k$ may be needed repeatedly at irregular intervals.

2. range-oriented:

$$k \circ m \equiv \bigcup_{z \in \uparrow m} [z {\downarrow} m \mapsto k(z)]$$

For evaluating this by a loop we only need one value of $k$ at a time to process a whole subset of $\downarrow m$. Hence we can avoid explicit representation of the complete $k$, which is particularly important in garbage collection, where storage is almost exhausted. Moreover, the repeated lookups are avoided and thus also time-efficiency is improved. Of course, this latter aspect is interesting only if $m$ is highly non-injective so that the inverse images $z\downarrow m$ are large.

We follow now the range-oriented variant. We need a way of representing the component maps $[z\downarrow m \ \mapsto\ y]$ suitably. For this we use an idea that is presented e.g. in [Dewar, McCann 77]: All elements of $z\downarrow m$ are chained into a linked list; then $[z\downarrow m \mapsto y]$ can be formed following the chain as $\bigcup_{x\in z\downarrow m} [x \mapsto y]$.

## 8.3.1  Chained Representation of Sets

Based on our notion of representation for sequences from section 4 we now introduce a chained representation of sets of proper cells: We represent such a set by a repetition-free sequence and this, in turn, by an anchored chain. Formally, a pair $(x, m)$ **represents** a set $s$ of proper cells if $set(sequ(x, m)) \equiv s$. We define

$$from(x, m) \stackrel{\text{def}}{\equiv} chain(sequ(x, m)) .$$

This is the subchain started by $x$ in $m$. Furthermore, we set

$$ischain(x, m) \stackrel{\text{def}}{\equiv} m = from(x, m) .$$

Now a representation function for sets is specified by

**funct** $chainset \ \equiv\ ($**cellset** $s)$ $($**cell**, **state**$)$ :
    **some cell** $x$, **state** $s : ischain(x, m)$ **cand** $set(sequ(x, m)) \ =\ s$ .

We want to develop an incrementation function

$$add(y, x, m) \stackrel{\text{def}}{\equiv} ischain(x, m) \textbf{ cand } y \notin set(sequ(x, m)) \cup \{\square\} \ \triangleright$$
$$chainset(set(sequ(x, m)) \cup \{y\}) .$$

To this end we observe that

$$y \notin set(sequ(x, m)) \cup \{\square\} \ \Rightarrow\ ischainable(<y> + sequ(x, m)) ,$$

and hence

$$set(sequ(chainrep(<y> + sequ(x, m))))$$
$$\equiv \quad \text{(by Lemma 4.3(1))}$$
$$set(<y> + sequ(x, m))$$
$$\equiv \quad \{y\} \cup set(sequ(x, m)) ,$$

so that $(z, l) \stackrel{\text{def}}{\equiv} chainrep(<y> + sequ(x, m))$ satisfies the second conjunct of the specification of $chainset(set(sequ(x, m)) \cup \{y\})$. Using Lemma 4.1(1), Lemma 4.2(2), and $ischain(x, m)$ this simplifies to

$$z \equiv y ,$$
$$l \equiv [y \mapsto x] \cup m .$$

Now $ischain(z, l)$ is easily checked, so that

$$add(y, x, m) \;\equiv\; ischain(x, m) \,\wedge\, y \notin set(sequ(x, m)) \cup \{\square\} \;\triangleright$$
$$(y, [y \mapsto x] \cup m)$$

is a correct refinement (i.e., a descendant) of our specification. Often we are interested just in the second component of the result of add. Hence we define

$$prefix(y, x, m) \;\overset{\text{def}}{\equiv}\; ischain(x, m) \,\wedge\, y \notin set(sequ(x, m)) \cup \{\square\} \;\triangleright$$
$$[y \mapsto x] \cup m \;.$$

**Corollary 8.2**

Assume $ischain(x, m) \,\wedge\, y \notin set(sequ(x, m)) \cup \{\square\}$. Then
(1) $prefix(y, x, m)(y) \equiv x$.
(2) $sequ(x, prefix(y, x, m)) \equiv sequ(x, m)$.

**Proof:** (1) is immediate from the definition.

(2) follows from Corollary 4.5, since $prefix(y, x, m) \equiv m \twoheadleftarrow [y \mapsto x]$.

■

### 8.3.2 Chained Representation of States

Let now $m$ be a state. From the range-oriented decomposition $m \equiv \bigcup\limits_{z \in \uparrow n} [z{\downarrow}m \mapsto z]$ we obtain

the partition $\downarrow m \equiv \bigcup\limits_{z \in \uparrow m} z{\downarrow}m$. We represent $m$ by a union of chains each of which represents

one of the sets $z{\downarrow}m$; the cell $z$ is prefixed as a header cell to the respective chain. To avoid confusion between these chains we require that

$$ischainable(m) \;\overset{\text{def}}{\Leftrightarrow}\; \downarrow m \cap \uparrow m \equiv \emptyset$$

holds; otherwise there would be a link from one chain to the beginning of another and the partition would be lost. We now define

$$\textbf{funct } chainmap \;\equiv\; (\textbf{state } m : ischainable(m)) \textbf{ state} :$$
$$\bigcup\limits_{z \in \uparrow m} prefix(z, chainset(z{\downarrow}m)) \;.$$

Since the elements of $z{\downarrow}m$ are the starting points of mutually unconnected chains, they are also sources (in the graph-theoretic sense) of the chained map. We set, for arbitrary state $n$,

$$src(n) \;\overset{\text{def}}{\equiv}\; \downarrow n \backslash \uparrow n \;;$$

this is the set of cells to which no pointer exists in $n$. With the help of this notion we can characterize chainings of maps by the following predicate $ischaining$ :

$$ischaining(l) \;\overset{\text{def}}{\equiv}\; \forall\, z_1, z_2 \in src(l) : z_1 \neq z_2 \;\Rightarrow\; disjoint(z_1, z_2, l)$$
$$\wedge \quad l = \bigcup\limits_{z \in src(l)} from(z, l)$$

Moreover, we can define the inverse operation to chaining:

$$unchain(l) \;\overset{\text{def}}{\equiv}\; ischaining(l) \;\triangleright\; \bigcup\limits_{z \in src(l)} [set(sequ(l(z), l)) \mapsto z] \;.$$

**Lemma 8.3**

$$ischainable(m) \implies unchain(chainmap(m)) \equiv m.$$

**Proof:** Set $l \stackrel{\text{def}}{\equiv} chainmap(m)$. Since $\downarrow m \cap \uparrow m \equiv \emptyset$ by the assumption, we have $src(l) \equiv \uparrow m$. Consider $z \in \uparrow m$ and $l_z \stackrel{\text{def}}{\equiv} chain(sequ(z, l))$. Then

$$l^*(l(z)) \equiv set(sequ(l(z)) \cup \{\square\} \equiv z{\downarrow}m \cup \{\square\}$$

and thus $l^*(l(z)) \cap \downarrow l_u \equiv \emptyset$ for all $u \in \uparrow m \setminus \{z\}$. Now we calculate

$$
\begin{aligned}
&\quad unchain(l) \\
\equiv\;& \bigcup_{z \in \uparrow m} [set(sequ(l(z), l)) \mapsto z] \\
\equiv\;& \quad \text{(by Lemma 4.4, since } l \equiv \bigcup_{u \in \uparrow m} l_u \text{)} \\
&\bigcup_{z \in \uparrow m} [set(sequ(l(z), l_z)) \mapsto z] \\
\equiv\;& \quad \text{(by Corollary 8.2(2))} \\
&\bigcup_{z \in \uparrow m} [z{\downarrow}m \mapsto z] \\
\equiv\;& \quad \text{(by Lemma 8.1(2))} \\
& m\;.
\end{aligned}
$$

∎

As in the case of set representations, we want to develop an incrementation function for extending a chained representation of a map into one of a larger map. Let therefore $l$ be a map satisfying $ischaining(l)$ and $unchain(l) \equiv m$, and let $x, y$ be cells such that $x \notin \downarrow m \cup \{\square\}$ and $ischainable(n)$ holds for $n \stackrel{\text{def}}{\equiv} m \cup [x \mapsto y]$. First we calculate

$$
\begin{aligned}
&\quad ischainable(n) \\
\Leftrightarrow\;& \downarrow(m \cup [x \mapsto y]) \cap \uparrow(m \cup [x \mapsto y]) \equiv \emptyset \\
\Leftrightarrow\;& (\downarrow m \cup \{x\}) \cap (\uparrow m \cup \{y\}) \equiv \emptyset \\
\Leftrightarrow\;& (\downarrow m \cap \uparrow m) \cup (\downarrow m \cap \{y\}) \cup (\{x\} \cap \uparrow m) \cup (\{x\} \cap \{y\}) \equiv \emptyset \\
\Leftrightarrow\;& \downarrow m \cap \uparrow m \equiv \emptyset \wedge \downarrow m \cap \{y\} \equiv \emptyset \wedge \{x\} \cap \uparrow m \equiv \emptyset \wedge \{x\} \cap \{y\} \equiv \emptyset \\
\Leftrightarrow\;& ischainable(m) \wedge y \notin \downarrow m \wedge x \notin \uparrow m \wedge x \not\equiv y \\
\Leftrightarrow\;& y \notin \downarrow m \wedge x \notin \uparrow m \wedge x \not\equiv y \\
\Leftrightarrow\;& y \notin set(l) \setminus src(l) \wedge x \notin src(l) \wedge x \not\equiv y\;.
\end{aligned}
$$

According to the definition of *chainmap* we have $l \equiv \bigcup_{z \in \uparrow m} l_z$ for certain chains $l_z$. To achieve a more uniform calculation we set $l_u \stackrel{\text{def}}{\equiv} \emptyset$ if $u \notin \uparrow m$. Moreover, we define

$$l(\!(y)\!) \stackrel{\text{def}}{\equiv} \text{if } y \in \downarrow l \text{ then } l(y) \text{ else } \square \text{ fi}\;.$$

Then for each $z$ the pair $rest(l_z) \stackrel{\text{def}}{\equiv} (l(\!(z)\!), l_z \ominus z)$ represents $z{\downarrow}m$ and hence $chainset(z{\downarrow}m) \sqsupseteq_D rest(l_z)$. We have

$$\downarrow l \equiv \bigcup_{z \in \uparrow m} (\{z\} \cup z{\downarrow}m)\;,$$

and hence $x \notin \downarrow l$. Consider now a $z \in \uparrow n$.

**Case 1:** $z \not\equiv y$. Then $z{\downarrow}n \equiv z{\downarrow}m$, and hence $prefix(z, chainset(z{\downarrow}n)) \sqsupseteq_D l_z$.

**Case 2:** $z \equiv y$. Then

$$
\begin{aligned}
& prefix(z, chainset(z{\downarrow}n)) \\
\equiv\ & prefix(z, chainset(z{\downarrow}m \cup \{x\})) \\
\sqsupseteq_D\ & prefix(z, add(x, chainset(z{\downarrow}m))) \\
\sqsupseteq_D\ & prefix(z, add(x, rest(l_z))) \ .
\end{aligned}
$$

Hence

$$
\begin{aligned}
& chain(n) \\
\equiv\ & \bigcup_{z \in \uparrow n} prefix(z, chainset(z{\downarrow}n)) \\
\equiv\ & \bigcup_{z \in \uparrow m \cup \{y\}} prefix(z, chainset(z{\downarrow}n)) \\
\equiv\ & \bigcup_{z \in \uparrow m \setminus \{y\} \cup \{y\}} prefix(z, chainset(z{\downarrow}n)) \\
\equiv\ & prefix(y, chainset(y{\downarrow}n)) \cup \bigcup_{z \in \uparrow m \setminus \{y\}} prefix(z, chainset(z{\downarrow}n)) \\
\sqsupseteq_D\ & prefix(y, add(x, rest(l_y))) \cup \bigcup_{z \in \uparrow m \setminus \{y\}} l_z \\
\equiv\ & prefix(y, add(x, rest(l_y))) \cup l \setminus l_y \\
\equiv\ & prefix(y, add(x, rest(l_y))) \cup l \ominus \downarrow l_y \\
\equiv\ & \quad (\text{since } x \notin \downarrow l) \\
& prefix(y, add(x, rest(l_y))) \cup l \ominus \downarrow prefix(y, add(x, rest(l_y))) \\
\equiv\ & l \twoheadleftarrow prefix(y, add(x, rest(l_y))) \\
\equiv\ & l \twoheadleftarrow ([y \mapsto x] \cup [x \mapsto l(\!(y)\!)] \cup l_y \ominus y) \\
\equiv\ & \quad (\text{by Lemma 3.8 (Annihilation), since } l_y \ominus y \subseteq l) \\
& l \twoheadleftarrow ([y \mapsto x] \cup [x \mapsto l(\!(y)\!)]) \ .
\end{aligned}
$$

Hence we define

$$
\begin{aligned}
& insert(l, y, x) \\
\stackrel{\text{def}}{\equiv}\ & ischaining(l) \ \wedge\ x \notin set(l) \ \wedge\ y \notin set(l) \setminus src(l) \ \wedge\ x \neq y \rhd \\
& l \twoheadleftarrow ([y \mapsto x] \cup [x \mapsto l(\!(y)\!)]) \ .
\end{aligned}
$$

The results of the above development then are summarized by

**Lemma 8.4**
Assume $ischaining(l) \ \wedge\ x \notin set(l) \ \wedge\ y \notin set(l) \setminus src(l) \ \wedge\ x \neq y$.
Then $chainmap(unchain(l) \cup [x \mapsto y]) \sqsupseteq_D insert(l, y, x)$.

### 8.3.3  Pointer Relocation Completed

We are now in the position to describe our efficient algorithm for computing $k \circ m$: We first construct a chained representation $l \equiv \bigcup_{z \in \uparrow m} l_z$ of $m$. Now we define

$$
relocate(l, k) \stackrel{\text{def}}{\equiv} ischaining(l) \ \rhd\ k \circ unchain(l) \ .
$$

We have

$$\begin{aligned}
&relocate(l,k) \\
\equiv\ &k \circ \bigcup_{z \in src(l)} [set(sequ(l(z), l_z)) \mapsto z] \\
\equiv\ &\bigcup_{z \in src(l)} [set(sequ(l(z), l_z)) \mapsto k(z)] \ .
\end{aligned}$$

This is the main loop of our algorithm. We now want to develop a more direct version of the inner loops that form the maps $[set(sequ(l(z), l_z)) \mapsto k(z)]$ for $z \in src(l)$. To this end we define

$$fibre(l, x, y) \stackrel{\text{def}}{\equiv} isanchored(x, l) \ \wedge \ x \in {\downarrow}l \ \triangleright \ [set(sequ(l(x), l)) \mapsto y] \ .$$

We have

$$\begin{aligned}
&fibre(l, x, y) \\
\equiv\ &\textbf{if } l(x) = \square \textbf{ then } [\emptyset \mapsto y] \\
&\qquad\quad \textbf{else } [\{l(x)\} \cup set(sequ(l(l(x)), l)) \mapsto y] \ \textbf{fi} \\
\equiv\ &\textbf{if } l(x) = \square \textbf{ then } \emptyset \\
&\qquad\quad \textbf{else } [l(x) \mapsto y] \cup [set(sequ(l(l(x)), l)) \mapsto y] \ \textbf{fi} \\
\equiv\ &\textbf{if } l(x) = \square \textbf{ then } \emptyset \\
&\qquad\quad \textbf{else } [l(x) \mapsto y] \cup fibre(l, l(x), y) \ \textbf{fi} \ .
\end{aligned}$$

Hence we have the recursion (termination is obvious)

$$\begin{aligned}
\textbf{funct } fibre \ \equiv\ &(\textbf{state } l, \textbf{cell } x, y : isanchored(x, l) \ \wedge \ x \in {\downarrow}l) \ \textbf{state}: \\
&\lceil \textbf{ cell } z \ \equiv\ l(x) \ ; \textbf{if } z = \square \textbf{ then } \emptyset \\
&\qquad\qquad\qquad\qquad \textbf{else } [z \mapsto y] \cup fibre(l, z, y) \ \textbf{fi} \ \rfloor \ .
\end{aligned}$$

Finally, we obtain

$$\begin{aligned}
&relocate(l, k) \\
\equiv\ &\bigcup_{x \in src(l)} [set(sequ(l(x), l_x)) \mapsto k(x)] \\
\equiv\ &\bigcup_{x \in src(l)} fibre(l, x, k(x)) \ .
\end{aligned}$$

## 8.4  Combining Relocation and Copying

Suppose that $unchain(l) \equiv m$. Then

$$\begin{aligned}
&k \circ m \circ k^{-1} \\
\equiv\ &k \circ unchain(l) \circ k^{-1} \\
\equiv\ &relocate(l, k) \circ k^{-1} \\
\equiv\ &copass(relocate(l, k), k) \ .
\end{aligned}$$

Therefore we define

$$\begin{aligned}
copy(l, k) \stackrel{\text{def}}{\equiv}\ &ischaining(l) \ \wedge \ isinjective(k) \ \wedge \ {\downarrow}k = set(unchain(l)) \ \triangleright \\
&copass(relocate(l, k), k) \ .
\end{aligned}$$

Then the following diagram commutes:

$$
\begin{array}{ccc}
set(m) & \xrightarrow{\;unchain(l)\;} & set(m) \\
\Big\downarrow{\scriptstyle k} & & \Big\downarrow{\scriptstyle k} \\
set(m){\uparrow}k & \xrightarrow{\;copy(l,k)\;} & set(m){\uparrow}k
\end{array}
$$

This concludes our treatment of the pointer relocation pass.

# 9   A Survey of the Further Development

## 9.1   Compressing Chained Graph States

The copying algorithm of the previous section can be specialized to the following problem: Given a graph $G$ and an overlap-free and order-preserving (but not necessarily gap-free) allocation $g$, compute from

$$
n \stackrel{\text{def}}{\equiv} blockrep(G, g)
$$

the compressed state

$$
n_c \stackrel{\text{def}}{\equiv} blockrep(G, g_c)
$$

for the unique compressing allocation $g_c$ of $G$.

In Section 7.4 we have already seen that $n_c$ is the copy of $n$ via the collapsing map $k$; i.e., $n_c \equiv k \circ n \circ k^{-1}$, where $k$ now is defined by

$$
(\text{K}) \qquad\qquad k(g(x) + i) \stackrel{\text{def}}{\equiv} g_c(x) + i
$$

for $x \in {\downarrow}G$ and $i \in [0 : |G(x)|]$.

A central assumption for chainable maps was that their domains should be disjoint from their ranges. Since, however, pseudo-graph states do not have this property, we cannot chain whole substates, but only their *arcs*, where

$$
arcs(m) \stackrel{\text{def}}{\equiv} m \ominus keys(m) \ .
$$

Assuming now that we have a map $l$ such that

$$
m \stackrel{\text{def}}{\equiv} l\big|\bigcup_{z \in keys(n)} set(sequ(z, l))
$$

is a chained representation of $arcs(n)$, we have the decomposition

$$
\begin{aligned}
& n \\
\equiv\ & n_\square \cup arcs(n) \\
\equiv\ & n_\square \cup unchain(m)
\end{aligned}
$$

where

$$
n_\square \stackrel{\text{def}}{\equiv} n|keys(n) \ (\ \equiv\ n \backslash arcs(n)) \ .
$$

Now we can employ the functions derived in Chapter 8 to calculate

151

$$
\begin{aligned}
& n_c \\
\equiv\ & k \circ n \circ k^{-1} \\
\equiv\ & k \circ (n_\square \cup unchain(m)) \circ k^{-1} \\
\equiv\ & (k \circ n_\square \circ k^{-1}) \cup (k \circ unchain(m) \circ k^{-1}) \\
\equiv\ & \quad (\text{since } k(\square) \equiv \square) \\
& (n_\square \circ k^{-1}) \cup copass(relocate(m,k),k) \\
\equiv\ & copass(n_\square, k) \cup copy(m,k) \ .
\end{aligned}
$$

By the definition of *copass* we have

$$
\begin{aligned}
& copass(n_\square, k) \\
\equiv\ & \bigcup_{z \in \downarrow n_\square} [k(z) \mapsto n_\square(z)] \\
\equiv\ & \bigcup_{z \in keys(n)} [k(z) \mapsto \square] \ .
\end{aligned}
$$

Furthermore we define

$$
p \overset{\text{def}}{\equiv} relocate(m,k) \equiv k \circ arcs(n) \ ,
$$
$$
size(z) \overset{\text{def}}{\equiv} succ_{\mathsf{cell} \setminus \downarrow arcs(n)}(z) - z \ .
$$

Hence $size(z)$ gives the number of cells in the block headed by $z$. Using that

$$
\begin{aligned}
& \downarrow p \\
\equiv\ & \downarrow arcs(n) \\
\equiv\ & \bigcup_{z \in keys(n)} [z+1 : z + size(z)]
\end{aligned}
$$

we obtain

$$
\begin{aligned}
& copass(p, k) \\
\equiv\ & \bigcup_{x \in \downarrow p} [k(x) \mapsto p(x)] \\
\equiv\ & \bigcup_{z \in keys(n)} \bigcup_{i \in [1:size(z)]} [k(z+i) \mapsto p(z+i)] \ .
\end{aligned}
$$

Therefore

$$
\begin{aligned}
& n_o \\
\equiv\ & copass(n_\square, k) \cup copass(p, k) \\
\equiv\ & \bigcup_{z \in keys(n)} \left([k(z) \mapsto \square] \cup \left( \bigcup_{i \in [1:size(z)]} [k(z)+i \mapsto p(z+i)] \right)\right)
\end{aligned}
$$

since $k(z+i) \equiv k(z) + i$ for $i \in [1 : size(z)]$.

Thus our problem divides into two parts:

1. Compute $p \equiv relocate(m,k)$ from $l$.

2. Compute the union above.

By the definition of relocate (cf. Section 8.3.3) it is immediate that

$$
relocate(m,k) \equiv \bigcup_{z \in keys(n)} fibre(m,z,k(z))
$$

where

$$fibre(m, z, y) \equiv [set(sequ(m(z), m)) \mapsto y] \,,$$

because $src(m) \equiv keys(n)$. From this specification we develop in [Berger et al. 89] a loop that traverses the set $keys(n)$ in ascending order.

The body of the function $copass$ is, like the body of $relocate$, a union over the index set $keys(n)$. Thus, forming $copass$ requires a second traversal of $keys(n)$. However, the successor function on $keys(n)$ can be $computed$ (as a map) simultaneously with $relocate$ and overwritten onto the state; afterwards we can use it to traverse $keys(n)$, thus improving speed efficiency considerably.

The relocation pass thus returns a union $lp$ of the state $p \stackrel{\mathrm{def}}{\equiv} relocate(m, k)$ and the map $l$ which is a chained representation of $keys(n)$. Now we have to construct $n_c$ from this union based on the values $size(z)$ for $z \in keys(n)$. As stated before, $n_c$ is represented by

$$n_c \equiv \bigcup_{z \in keys(n)} \left( [k(z) \mapsto \square] \cup \bigcup_{i \in [1:size(z)]} [k(z) + i \mapsto p(z + i)] \right),$$

where now $lp$ can be substituted for $p$. The outer union represents the main loop of the algorithm. In evaluating it we only need one value of $k$ at a time; the value for the next cycle can be computed using the recursion relation (cf. Section 7.2)

$$k(next_n(y)) \equiv k(y) + size(y)$$

for $y \in keys(n) \backslash \{max(keys(n))\}$. Hence we can eliminate $k$ and use its values on the single cells instead. So no extra space for $k$ is necessary.


## 9.2 Determining the Reachable Subgraph

We now turn to the problem of determining the reachable part of the store. We prepare this step at the level of pseudo-graphs. Let **pgraph** be the set of pseudo-graphs and **nodeset** be the set of finite subsets of **node**. The specification of the reachability problem now reads

> **funct** $reach \equiv$ (**pgraph** $G$, **nodeset** $s : s \subseteq \downarrow G$)**pgraph** :
> $\quad G | rnset_G(s)$ .

From this specification the following algorithm is derived in [Berger et al. 89]:

> **funct** $reach \equiv$ (**pgraph** $G$, **nodeset** $s : s \subseteq \downarrow G$)**pgraph** :
> $\quad$ **if** $s = \emptyset$ **then** $\emptyset$
> $\quad\quad\quad$ **else** **node** $z \equiv elem(s)$ ;
> $\quad\quad\quad\quad$ **pgraph** $G_1 \equiv G \ominus z$ ;
> $\quad\quad\quad\quad$ **nodeset** $s_1 \equiv (s \cup set(G(z))) \cap \downarrow G_1$ ;
> $\quad\quad\quad\quad$ $[z \mapsto G(z)] \cup reach(G_1, s_1)$ $\quad\quad\quad\quad$ **fi** ,

where

$$elem(s) \stackrel{\mathrm{def}}{\equiv} s \neq \emptyset \; \triangleright \; \textbf{some node } x : x \in s \,.$$

Termination is guaranteed, since $|\downarrow G_1| < |\downarrow G|$.

## 9.3 Merging Reachability and Chaining

In Section 9.1 we have described an efficient compressing algorithm based on a chaining of the arcs of the state to be copied. We now discuss the integration of the construction of such a representation with the computation of the reachable part.

The first step consists in transforming the reachability algorithm on pseudo-graphs into a corresponding algorithm for their state representations. An additional requirement for this algorithm is that it should not use a separate parameter for the set $t$ of cells already visited (corresponding to the parameter $s$ of *reach*), since this would occupy a lot of storage space. Everything should be done on the store and on some auxiliary cells.

A first idea how to realize this would be to represent the set $t$ as a chain and to overwrite the store with this chain. However, one sees immediately that this is in conflict with the chaining for representing the arcs which requires a different overwriting of the store. These difficulties are overcome by not storing $t$ itself but (a code of) a set of cells pointing to the elements of $t$ and, when adding a piece $block(x, g)$, by chaining the cells not in their original order but in a different one arising during the traversal of the reachable part. In fact, we use a chaining of the heads of the block fragments that have not been visited yet. This chaining arises from a specialization of the reachability algorithm above: The set $s$ is represented by a sequence, and *elem* is refined to *first*. In this way we obtain a depth-first traversal of the storage structure in which the representation of $s$ acts as a stack. This stack is then compressed considerably by retaining only the leading cells of the unprocessed block fragments.

Next we add a parameter that accumulates a chaining of the map that is the already visited storage part. The resulting algorithm still uses the three "large" parameters, one for the not yet visited part of the store, one for the stack, and one for the chaining of the visited part of the store. In the case of garbage collection, however, there is no space available for three separate parameters. So we need to represent them by *one* map parameter, viz. by the overall store. This is possible since the three original parameters are maps with pairwise disjoint domains which can be united into a single one. The complete algorithm (cf. [Berger et al. 89]) is linear in the size of the store in which garbage collection takes place.

# 10 Conclusion

We have shown with several examples how to derive algorithms involving pointers and selective updating from formal specifications using standard transformation techniques. The key to the method consists in considering the store as an explicit parameter, since then one has complete information about sharing and therefore complete control about side effects. We deem this approach much clearer (and much more convenient) than the idea of hiding the store and coming up with special logics (see e.g. [Mason 88, Kausche 89]) that capture the side-effects indirectly, as needs to be done in the field of verification of procedural programs.

Staying at the applicative level almost to the very end of the derivations has allowed us to take full advantage of the powerful algebra of partial maps. Using this algebra one saves an enormous amount of quantifiers (as compared e.g. with [Bijlsma 88]). Moreover, the operations of that algebra are expressive enough that we did not need to explain anything with the help of diagrams. Even when developing the intricate garbage collection algorithm described above we quite soon stopped drawing diagrams because the algebraic formulation was clearer and much more modular. Another advantage of the applicative treatment is that if additional predicates or operations on maps are needed, they are much more easily added at the applicative than at the procedural level. Finally, if pointer algorithms are developed in a systematic way at the applicative language level, there is no need for introducing additional imperative language concepts such as the highly imperspicuous pointer rotation [Suzuki 80].

We are convinced that our approach can be extended into a convenient method for constructing systems software with guaranteed correctness.

**Acknowledgement**
The idea of an algebraic treatment of pointers was stimulated by discussions within IFIP WG 2.1, notably by the algebraic way in which R. Bird and L. Meertens develop tree and list algorithms. I gratefully acknowledge helpful remarks from F.L. Bauer, U. Berger, R. Berghammer, R. Dewar, W. Dosch, F. Erhard, M. Lichtmannegger, W. Meixner, H. Partsch, P. Pepper, M. Sintzoff. and, particularly, H. Ehler. C. Karpf has pointed out a significant simplification in the derivation of the concatenation algorithm.

# 11   References


[Bauer, Wössner 82]
F.L. Bauer, H. Wössner: Algorithmic language and program development. New York: Springer 1982

[Bauer et al. 85]
F.L. Bauer et al.: The Munich project CIP. Volume I: The wide spectrum language CIP-L. Lecture Notes in Computer Science **183**. New York: Springer 1985

[Bauer et al. 89]
F.L. Bauer, B. Möller, H. Partsch, P. Pepper: Formal program construction by transformations — Computer-aided, Intuition-guided Programming. IEEE Transactions on Software Engineering **15**. 165–180 (1989)

[Berger et al. 89]
U. Berger, W. Meixner, B. Möller: Calculating a garbage collector. In: M. Broy M. Wirsing (ed.): Methodik des Programmierens. Fakultät für Mathematik und Informatik der Universität Passau, MIP-8915, 1989, 1–52. Also in: M. Broy, M. Wirsing (eds.): Programming methodology — The CIP approach. To appear in Lecture Notes in Computer Science. Berlin: Springer

[Bijlsma 88]
A. Bijlsma: Calculating with pointers. Science of Computer Programming **12**, 191–205 (1988)

[Broy, Pepper 82]
M. Broy, P. Pepper: Combining algebraic and algorithmic reasoning: An approach to the Schorr-Waite-Algorithm. ACM TOPLAS **4**, 362–381 (1982)

[Burstall 72]
R. Burstall: Some techniques for proving correctness of programs which alter data structures. In: B. Meltzer, D. Mitchie (eds.): Machine Intelligence **7**. Edinburgh University Press 1972, 23–50

[Dewar, McCann 77]
R. Dewar, A. McCann: MACRO SPITBOL — a SNOBOL4 compiler. Software — Practice and Experience **7**, 95–113 (1977)

[Dewar et al. 82]
R. Dewar, M. Sharir, E. Weixelbaum: Transformational derivation of a garbage collection algorithm. ACM TOPLAS **4**, 650–667 (1982)

[van Diepen, de Roever 86]
N. van Diepen, W. de Roever: Program derivation through transformations: The evolution of list-copying algorithms. Science of Computer Programming **6**, 213–272 (1986)

[Jones 80]
C.B. Jones: Software development: A rigorous approach. Englewood Cliffs: Prentice-Hall 1980

[Kausche 89]
A. Kausche: Modale Logiken von geflechtartigen Datenstrukturen und ihre Kombination mit temporaler Programmlogik. Fakultät für Mathematik und Informatik der TU München, Dissertation, 1989

[Levy 78]
M. Levy: Verification of programs with data referencing. Proc. 3me Colloque sur la Programmation 1978, 413–426

[Manna 74]
Z. Manna: Mathematical theory of computation. New York: McGraw-Hill 1974

[Mason 88]
I. Mason: Verification of programs that destructively manipulate data. Science of Computer Programming **10**, 177–210 (1988)

[Möller 89]
B. Möller: Applicative assertions. In : J.L.A. van de Snepscheut (ed.): Mathematics of Program Construction, Groningen, 26–30 June 1989. Lecture Notes in Computer Science **375**. Berlin: Springer 1989, 348–362

[Pepper, Möller 89]
P. Pepper, B. Möller: Programming with (finite) mappings. In: M. Broy (ed.): Informatik im Kreuzungspunkt von Numerischer Mathematik, Rechnerentwurf, Programmierung, Algebra und Logik. Festkolloquium für F.L. Bauer, Juni 1989. To appear in Lecture Notes in Computer Science. Berlin: Springer

[Reynolds 79]
J. Reynolds: Reasoning about arrays. Commun. ACM **22**, 290–299 (1979)

[Suzuki 80]
N. Suzuki: Analysis of pointer rotation. Conf. Record 7th POPL, 1980, 1–11. Revised version: Commun. ACM **25**, 330–335 (1982)

Some generalizations and applications of
Dijkstra's guarded commands

Greg Nelson
Digital Equipment Corporation
Systems Research Center
130 Lytton Avenue
Palo Alto, CA 94301
USA

This paper presents a series of topics in which Dijkstra's calculus
of guarded commands is generalized and applied in various ways.
Many proofs are left to the reader, in order to survey the topics
without getting bogged down in details.

In the first section, Dijkstra's calculus is simplified and generalized
by dropping the law of the excluded miracle, introducing "partial
commands".  Recursion is treated by the fixpoint method using a variant
of the Egli-Milner order.

In the second section, the value of partial commands is illustrated
by presenting a proof of correctness for a set of equations that
transform high-level control structures into semantically equivalent
sequences of low-level tests and branches.  Both the source and target
languages are specified semantically using guarded commands.  This
is joint work with Mark Manasse.

The third section discusses fair choice.  Fair choice can be
incorporated into the calculus of guarded commands by adding a
"dovetail" operator, but the operator is not monotonic with respect
to any of the usual orders for proving the existence of least fixpoints
for recursive definitions.  Nevertheless, we can prove that many
important classes of recursions that include the operator do have
fixpoints.  This is joint work with Manfred Broy.

The fourth section describes a theory of games.  Ordinarily the
nondeterministic choices in guarded commands are made by a single agent
(often called "the daemon").  But if some of the choices are made by

one agent and some by another, then the computation becomes a game
between the two agents.  We describe the semantics of these games using
predicate transformers.  As the ordinary calculus of commands gives
all conjunctive predicate transformers, the calculus of games gives
all monotonic predicate transformers.  This section generalizes the
work of Back and von Wright [0].


## 0. Notation

We will use a left-associative infix dot to denote function application.
That is, we write f.x instead of f(x), and g.x.y instead of g(x,y),
and g.x instead of $(\lambda y.\ g(x, y))$.

We use square brackets for Dijkstra's everywhere operator: [<u>true</u>] = <u>true</u>,
and [P] = <u>false</u> for all predicates P other than <u>true</u>.

We write

   (operation dummies: range: term)

to denote the combination via the given operation of the values assumed
by the given term as the dummies vary over the given range.  If the
range is obvious from the context, it will be omitted.

We write $A \equiv B$ with the same meaning as A = B, but give $\equiv$ a lower
binding power than any other operator.

We strive to write proofs as a chain of equivalences or implications,
with justifications for the individual steps inserted in curly braces.


## 1. Partial commands and recursive definitions

We begin by taking a relational approach.  In this approach, a "command"
is defined to be a relation between states and outcomes.  The
operational motivation is that a command A relates a state s to an
outcome o if o is a possible outcome of activating A in initial state
s.  An outcome is either a state, or the special outcome $\perp$ that
represents infinite looping.

Here are relational definitions of some fundamental commands:

$$\text{Skip}.s.o \;\equiv\; o = s$$
$$\text{Loop}.s.o \;\equiv\; o = \bot$$
$$\text{Fail}.s.o \;\equiv\; \underline{false}$$
$$\text{Havoc}.s.o \;\equiv\; o \neq \bot$$

Skip is a no-op; Loop always loops; Fail has no outcomes from any initial state; and Havoc is guaranteed not to loop, but is otherwise unpredicatable.

Notice that partial relations are allowed.  In fact, we define the "guard" of a command A (written grd.A) to be the domain of the relation considered as a predicate on initial states.  That is

$$\text{grd}.A.s \;\equiv\; (\exists\, o :: A.s.o)$$

For example, grd.Loop = $\underline{true}$ and grd.Fail = $\underline{false}$.

We define the following operations on commands, called the fundamental operations:

$$(A \;\square\; B).s.o \;\equiv\; A.s.o \;\vee\; B.s.o$$
$$(A \;\boxtimes\; B).s.o \;\equiv\; A.s.o \;\vee\; (\neg\, \text{grd}.A.s \;\wedge\; B.s.o)$$
$$(P \rightarrow A).s.o \;\equiv\; P.s \;\wedge\; A.s.o$$
$$(A \;;\; B).s.o \;\equiv\; (\exists\, t : A.s.t \;\wedge\; B.t.o) \;\vee\; (A.s.\bot \;\wedge\; o = \bot)$$

These are called demonic choice, deterministic choice, guard, and sequential composition.  A $\boxtimes$ B is read "A else B"; it can also be defined as A $\square \neg$ grd.A $\rightarrow$ B.  Sequential composition has greater binding power than the guard arrow, which has greater binding power than the two choice operators.  The relative binding power of the two choice operators is not specified.

We also define P?, called "test", to be short for P $\rightarrow$ Skip. Alternatively we could take test to be primitive and define the guard arrow by the equation P $\rightarrow$ A $\equiv$ P? ; A.

For example, working out the relational definitions, we compute that

$$(x := 0 \;\square\; x := 1); \; (x = 1 \rightarrow Skip) \;\equiv\; x := 1$$
$$(x := 0 \;\boxtimes\; x := 1); \; (x = 1 \rightarrow Skip) \;\equiv\; Fail$$
$$(\; Loop \;\square\; x := 1); \; (x = 1 \rightarrow Skip) \;\equiv\; Loop \;\square\; x := 1$$

The first example makes it clear that the operational meaning of
executing a command in a state where its guard is false is to backtrack
(or "fail").  In fact, operationally we have:

$$A \;\square\; B \;\equiv\; \text{execute either A or B}$$
$$A \;\boxtimes\; B \;\equiv\; \text{execute A; if this backtracks, execute B}$$
$$P \rightarrow A \;\equiv\; \text{backtrack if P is false, otherwise execute A}$$
$$A \; ; \; B \;\equiv\; \text{execute A, then execute B.}$$

So much for the relational and operational approaches.  On to predicate
transformers.

For any command A and predicate R, we define the predicates
wp.A.R and wlp.A.R by the rules that for any state s,

$$wp.A.R.s \;\equiv\; (\forall \, t: A.s.t: t \neq \perp \wedge R.t)$$
$$wlp.A.R.s \;\equiv\; (\forall \, t: A.s.t \wedge t \neq \perp: R.t)$$

It is left to the reader to show that for any command A, the predicate
transformers wp.A and wlp.A satisfy:

$$wlp.A.(\wedge \, R:: R) \;\equiv\; (\wedge \, R:: wlp.A.R)$$
$$wp.A.R \;\equiv\; wp.A.\underline{true} \wedge wlp.A.R$$

where in the first formula the dummy R ranges over any set of
predicates.  The first formula states that wlp is universally
conjunctive; the second is called the "pairing condition".  Together
they are called the "healthiness conditions" for commands.

It is not difficult to show that if f and g are two predicate
transformers that satisfy the healthiness conditions, then there exists
a relation A from states to outcomes such that f = wp.A and g = wlp.A.
This suggests that we dispense with relations and simply define a
command to be a pair of predicate transformers that satisfy the
healthiness conditions, which in fact is the view that we will take

in the remainder of the paper.  In this view, the constant commands
and fundamental operations are defined as follows:

$$wlp.Loop.R \equiv \underline{true}$$
$$wp.Loop.R \equiv \underline{false}$$
$$w.Skip.R \equiv R$$
$$w.Fail.R \equiv \underline{true}$$
$$w.Havoc.R \equiv [R]$$
$$w.(A;B).R \equiv w.A.(w.B.R)$$
$$w.(A \;\square\; B).R \equiv w.A.R \wedge w.B.R$$
$$w.(P \rightarrow A).R \equiv \neg P \vee w.A.R$$
$$w.(A \;\boxtimes\; B).R \equiv w.A.R \wedge (grd.A \vee w.B.R)$$

in which a single equation in w stands for two identical equations
in wp and wlp.  It is left to the reader to verify that the operators
so defined produce commands (that is, that the pairs of predicate
transformers they define satisfy the healthiness conditions).

We also define grd.A to be the predicate $\neg$ wp.A.$\underline{false}$.  This is
consistent with the previous relational definition of grd:

$$\begin{aligned}
&\quad\ grd.A.s \text{ (new definition)}\\
\equiv\ & (\neg\ wp.A.false).s\\
\equiv\ & \neg\ (\forall\ o:\ A.s.o:\ \underline{false})\\
\equiv\ & (\exists\ o::\ A.s.o)\\
\equiv\ & grd.A.s \text{ (old definition)}
\end{aligned}$$

Dijkstra's law of the excluded miracle states that wp.A.$\underline{false} \equiv \underline{false}$,
that is, that grd.A $\equiv \underline{true}$.  We call a command "total" if it satisfies
this law and "partial" otherwise.

We also define hlt.A to be the predicate wp.A.$\underline{true}$.

Next we turn to the solution of recursion equations.

Operationally, <u>do</u> A <u>od</u> means to execute A repeatedly until it fails;
that is, we have the equation

$$\underline{do}\ A\ \underline{od} \equiv A;\ \underline{do}\ A\ \underline{od} \;\boxtimes\; Skip\ .$$

This suggests that we define <u>do</u> A <u>od</u> to be a solution for X of the equation

$$X \equiv A;X \boxtimes Skip .$$

However, the solution (supposing that we can prove that one exists) need not be unique. For example, if A is Skip, then both Skip and Loop solve the equation for X.

The way to proceed is the famous "fixpoint method", in which an order is introduced on the commands and recursive definitions are taken to denote the least fixpoint of the corresponding function from commands to commands, which is proved to exist by appealing to any of a variety of theorems. Here are three theorems that are useful in various circumstances.

Three fixpoint theorems: Let $f: S \to S$ be a map from a set S to itself, and let $\preceq$ be a reflexive partial order on S. Let $\sqcup$ and $\sqcap$ denote join and meet on S with respect to $\preceq$. Define:

$f$ monotonic $\equiv [x \preceq y \Rightarrow f.x \preceq f.y]$
$f$ continuous $\equiv [f.(\sqcup x: x \in C: x) = (\sqcup x: x \in C: f.x)]$ for
   all non-empty chains C whose join exists.

Then $f$ has a least fixpoint $p$ if

$f$ is monotonic and $p = (\sqcap x: f.x \preceq x: x)$, or
$f$ is monotonic and $p = (\sqcap x: f.x = x: x)$
(Knaster-Tarski 1928)

$f$ is continuous and $p = (\sqcup n: n \geq 0: f^n(\min S))$
(Limit Theorem, Kleene 1952)

$f$ is monotonic and $p = (\sqcup i: i \text{ an ordinal}: f^i(\min S))$
(Generalized Limit Theorem, Hitchcock and Park 1972)

In these theorems min S is the minimum element of S. See Hitchcock and Park's original paper [7] or the author's paper [11] for the formal definition of $f^i(\min S)$ when i is an ordinal. The attributions given

may not be the earliest, but they are the earliest that I have actually
checked.  For a thorough survey of the early literature on fixpoint
theorems, see [8].

To apply the fixpoint method, we need a suitable partial order on
our commands.  What order should we use?

Two important orders on commands are defined by:

$$A \sqsubseteq_{wp} B = (\forall R:: [wp.A.R \Rightarrow wp.B.R])$$
$$A \sqsubseteq_{wlp} B = (\forall R:: [wlp.B.R \Rightarrow wlp.A.R])$$

(Notice that the antecedent and consequent are reversed in the right
hand side of the second definition.)

The order $\sqsubseteq_{wp}$ is the "as good as" order used in program refinement:
B is as good as A if B establishes every postcondition that A does,
given the same precondition. The order $\sqsubseteq_{wlp}$ is the subset order on
commands considered as relations on states; that is,

$$A \sqsubseteq_{wlp} B \equiv (\forall s, o: o \neq \bot: A.s.o \Rightarrow B.s.o)$$

which the reader can easily verify is equivalent to the other definition.

Since $\sqsubseteq_{wp}$ depends only on wp, it cannot distinguish commands
that have the same wp but different wlps.  Similarly, $\sqsubseteq_{wlp}$ cannot
distinguish commands that have the same wlp but different wps.
Therefore, neither of these is a partial order, and therefore
neither of them is suitable as a basis for the fixpoint method.

It doesn't help to identify commands with the same wp and use $\sqsubseteq_{wp}$
on the equivalence classes, or to identify commands with the same wlp
and use $\sqsubseteq_{wlp}$ on the equivalence classes, since neither <u>do-od</u>
nor ⊠ are monotonic with respect to $\sqsubseteq_{wp}$ or $\sqsubseteq_{wlp}$.

The solution is to define

$$A \sqsubseteq B \equiv (A \sqsubseteq_{wp} B) \wedge (A \sqsubseteq_{wlp} B)$$

and carry out the fixpoint method using the relation $\sqsubseteq$.  This is a version of the Egli-Milner order [6]; we will call it the "approximation" order. We will state without proof the theorems necessary to apply the fixpoint method in the approximation order; see the author's paper [11] for proofs.

Theorem 1.1. The fundamental operators on commands satisfy the following continuity and monotonicity properties with respect to approximation:

$$P \rightarrow A \quad \text{continuous in A}$$
$$A \;\square\; B \quad \text{continuous in A and B}$$
$$A \;\boxtimes\; B \quad \text{continuous in A and B}$$
$$A \;;\; B \quad \text{continous in A, monotonic in B. Also}$$
$$\text{continuous in B if wp.A is } \vee\text{-continuous}$$

In particular, since continuity implies monotonicity, all of the operators are monotonic.

To apply the two limit theorems, it is necessary to establish that chains of commands have joins:

Theorem 1.2. Every chain of commands has a $\sqsubseteq$-join.

(Sketch of proof: Let C be a $\sqsubseteq$-chain of commands, and define the command J by the equations

$$\text{wp.J.R} = (\wedge\; A: \quad A \in C: \quad \text{wp.A.R})$$
$$\text{wlp.J.R} = (\vee\; A: \quad A \in C: \quad \text{wlp.A.R})$$

Then wp.J and wlp.J satisfy the healthiness conditions, and hence J is a well-defined command; furthermore J is the $\sqsubseteq$-least upper bound of the A's.)

To apply the Knaster-Tarski theorem, it is necessary to establish that the $\sqsubseteq$-meets that occurs in the theorem exist.

Theorem 1.3. Every non-empty set of commands has a $\sqsubseteq$-meet.

(The obvious strategy for proving Theorem 1.3 is to exchange $\wedge$ and $\vee$

in the proof of Theorem 1.2.  Unfortunately this doesn't work; the
resulting pair of predicate transformers fails to satisfy the healthiness
conditions.  For a proof, see reference [11].)

Combining all these theorems, we have:

Theorem 1.4.  Let f be a map from commands to commands defined by an
expression of the form f(X) = E, where E is a expression built from
the fundamental operators, the command parameter X, and any number
of fixed commands and predicates.  Then f has a least fixpoint under
the approximation relation, given by

$$(\sqcup\ i:\ i\ \text{ordinal}:\ f^i(\text{Loop}))\ \text{and also by}$$
$$(\sqcap\ X:\ f(X) \sqsubseteq X:\ X)\ \text{and finally by}$$
$$(\sqcap\ X:\ f(X) \equiv X:\ X)$$

Proof.  Since the operators are monotonic (Theorem 1.1), f is monotonic.
Since every chain has a join (Theorem 1.2), the join in the first
formula for the fixpoint exists, and equals f's least fixpoint by the
generalized limit theorem.  Thus f has a fixpoint, and therefore the
meets in the other formulas for the fixpoint are non-empty, and
therefore exist (Theorem 1.3), and equal f's least fixpoint by the
Knaster-Tarski theorem.


## 2. Compiler verification

Our goal in this section is to sketch the proof of validity of a set
of compilation equations that transform high-level control structure
(do-od, $\square$, $\rightarrow$, etc.) into low-level tests and branches.

The input to the compiler will be a total command with the following
syntax:

```
<total command> ::=
  <atomic command>
| <total command> ; <total command>
| "if" <partial command> "fi"
| "do" <partial command> "od"
```

```
<partial command> ::=
  <predicate> "→" <total command>
| <partial command> "▨" <partial command>
```

where the exact nature of atomic commands and predicates need not
concern us.  Notice that we have restricted the input to be
deterministic.

The output of the compiler is an instruction list with the syntax:

```
<instruction list> ::=
  <instruction>
| <instruction> "□" <instruction list>

<instruction> ::=
  pc "=" <number> "→" <Action>
```

where the exact nature of <Action> need not concern us.  E.g., an
instruction list might be:

```
  pc = 32 → Push x; pc := 36
□ pc = 36 → Push y; pc := 40
□ pc = 40 → Add; pc := 44
□ ...
```

The idea is that an instruction list is "executed" by bracketing it
with do-od and preceding it with an initialization of pc.  In addition
to the <total command> to be translated, the compiler takes as an
argument the initial value of the program counter, the final value
of the program counter, and the allowable intermediate values for
the program counter.

For example, suppose that we are compiling for a machine that includes
the following actions:

```
Compare.(n, m, L1, L2) ≡
  if n ≥ m → pc := L1 □ n < m → pc := L2 fi

Subtract.(n, m, L) ≡ m := m - n; pc := L
```

and that we direct the compiler to compile the source:

$$\underline{do}\ n < m \rightarrow m := m - n\ \square\ m < n \rightarrow n := n - m\ \underline{od}$$

under the side conditions that initially pc = 2, finally pc = 17, and only prime values are to be allowed for the program counter. Then the output might be:

$$pc = 2 \rightarrow Compare.(n, m, 7, 3)$$
$$\square\ pc = 3 \rightarrow Subtract.(n, m, 2)$$
$$\square\ pc = 7 \rightarrow Compare.(m, n, 17, 11)$$
$$\square\ pc = 11 \rightarrow Subtract.(m, n, 2)$$

The informal specification for the compiler $\Gamma$ is

$$\Gamma.(A, s, h, g, L)$$
= Compile the total or partial command A with
    start label s
    halt label h
    guard failure label g
    label pool L

This means that executing the instruction list $\Gamma.(A, s, h, g, L)$ from an initial state with pc=s will perform a computation equivalent to A and halt with pc=h, or will halt with pc=g if started in a state where A's guard is false; and that during the computation pc will take on values only from the set of labels L.

We use numbers both as labels and as sets of labels, according to the rule

$$n \in L$$
= {definition}
    L's binary representation is a prefix of n's binary representation.

We also write $L^0$ for 2*L and $L^1$ for 2*L+1. It follows that

$$n \in L\ \equiv\ n = L \lor n \in L^0 \lor n \in L^1$$

That is, each label pool L is partitioned into the singleton {L} and the two infinite sets $L^0$ and $L^1$.

We are now ready for the formal specification: the compiler $\Gamma$ satisfies (2.1), (2.2), and (2.3):

$$\text{pc} := s; \;\underline{\text{do}}\; \Gamma.(A, s, h, g, L) \;\underline{\text{od}} \tag{2.1}$$
$$\equiv A; \; \text{pc} := h \;\square\; \text{pc} := g$$

$$[\text{grd}.(\Gamma.(A, s, h, g, L)) \Rightarrow \text{pc} \in L \vee \text{pc} = s] \tag{2.2}$$

$$[\text{wlp}.(\Gamma.(A, s, h, g, L)).(\text{pc} \in L \vee \text{pc} \in \{s,h,g\})] \tag{2.3}$$

provided that

    A is independent of pc;
    s, h, and g are not in L; and        (2.4)
    s is not in {h, g}.

Condition (2.1) defines the meaning of the instruction list. Condition (2.2) specifies that the target program's instructions are all loaded into location s or into locations in L. Condition (2.3) specifies that the only locations outside L to which target instructions can branch are s, h, and g. The proviso that A be independent of the pc means that A neither tests nor sets the pc (for an axiomatic definition of this condition, see [11]).

So much for the specification. The definition of $\Gamma$ is by structural induction over the syntax of its first argument. For example,

$$\Gamma.(A;B, s, h, g, L)$$
$$\equiv \Gamma.(A, s, L, g, L^0) \;\square\; \Gamma.(B, L, h, g, L^1)$$

That is, to compile A;B so as to start at s and halt at h, compile A starting at s and halting at the fresh label L, combining this with the result of compiling B starting at L and halting at h. The label pool L provides the intermediate label as well as the two infinite label pools for the recursive subcompilations.

Notice that this equation is only valid if B is total, in which case
the g argument to $\Gamma.(B, \ldots)$ is irrelevant. Our source syntax
ensures that the second argument to semicolon is total; without this
restriction the compilation of semicolon would be more difficult.

Similarly we have:

$$\Gamma.(A \boxtimes B, s, h, g, L)$$
$$\equiv \Gamma.(A, s, h, L, L^0) \square \Gamma.(B, L, h, g, L^1)$$

$$\Gamma.(\underline{do}\ A\ \underline{od}, s, h, g, L)$$
$$\equiv \Gamma.(A, s, L, h, L^0) \square pc = L \rightarrow pc := s$$

$$\Gamma.(\underline{if}\ A\ \underline{fi}, s, h, g, L)$$
$$\equiv \Gamma.(A, s, h, L, L^0) \square pc = L \rightarrow pc := s$$

$$\Gamma.(P \rightarrow A, s, h, g, L)$$
$$\equiv pc = s \rightarrow (P \rightarrow pc := L \boxtimes pc := g) \square \Gamma.(A, L, h, g, L^0)$$

(The Manasse-Nelson paper [10] gives a more satisfactory treatment for
the guard arrow, which allows a structural induction over the boolean
structure of the predicate P, but for brevity's sake we omit it.)

It remains to prove that $\Gamma$ satisfies its specification. We will
concentrate on the equation for $\Gamma.(A;B, \ldots)$. Launching into the
algebra, we find

$$pc := s;\ \underline{do}\ \Gamma.(A;B, s, h, g, L)\ \underline{od}$$
$$=\ \{\text{definition of}\ \Gamma.(A;B, \ldots)\}$$
$$pc := s;\ \underline{do}\ \Gamma_A\ \square\ \Gamma_B\ \underline{od}$$

where

$$\Gamma_A = \Gamma.(A, s, L, g, L^0)\ \text{and}\ \Gamma_B = \Gamma.(B, L, h, g, L^1).$$

To continue the derivation, we need to distribute $\underline{do}$-$\underline{od}$ over $\square$, in
order to introduce the subexpressions $\underline{do}\ \Gamma_A\ \underline{od}$ and $\underline{do}\ \Gamma_B\ \underline{od}$.
That is, the "shape of the formula" dictates that we investigate
the distribution properties of $\underline{do}$-$\underline{od}$ over $\square$.

There is a well-known identity for distributing * over |
in regular expressions:

$$(A \mid B)^* = A^*(BA^*)^*$$

(Here * denotes "any finite number of repetitions" and "|" denotes
non-deterministic choice.  There is a similar identity for
distributing <u>do</u>-<u>od</u> over □, but it requires that the guards of
the two commands be disjoint:

Lemma 2.5 (the do-box lemma): if [grd.A $\wedge$ grd.B $\equiv$ <u>false</u>], then

$$\underline{do}\ A\ \square\ B\ \underline{od} = \underline{do}\ A\ \underline{od};\ \underline{do}\ B;\ \underline{do}\ A\ \underline{od}\ \underline{od}$$

Before proving this lemma, we will complete the proof of the
compilation equation for semicolon:

$$pc := s;\ \underline{do}\ \Gamma.(A;B,\ s,\ h,\ g,\ L)\ \underline{od}$$
= {derivation above}
$$pc := s;\ \underline{do}\ \Gamma_A\ \square\ \Gamma_B\ \underline{od}$$
= {do-box lemma; (2.2), (2.4)}
$$pc := s;\ \underline{do}\ \Gamma_A\ \underline{od};\ \underline{do}\ \Gamma_B;\ \underline{do}\ \Gamma_A\ \underline{od}\ \underline{od}$$
= {(2.2), (2.3), (2.4) imply [wp.$\Gamma_B$.($\neg$ grd.$\Gamma_A$)]}
$$pc := s;\ \underline{do}\ \Gamma_A\ \underline{od};\ \underline{do}\ \Gamma_B\ \underline{od}$$
= {structural induction}
$$(A\ ;\ pc := L \boxtimes pc := g)\ ;\ \underline{do}\ \Gamma_B\ \underline{od}$$
= {(X $\boxtimes$ Y);Z = X;Z $\boxtimes$ Y;Z if Z is total}
$$A;\ pc := L;\ \underline{do}\ \Gamma_B\ \underline{od} \boxtimes pc := g;\ \underline{do}\ \Gamma_B\ \underline{od}$$
= {(2.2), (2.4) $\Rightarrow$ [pc = g $\Rightarrow$ $\neg$ grd.$\Gamma_B$]}
$$A;\ pc := L;\ \underline{do}\ \Gamma_B\ \underline{od} \boxtimes pc := g$$
= {structural induction}
$$A;\ (B;\ pc := h \boxtimes pc := g) \boxtimes pc := g$$
= {B is total}
$$A;\ (B;\ pc := h) \boxtimes pc := g$$
= {associativity of semicolon}
$$(A\ ;\ B)\ ;\ pc := h \boxtimes pc := g$$

which is what was to be proved.

The proofs of the $\Gamma$-equations for the other control structures
all similar, and we will not present them.  We note, however, that
our proof required the proviso $s \neq h$; and it is because of this proviso
that we were forced to use the equation

$$\Gamma.(\underline{do}\ A\ \underline{od},\ s,\ h,\ g,\ L)$$
$$\equiv \Gamma.(A,\ s,\ L,\ h,\ L^0)\ \square\ pc = L \rightarrow pc := s$$

instead of the more efficient and natural

$$\Gamma.(\underline{do}\ A\ \underline{od},\ s,\ h,\ g,\ L) \equiv \Gamma.(A,\ s,\ s,\ h,\ L)\ .$$

Luc Rooijakkers has shown how to remove the proviso $s \neq h$ by strenthening
(2.1), but we will not present his improvement here.

We conclude this section with the proof of the do-box lemma.  First
we need an equation for $wp.(\underline{do}\ A\ \underline{od})$.

From the fact that $\underline{do}\ A\ \underline{od}$ is a solution for X of the equation in
commands

$$X = A;\ X\ \boxtimes\ Skip$$

we conclude that $wp.(\underline{do}\ A\ \underline{od}).R$ is a solution for X of the equation
in predicates

$$X = wp.A.X \wedge (grd.A \vee R)$$

Actually it can be shown that $wp.(\underline{do}\ A\ \underline{od}).R$ is the strongest such
X, and the $wlp.(\underline{do}\ A\ \underline{od}).R$ is the weakest solution for X of the similar
equation

$$X = wlp.A.X \wedge (grd.A \vee R).$$

(For a proof, see the author's paper [11].)

To prove the do-box lemma, we will show that the left and right sides
of the equation in the lemma have the same wp for an arbitrary
postcondition.  The proof that they have the same wlp is similar and
will be left to the reader.

Starting with the left hand side, we observe that

$$wp.(\underline{do}\ A\ \square\ B\ \underline{od}).R$$

is the strongest X such that

$$[X = wp.A.X \land wp.B.X \land (grd.A \lor grd.B \lor R)] \qquad (2.6)$$

The right hand side requires that we compute

$$wp.(\underline{do}\ A\ \underline{od};\ \underline{do}\ B;\ \underline{do}\ A\ \underline{od}\ \underline{od}).R \qquad\qquad (2.7)$$

Applying the wp equation for $\underline{do\text{-}od}$ three times, we find that (2.7)
is the X component of the strongest predicate triple (X, Y, Z) that
satisfies

$$
\begin{aligned}
&[X = wp.A.X \land (grd.A \lor Y)]\ \land \\
&[Y = wp.B.Z \land (grd.B \lor R)]\ \land \qquad\qquad (2.8) \\
&[Z = wp.A.Z \land (grd.A \lor Y)]
\end{aligned}
$$

Here we have used some simple theory about extreme solutions of
simultaneous predicate equations; see for example Chapter 8 of Dijkstra
and Scholten's recent book, [5].  The passage from (2.7) to (2.8)
can be checked mentally by thinking of X, Y, and Z as the invariants
of the three loops, and verifying that the three constraints in
(2.8) correspond to the control flow from invariant to invariant
in (2.7).

Next, substitute the value of Y from the second equation in (2.8)
for the occurrences of Y in the first and third equations, producing:

$$
\begin{aligned}
&[X = wp.A.X \land (grd.A \lor wp.B.Z \land (grd.B \lor R))]\ \land \\
&[Y = wp.B.Z \land (grd.B \lor R)]\ \land \qquad\qquad (2.9) \\
&[Z = wp.A.Z \land (grd.A \lor wp.B.Z \land (grd.B \lor R))]
\end{aligned}
$$

Next, observe that under the assumptions of the do-box lemma,
grd.A implies $\neg$ grd.B, which in turn implies wp.B.Q for
any Q. In particular, [grd.A $\Rightarrow$ wp.B.Z]. This fact justifies
a transformation of the first and third equations, producing:

$$[X = wp.A.X \wedge wp.B.Z \wedge (grd.A \vee grd.B \vee R)] \wedge$$
$$[Y = wp.B.Z \wedge (grd.B \vee R)] \wedge \qquad\qquad (2.10)$$
$$[Z = wp.A.Z \wedge wp.B.Z \wedge (grd.A \vee grd.B \vee R)]$$

Notice that the equation for Z in (2.10) is identical to the equation
for X in (2.6).  Therefore the strongest Z satisfying (2.10) agrees
with the strongest X satisfying (2.6).  Since our goal is to establish
that the strongest X satisfying (2.10) is the same as the strongest
X satisfying (2.6), we will be done if we can show that in the
strongest solution to (2.10), X and Z have the same value.

By the Knaster-Tarski theorem, the strongest (X, Y, Z)
satisfying (2.10) is the same as the strongest (X, Y, Z) satisfying

$$[X \Leftarrow wp.A.X \wedge wp.B.Z \wedge (grd.A \vee grd.B \vee R)] \wedge$$
$$[Y \Leftarrow wp.B.Z \wedge (grd.B \vee R)] \wedge \qquad\qquad (2.11)$$
$$[Z \Leftarrow wp.A.Z \wedge wp.B.Z \wedge (grd.A \vee grd.B \vee R)]$$

If (X, Y, Z) is a solution to (2.11), calculation shows that

$$(X \wedge Z, \; wp.B.(X \wedge Z) \wedge (grd.B \vee R), \; X \wedge Z)$$

is also a solution, and is not weaker than (X, Y, Z).  It follows
that in the strongest solution for (X, Y, Z), the predicates X and
Z coincide.  This completes the proof of the do-box lemma.

This concludes our sketch of the compiler correctness proof.  Notice
the basic algebraic idea of the proof was to distribute <u>do-od</u>
over □, a notion that is more easily conceived of in the presence
of partial commands.  The equations for $\Gamma$ were also simplified
by treating $P \rightarrow A$ and $A \;\square\; B$ as semantic entities in their own right,
which is only possible using partial commands.


3. The Dovetail operator

In this section we present an equational characterization of fairness
by studying the dovetail operator.  This is joint work with Manfred Broy.

The operational definition of dovetail (which we shall write $\nabla$) is

      A ▽ B

= Execute the command A and B in parallel, on separate copies of
  the state, interleaving the two computations non-deterministically
  but fairly, accepting as an outcome any proper (that is, non-looping)
  outcome of either computation.

By "fairly" it is meant that neither computation is starved: if the
computation is infinite, then each of the A and B parts are either
infinite, or else run to completion without producing an outcome,
as can happen in the case of partial commands.

For example, the recursion

    X = (n := 0 ▽ (X; n := n + 1))

defines X = "set n to any natural number".  This is in contrast to
the recursion

    Y = (n := 0 ☐ (Y; n := n +1))

which defines Y = "set n to any natural number, or loop".  Thus the
dovetail operator is sufficiently powerful to express unbounded
non-determinism.

The dovetail operator is the imperative counterpart of the ambiguity
operator introduced by McCarthy in 1963 [9].  The ambiguity operator
is not monotonic in the ordering of either the Smyth or the Plotkin
powerdomains.  Therefore its fixpoint theory, presented by Broy in
1986, is far from straightforward [1].  The dovetail operation also
is not monotonic, and to treat it by the fixpoint method requires
some of Broy's techniques.  But the presence of partial commands
introduces even more difficulties.

Dovetail is powerful enough to model classical weak fairness.  Define
FairDo(A, B) to be like do A ☐ B od but with the extra requirement
that neither A nor B be starved; that is, the scheduler must not loop
executing A infinitely if grd.B remains true throughout the infinite
loop, nor vice versa.  To implement FairDo using Dovetail, we first

implement SEQ(X) (meaning "execute X some finite number of times, without looping") and $X^+$ (meaning "execute X a finite non-zero number of times"), and $X^*$ (meaning "execute X a finite number of times"). These are defined as follows:

$$\text{SEQ}(X) = (X \; ; \; \text{SEQ}(X)) \; \triangledown \; \text{Skip}$$
$$X^+ = \text{SEQ}(X) \; ; \; X$$
$$X^* = X^+ \; \Box \; \text{Skip}$$

Notice that SEQ(X) cannot loop, even if X can; but $X^+$ and $X^*$ can loop if X can. Finally we define FairDo(A, B) as follows:

$$\text{FairDo}(A, B) = \underline{\text{do}} \; A^+;(B \boxtimes \text{Skip}) \; \Box \; B^+;(A \boxtimes \text{Skip}) \; \underline{\text{od}}$$

It will be left to the reader to persuade himself of the appropriateness of this formula, and to generalize to FairDo(A, B, C).

Finally, to connect Dovetail to an existing formalism for reasoning about fair concurrent programs, we remark that the "leads to" relation of Chandy and Misra's Unity system [3] can now be defined: P leads to Q with respect to a Unity program whose body consists of the commands A and B if

$$P \Rightarrow \text{hlt.}(\text{FairDo}(\neg \, Q \rightarrow A, \; \neg \, Q \rightarrow B))$$

So much for the motivation of dovetail. Our goal is to reason about it in the same axiomatic equational style used to reason about the fundamental operators. We begin with the precondition equations, which are somewhat subtle:

$$\text{wlp.}(A \; \triangledown \; B).R = \text{wlp.}A.R \; \wedge \; \text{wlp.}B.R$$

$$\begin{aligned}
\text{hlt.}(A \; \triangledown \; B) = \; &(\text{hlt.}A \vee \text{hlt.}B) \; \wedge \\
&(\text{grd.}A \vee \text{hlt.}B) \; \wedge \\
&(\text{hlt.}A \vee \text{grd.}B)
\end{aligned}$$

That is, as far as wlp is concerned, $\triangledown$ is the same as $\Box$. It differs by having a more liberal wp equation: to ensure that $A \; \triangledown \; B$ halts,

it suffices to forbid A and B from both looping and to forbid either
from looping in a state where the other fails.  The value of wp for
postconditions other than <u>true</u> is determined by the pairing condition.

As is well-known, fairness leads to non-monotonicity with respect
to approximation.  For example,

    Loop $\sqsubseteq$ x := 0

but

    Loop $\nabla$ x := 1 $\equiv$ x := 1
    x := 0 $\nabla$ x := 1 $\equiv$ x := 0 $\square$ x := 1

and x := 1 does not approximate x := 1 $\square$ x := 2. This non-monotonicity
is a disaster for the fixpoint method.  There is however a glimmer
of hope: dovetail is monotonic with respect to approximation when
restricted to equivalence classes with respect to wlp.  That is:

Lemma 3.1 If wlp.A = wlp.B, then for any C,

    A $\sqsubseteq$ B $\Rightarrow$ A $\nabla$ C $\sqsubseteq$ B $\nabla$ C

(Proof left to the reader.)

This glimmer suggests the following strategy for finding a fixpoint
of $f(X) = E$. First, let $E^*$ be the result of replacing all occurrences
of $\nabla$ in E by $\square$, and define $f^*(X) = E^*$. By Theorem 1.4, $f^*$ has an
approximation-least fixpoint, say $X^*$.  Now let S be the set of all
commands Y such that $wlp.Y = wlp.(X^*)$ and $X^* \sqsubseteq Y$. Operationally we expect
the fixpoint of f to be in S, since changing $\square$ to $\nabla$ does not affect
wlp, and can only decrease the number of looping outcomes, which tends
to move the fixpoint upwards in the approximation relation.  Furthermore,
within Y, $\nabla$ is monotonic (Lemma 3.1), and therefore we should be
able to carry out a fixpoint contruction for f within Y.

In fact this strategy works just fine, provided that $f: S \rightarrow S$.
For this condition we have the following proof sketch:

$Y \in S$

$\Rightarrow \{\text{def. of } S\}$

$X^* \sqsubseteq Y \land \text{wlp.}(X^*) = \text{wlp.}Y$

$\Rightarrow \{\text{Proposition to be proved below}\}$

$f^*(X^*) \sqsubseteq f(Y) \land \text{wlp.}(f^*(X^*)) = \text{wlp.}(f(Y))$

$= \{X^* \text{ is a fixpoint of } f^*\}$

$X^* \sqsubseteq f(Y) \land \text{wlp.}(X^*) = \text{wlp.}(f(Y))$

$= \{\text{definition of } S\}$

$f(Y) \in S$

Thus we have a proof of the existence of fixpoints, provided we can prove the following proposition:

> If $f^*$ and $f$ are defined as in the proof sketch above,
> and $X \sqsubseteq Y$ and $\text{wlp.}X = \text{wlp.}Y$, then $f^*(X) \sqsubseteq f(Y)$ and
> $\text{wlp.}(f(X)) = \text{wlp.}(f(Y))$.

This can easily be proved by structural induction on the expression defining f, provided that all the operators in that expression preserve equivalence with respect to wlp.

But, unfortunately, the ⊠ operator does not preserve equivalence with respect to wlp.  For example,

> wlp.Loop is equal to wlp.Fail, but
> wlp.(Loop ⊠ x := 0) is not equal to
> wlp.(Fail ⊠ x := 0), since
> wlp.Loop is not equal to wlp.(x := 0).

We have failed to prove the theorem we want.  But we have proved that if deterministic choice is excluded, then recursions involving dovetail have solutions.  A closer analysis of the proof sketch above also gives a characterization of the fixpoint.  Here we state the theorem and refer the reader to the paper by Broy and Nelson [2] for the details of the proof:

Theorem 3.2.  If f(x) = E is defined by an expression E built from dovetail, the fundamental operators excluding ⊠, and any number of fixed commands and predicates, than f has a least fixpoint in the order $\preceq$ defined by

$$(A \preceq B) \equiv (A \sqsubseteq_{wlp} B) \wedge (wlp.A = wlp.B \Rightarrow A \sqsubseteq B)$$

The approximation order $\sqsubseteq$ is the intersection of $\sqsubseteq_{wlp}$ and $\sqsubseteq_{wp}$; the order $\preceq$ is a sort of lexicographic combination of $\sqsubseteq_{wlp}$ and $\sqsubseteq_{wp}$.

Notice that Theorem 3.2 is strong enough to justify the definition of FairDo in terms of Dovetail, since ⊠ does not appear in the recursions that define SEQ.

The reason that ⊠ does not respect wlp-equivalence is that ⊠ depends on the guard of its left argument, and the guard is determined by wp, not by wlp.  This suggests a new glimmer of hope: suppose we define two commands A and B to be "similar" (and write $A \simeq B$) if wlp.A = wlp.B and grd.A = grd.B.  Then we can carry out the same proof strategy above, but define the set S to be the set of Y such that X* $\simeq$ Y and X* $\sqsubseteq$ Y. Since $\nabla$ is monotonic on equivalence classes with respect to wlp, it is certainly monotonic on the new Y.  Thus we hope to handle ⊠ along with the other fundamental operators.

Unfortunately this doesn't work either: now ⊠ is fine, but the backtracking semicolon does not preserve $\simeq$.  For example,

$$(x=0 \rightarrow \text{Skip}) \quad \simeq \quad (x=0 \rightarrow (\text{Skip} \; \square \; \text{Loop})), \text{ but}$$

```
    (x=0 → Skip)           ; Fail  is not similar to
(x=0 → (Skip ☐ Loop))  ; Fail
```
since the first is Fail, the second is $x=0 \rightarrow$ Loop, and these have different guards.

However, it is possible to prove that ; respects $\simeq$ if its right argument is restricted to be total.  More precisely, define

$$A \;;; B \equiv A \; ; \; (B \; ⊠ \; \text{Loop})$$

Then A ; B and A ;; B are the same if B is total, and ;; respects $\simeq$.

This leads to:

Theorem 3.3.  If $f(x) = E$ is defined by an expression E built from
the operators ;; and $\nabla$ and the fundamental operators other than
semicolon, together with any number of fixed commands and predicates,
then f has a least fixpoint in the order $\preceq$ defined by

$$(A \preceq B) \equiv (A \sqsubseteq_{wlp} B) \wedge (A \simeq B \Rightarrow A \sqsubseteq B)$$

See reference [2] for the details of the proof.

Thus we have proved the existence of fixpoints for recursions that
involve any two of the operators $\boxtimes$, $\nabla$, and backtracking semicolon,
but not all three at once.  Perhaps by fiddling with the equivalence
relation further, we can handle all the operators at once?  We tried
and tried, and eventually found a proof sketch that seemed to work
for all recursions except those in which the left argument to a
backtracking semicolon contained an application of dovetail one of
whose arguments contained an application of $\boxtimes$, whose left argument
contained a recursive call.  At that point it was not difficult to
find the following example:

$$f.X = ((X \boxtimes b := 0) \nabla b := 1); (b = 0 \rightarrow \text{Loop})$$

Notice that f.Loop = Fail and f.Fail = Loop. More generally, the only
outcome of f.X is the looping outcome.  Therefore a fixpoint, if
it exists, has the form $P \rightarrow \text{Loop}$, for some predicate P.  But
direct calculation yields

$$f.(P \rightarrow \text{Loop}) = (\neg P) \rightarrow \text{Loop}$$

from which it is obvious that f has no fixpoint.

Operationally, if X is defined by the recursion f, then executing
a call to X searches an infinite tree.  At each level of the tree,
there is a fork caused by the dovetail.  The non-recursive branch
of the fork "evaporates", since it sets $b := 1$ and then runs into
the guard $b = 0$.  The other branch continues the infinite
search.

In the most obvious implementation, the call to X would loop forever. There is, however, another possible implementation.  In implementing X $\triangledown$ Y, if the computation of X goes into a detectable infinite loop (in the sense that computation from one of its states leads back to the exact same state, with all other alternatives exhausted) then a clever implementation could ignore the X computation and continue only the Y computation.  If this clever implementation were used, the call to the recursive procedure X would result in Fail, not Loop.

The fact that different rules for executing the recursively defined X lead to different results is consistent with the fact that the recursion has no fixpoint.  A similar situation hold in the theory of infinite series, where, if a series is not summable, it often can be summed to different results by parenthesizing it differently. Theorems 3.2 and 3.3 provide criteria to establish that a given recursion involving dovetail has a fixpoint, just as various tests are known for establishing that a given series is summable.

In principle the definition of FairDo in terms of dovetail together with the fixpoint characterization in Theorem 3.2 could serve as the basis of a proof rule for a system like Unity.  However, we have not yet tried to derive the proof rule.


4. A precondition calculus for games

In this section we generalize from commands to games.  Our results are similar to those of Back and J von Wright [0], but we deal with pairs of predicate transformers rather than single predicate transformers, thereby handling looping in its full generality.

An angel and a devil play a game by executing a program in which some of the non-deterministic choices are made by the angel, others by the devil.  The possible outcomes of the computation are

    victory for one of the two players
    non-termination (looping outcome)
    termination in some state

We shall develop the language in which the game-program is written
bit by bit throughout this note; for starters, consider the following
four operators:

A ; B              Execute A, then B.

A □ B              Execute A or B; the devil chooses which.

P?                 If P is true, this is a no-op; else the
                   angel wins.

A~                 Execute A, with the roles of the angel and
                   devil reversed.

The first three operators are sequential composition, demonic choice,
and test, as introduced in the first section.  The last operator is
called alternation.

For game A and predicate R, we define

wp.A.R             the predicate that holds of those initial
                   states from which the angel can win or force
                   the program to halt in a state satisfying R.

wlp.A.R            the predicate that holds of those states from
                   which the angel can win or force the program
                   to loop forever or halt in a state satisfying R.

Writing one equation in w instead of two identical equations in wp
and wlp, we have:

$$w.(A;B).R \equiv w.A.(w.B.R)$$

$$w.(A \,\square\, B).R \equiv w.A.R \wedge w.B.R$$

$$w.P?.R \equiv (\neg P) \vee R$$

$$wp.A^{\sim}.R \equiv \neg\, wlp.A.(\neg R)$$

$$wlp.A^{\sim}.R \equiv \neg\, wp.A.(\neg R)$$

The precondition equations for sequential composition, demonic choice, and test are the same for games as for commands.  Thus the technical difference between the calculus of games and the calculus of commands is mainly that the latter must deal with alternation.

Some additional games and operators:

$$A \ \Diamond\ B \ = \ (A^\sim \ \Box\ B^\sim)^\sim \qquad \text{execute A or B; the angel chooses.}$$

Skip = <u>true</u>?                    The no-op.

Win  = <u>false</u>?                   If this command is executed,
                               the angel wins.

Lose  =  Win$^\sim$                    If this command is executed,
                               the devil wins.

Loop                           infinite looping

Havoc                          The devil chooses any halting
                               outcome.

$$P \rightarrow A \ = \ P? \ ; \ A \qquad \text{If P is false, the angel wins,}$$
                                   else continue with A

Skip, Win, Havoc, Loop, and the guard arrow are familiar from the section 1 (but for the fact that Win was called Fail).  Their precondition equations are familiar:

$$\text{w.Skip.R} \ \equiv \ R$$
$$\text{w.Win.R} \ \equiv \ \underline{true}$$
$$\text{w.}(P \rightarrow A).R \ \equiv \ (\neg\ P) \lor R$$
$$\text{wp.Loop.R} \ \equiv \ \underline{false}$$
$$\text{wlp.Loop.R} \ \equiv \ \underline{true}$$
$$\text{w.Havoc.R} \ \equiv \ [R]$$

Angelic choice and Lose have new equations:

$$\text{w.}(A \ \Diamond\ B).R \ \equiv \ \text{w.A.R} \lor \text{w.B.R}$$
$$\text{w.Lose.R} \ \equiv \ \underline{false}$$

The game A $\Diamond$ B cannot be interpreted as a command, since
wp.(A $\Diamond$ B) is not conjunctive.  For example, let

$$A \ \equiv \ x := 0 \ \Diamond \ x := 1$$

Then

$$wp.A.(x=0) = \underline{true}$$
$$wp.A.(x=1) = \underline{true}$$
$$wp.A.\underline{false} = \underline{false}$$

which violates conjunctivity.

Lose also cannot be interpreted as a command, since wlp.Lose is not
universally conjunctive; that is, it does not distribute over the
empty conjunction.

We define a game A to be "deterministic" if A = A$^\sim$, in the sense
that wp.A = wp.A$^\sim$ and wlp.A = wlp.A$^\sim$.  For example, it is easy to
verify that

$$P \rightarrow A \ \Box \ \neg P \rightarrow B$$

is deterministic if A and B are.  The "choice" offered to the devil
in the construct above is Hobson's choice: to avoid losing immediately,
his choice is determined by the state.

We define for games (as previously for commands):

$$grd.A \ \equiv \ \neg \ wp.A.\underline{false}$$

The states satisfying wp.A.$\underline{false}$ are those from which the angel can
force a win, regardless of the postcondition.  In the model of commands
as partial relations, such a state corresponds to a state that is
related to no outcome.  Thus the predicate grd.A characterizes those
states from which the angel must accept some outcome other than
out-and-out victory for himself.

We also define deterministic choice for games:

$$A \boxtimes B \;\equiv\; A \;\square\; (\neg\, grd.A) \to B$$

The precondition equation for $\boxtimes$ is, as before:

$$w.(A \boxtimes B).R \;\equiv\; w.A.R \,\wedge\, (grd.A \,\vee\, w.B.R)$$

The effect of $A \boxtimes B$ is to execute A, unless this is a forced win for the angel regardless of the postcondition, in which case B is executed.  The demonic choice operator occurs in the definition of $A \boxtimes B$ , but the "choice" offered to the devil is again Hobson's choice:  $A \boxtimes B$ is deterministic if A and B are.

We now turn from the operational to the axiomatic approach. The healthiness conditions used for commands are too strict for games. We have already encountered games that violate conjunctivity, and they may violate the pairing condition, too: consider

$$A \;\equiv\; (\text{Loop}\ \square\ x := 1)\ \Diamond\ x := 0$$
$$R \;\equiv\; x = 1$$

The angel playing A can force the computation to halt, or can force all halting outcomes to satisfy $x = 1$, but cannot force both at once.

These are the correct healthiness conditions for games:

$$[wp.A.R \Rightarrow wlp.A.R]$$
$$[P \Rightarrow Q] \;\Rightarrow\; [wp.A.P \Rightarrow wp.A.Q]$$
$$[P \Rightarrow Q] \;\Rightarrow\; [wlp.A.P \Rightarrow wlp.A.Q]$$

We define a game A to be a pair of predicate transformers wp.A and wlp.A satisfying these three conditions.  All commands are games, but not all games are commands.

The fact that for any operational game A, wp.A and wlp.A satisfy the healthiness conditions is clear from the operational definitions of wp and wlp.  The next theorem shows the converse, that for any pair of predicate transformers f and g satisfying the conditions, there is an operational game A for which f = wp.A and g = wlp.A.

Theorem 4.1.  For any game A,

$$A = (\Diamond\, R::\ G.A.R)$$

where R ranges over all predicates and

$$G.A.R = \text{Havoc};\ R?\ \square\ \neg\ \text{wlp}.A.R \rightarrow \text{Lose}\ \square\ \neg\ \text{wp}.A.R \rightarrow \text{Loop}$$

Proof: First we establish the identity:

$$w.(G.A.R).Q\ =\ [R \Rightarrow Q]\ \wedge\ w.A.R$$

This requires two calculations:

$$
\begin{aligned}
&\text{wp}.(G.A.R).Q\\
=\ &[R \Rightarrow Q]\ \wedge\ \text{wlp}.A.R\ \wedge\ (\text{wp}.A.R\ \vee\ \text{wp}.\text{Loop}.R)\\
=\ &[R \Rightarrow Q]\ \wedge\ \text{wp}.A.R
\end{aligned}
$$

$$
\begin{aligned}
&\text{wlp}.(G.A.R).Q\\
=\ &[R \Rightarrow Q]\ \wedge\ \text{wlp}.A.R\ \wedge\ (\text{wp}.A.R\ \vee\ \text{wlp}.\text{Loop}.R)\\
=\ &[R \Rightarrow Q]\ \wedge\ \text{wlp}.A.R
\end{aligned}
$$

Now we compute

$$
\begin{aligned}
&w.(\Diamond\, R::\ G.A.R).Q\\
=\ &(\vee\ R::\ w.(G.A.R).Q)\\
=\ &(\vee\ R::\ [R \Rightarrow Q]\ \wedge\ w.A.R)\\
=\ &\{\text{Split the range}\}\\
=\ &(\vee\ R:\ [R \Rightarrow Q]:\ w.A.R)\ \vee\ (\vee\ R:\ \neg\ [R \Rightarrow Q]:\ \underline{false})\\
=\ &\{\ w\ \text{is monotonic}\ \}\\
&w.A.Q
\end{aligned}
$$

This completes the proof of Theorem 4.1.

The theorem shows that any game defined axiomatically as a pair of
predicate transformers can be realized by an explicit formula involving
angelic and demonic choice.  Hence our axiomatic definition of a game
is not too general.

Notice that the game $(\Diamond\ R::\ G.A.R)$ begins with a giant choice for
the angel, after which all remaining choices are made by the devil.
From the point of view of classical game theory, there is one disjunct
for each "strategy" for the angel.  Once the angel has chosen his
strategy, his moves are determined, and only demonic non-determinism
remains.  Thus every game can be imagined to have two moves: the
angel chooses a strategy, and then the devil chooses a response.

We would like to allow games to be defined recursively.  This
essentially requires that we redo the theory outline in section 1,
using monotonicity only, not conjunctively.  In fact, the theory is
slightly simpler for games than for commands.

For predicate transformers f and g, we write $(f \Rightarrow g)$ to mean that
$[f.R \Rightarrow g.R]$ for all predicates R.  In this notation, the familiar
approximation order $A \sqsubseteq B$ can be defined as

$$(wp.A \Rightarrow wp.B) \wedge (wlp.B \Rightarrow wlp.A).$$

We need to establish the existence of joins and meets of games in the
approximation order.

Theorem 4.2.  Every chain of games has a join in the approximation
order.  If A ranges over the chain then the join J is defined by

$$wp.J\ =\ (\sqcup\ A::\ wp.A)$$
$$wlp.J\ =\ (\sqcap\ A::\ wlp.A)$$

Here $\sqcup$ and $\sqcap$ refer to predicate transformers under the order $(? \Rightarrow ?)$.

Proof.  First we show that J satisfies the healthiness conditions.
Since the join and meet of monotonic predicate transformers are
monotonic, both wp.J and wlp.J are monotonic. Furthermore:

$$(wp.J \Rightarrow wlp.J)$$
$$= ((\sqcup A:: wp.A) \Rightarrow (\sqcap A:: wlp.A))$$
$$= (\forall A, A':: (wp.A \Rightarrow wlp.A'))$$
$$= \{\text{The A's range over a chain}\}$$
$$(\forall A, A': A \sqsubseteq A': (wp.A \Rightarrow wlp.A'))$$
$$= \{(wp.A \Rightarrow wp.A') \text{ since } A \sqsubseteq A'\}$$
$$\{(wp.A' \Rightarrow wlp.A') \text{ since } A' \text{ is a game}\}$$
$$\underline{true}$$

Thus J is a game.

To prove that this game is the least upper bound of the A's, let B be any game, and compute

$$J \sqsubseteq B$$
$$= (wp.J \Rightarrow wp.B) \wedge (wlp.B \Rightarrow wlp.J)$$
$$= ((\sqcup A:: wp.A) \Rightarrow wp.B) \wedge (wlp.B \Rightarrow (\sqcap A:: wlp.A))$$
$$= \{\text{definition of } \sqcup \text{ and } \sqcap\}$$
$$(\forall A:: (wp.A \Rightarrow wp.B)) \wedge (\forall A:: (wlp.B \Rightarrow wlp.A))$$
$$= (\forall A:: A \sqsubseteq B)$$

Thus J approximates exactly those games that are approximated by every A, which is what it means to be the join. Q.E.D

The formula in Theorem 4.2 is the same as the formula in Theorem 1.2. That is, the game-join of a chain of commands is a command.

Theorem 4.3. Every non-empty set of games has a meet in the approximation order. If A ranges over the set of games, then the meet M is defined by:

$$wp.M \ = (\sqcap A:: wp.A)$$
$$wlp.M = (\sqcup A:: wlp.A)$$

Proof. As wp.M and wlp.M are the join and meet of monotonic functions, they are monotonic. So much for the first two healthiness conditions. The proof of the third healthiness condition is:

$$(\text{wp}.M \Rightarrow \text{wlp}.M)$$
$$= ((\wedge\ A::\ \text{wp}.A) \Rightarrow (\vee\ A::\ \text{wlp}.A))$$
$$= \{A \text{ ranges over a non-empty set}\}$$
$$(\forall\ A::\ (\exists\ A'::\ (\text{wp}.A \Rightarrow \text{wlp}.A')))$$
$$\Leftarrow (\forall\ A::\ (\text{wp}.A \Rightarrow \text{wlp}.A))$$
$$= \{\text{each } A \text{ is a game}\}$$
$$\underline{\text{true}}$$

To prove that M is the greatest lower bound of the A's, let
B be any command, and compute:

$$B \sqsubseteq M$$
$$= (\text{wp}.B \Rightarrow \text{wp}.M) \wedge (\text{wlp}.M \Rightarrow \text{wlp}.B)$$
$$= (\text{wp}.B \Rightarrow (\sqcap\ A::\ \text{wp}.A)) \wedge ((\sqcup\ A::\ \text{wlp}.A) \Rightarrow \text{wlp}.B)$$
$$= (\wedge\ A::\ (\text{wp}.B \Rightarrow \text{wp}.A) \wedge (\text{wlp}.A \Rightarrow \text{wlp}.B))$$
$$= (\wedge\ A::\ B \sqsubseteq A)$$

This completes the proof of Theorem 4.3.

Theorem 4.3 is easier to prove than Theorem 1.3, since the obvious
formula for the meet works. As a consequence, the game-meet of two
commands may not be a command. For example, in a state space with
a single variable x ranging over $\{0, 1\}$, consider the meet of the two
commands x := 0 and x := 1. As commands, their meet is Loop. As games,
we compute their meet M by

$$\text{wp}.M.R = R(x:0) \wedge R(x:1) = [R]$$
$$\text{wlp}.M.R = R(x:0) \vee R(x:1) = \neg\ [\neg\ R].$$

One formula for M is

$$\text{Havoc} \diamond (\text{Loop} \square \text{Havoc}^\sim)$$

The angel can avoid looping, at the cost of losing control over x;
or control x, at the cost of allowing looping.

The proofs that sequential composition, demonic choice, and test
are monotonic with respect to approximation are the same for games
as they were for commands. The only new operator to check is
alternation:

$$A^\sim \sqsubseteq B^\sim$$

$$= (\forall R:: [wp.A^\sim.R \Rightarrow wp.B^\sim.R] \wedge [wlp.B^\sim.R \Rightarrow wlp.A^\sim.R])$$

$$= (\forall R:: [wlp.B.(\neg R) \Rightarrow wlp.A.(\neg R)])$$
$$\wedge [wp.A.(\neg R) \Rightarrow wp.B.(\neg R)])$$

$$= A \sqsubseteq B$$

Thus all the operators are monotonic, and therefore, games can
be defined by recursions:

Theorem 4.4:  Let f be a map from games to games defined by an
expression of the form $f(X) = E$, where E is an expression built from
the fundamental operators, alternation, and any number of fixed games
and predicates.  Then f has a least fixpoint in the approximation
order $\sqsubseteq$, equal to

$$(\sqcup i: i \text{ an ordinal}: f^i(\text{Loop})) \text{ and to}$$
$$(\sqcap X: f(X) \sqsubseteq X: X) \text{ and to}$$
$$(\sqcap X: f(X) \equiv X: X)$$

The proof is essentially the same as the proof of Theorem 4.3.

So much for alternation.  Handling alternation and fair choice
simultaneously seems difficult, since alternation does not
preserve equivalence with respect to wlp.


Conclusion

Following Dijkstra, we have concentrated on total correctness---that
is, we have preserved the distinction between wp and wlp.  Also
following Dijkstra, we have taken non-determinism for granted.  But
we have generalized by dropping the law of the excluded miracle, and
extended the calculus by adding deterministic choice, fair choice,
and alternation to Dijkstra's sequential composition, domonic choice,
and test.  For some combinations of the operators, recursions are
guaranteed to have fixpoints, but not for all.

References

0. R.J.R. Back, J. von Wright. Combining Angels, Demons, and Miracles in Program Specifications. Inst. for Informationsbehandling, Lemminkainengatan 14, SF-20520 Abo, Finland. ISBN 952-649-626-1. 1989.

1. M. Broy. A theory for nondeterminism, parallelism, communication, and concurrency. TCS 45, 1986, pp. 1-61.

2. Manfred Broy and Greg Nelson. Can fair choice be added to Dijkstra's calculus? Research Report 38, Digital Systems Research Center, February 1989.

3. K. Mani Chandy and Jayadev Misra. Parallel Program Design: A Foundation. Addison-Wesley, 1989.

4. J. W. deBakker. Semantics and termination of nondeterministic recursive programs. Automata, Languages, and Programming, Edinburgh, 1976.

5. Edsger W. Dijkstra and Carel S. Scholten. Predicate Calculus and Program Semantics. Springer-Verlag, 1990.

6. H. Egli. A mathematical model for nondeterministic computations. Zuric, ETH (1975). Cited by deBakker [4].

7. Peter Hitchcock and David Park. Induction Rules and Termination Proofs. IRIA Conf. Autom. Lang. and Program. Theory. France, 1972.

8. J.L. Lassez, V.L. Nguyen, and E. A. Sonenberg. Fixed point theorems and semantics: a folk tale. Information Processing Letters vol. 14 no. 3, May 1982.

9. J. McCarthy. Towards a mathematical science of computation. In P. Braffort, D. Hirschberg (eds.): Computer Programming and Formal Systems, Amsterdam, North-Holland 1963. pp. 33-70.

10. Mark Manasse and Greg Nelson.  Correct compilation of control structures.  Bell Labs Technical Memorandum, September 9, 1984.

11. Greg Nelson. A generalization of Dijkstra's calculus.  TOPLAS 11, October 1989, pp. 517--61.

# Chapter 3

# Refinement and Program Composition

Refinement is the key issue to form a method of program design. The concepts of refinement allow to start from very high level specification leaving out a lot of details and to add more and more details to a program or system description until an appropriate level is reached. The simplicity and flexibility of the concept of refinement are decisive in the program development, because only if high flexibility is provided, then all the interesting steps of program development can be explained, specified, verified and finally carried out within a refinement calculus. This is especially important for more complicated examples, such as the development of for instance compilers by stepwise refinement.

K.M. Chandy

D. Gries

C.A.R. Hoare

# A Theory of Program Composition


K. Mani Chandy

California Institute of Technology


September 25, 1990

## Abstract


Program composition is defined using a notation called Program Composition Notation (PCN). The syntax of a PCN program is that of a formula in the predicate calculus, in a normal form. Given a program specification, we manipulate the specification into a predicate in the normal form, and thus derive a PCN program. If the efficiency of the program is not adequate when executed on a target (possibly parallel) computer, the program is made more efficient by the judicious introduction of imperative programming constructs: sequential composition and mutable variables that can be assigned values arbitrarily often. Fairness is introduced, where necessary, by predicates that specify UNITY programs.


## 1  Introduction

PCN is a notation for composing programs in a variety of languages such as Fortran, C and UNITY [2], so that the composed programs execute efficiently on a variety of architectures including sequential computers, shared-memory multiprocessors, and message-passing multicomputers. For the time being we do not consider sequential composition, and we restrict the syntax of a program to be an equation in the predicate calculus, in a normal form.

# 2  Syntax

In the syntax, the notation $\prec su \succ$, where $su$ is a syntactic unit, is a list of zero or more instances of $su$. The syntax for some nonterminals is discussed later.

---

| | | |
|---|---|---|
| *program* | :: | *program-name*(*variable* $\prec$, *variable* $\succ$) <br> = <br> *block* |
| *block* | :: | *true* \| <br> *directed-equation* \| <br> *program-name*(*expression* $\prec$, *expression* $\succ$) \| <br> *conjunctive-form* \| <br> *disjunctive-form* |
| *directed-equation* | :: | *variable* $\overset{\text{def}}{=}$ *term* |
| *conjunctive-form* | :: | (*block*) $\wedge$ (*block*) $\prec \wedge$ (*block*) $\succ$ \| <br> ($\exists$ *variable* $\prec$, *variable* $\succ$ :: <br>     (*block*) $\wedge$ (*block*) $\prec \wedge$ (*block*) $\succ$) |
| *disjunctive-form* | :: | (*guard* & *block*) $\vee$ (*guard* & *block*) <br> $\prec \vee$ (*guard* & *block*) $\succ$ |

---

## 2.1  The Predicate Corresponding to a Block

Before we discuss the syntax of a program, we define an important concept in PCN: *the safety-predicate corresponding to a block*. The safety-predicate corresponding to a block is the predicate with the same syntax as a block and in which & is replaced by $\wedge$, and $\stackrel{\text{def}}{=}$ is replaced by $=$. The safety-predicate corresponding to a block $b$ is denoted by $\mathcal{SAFETY}.b$ . In a program block, $\&$ and $\stackrel{\text{def}}{=}$ are asymmetric, for reasons that will become clear shortly.

A predicate can be defined in terms of itself. The predicate defined by a recursive equation is the weakest predicate satisfying the equation. For example Hoare's clock[9] can be specified as:

$$clock(x) \;\; = \;\; ((x = [\texttt{tick}|y]) \wedge clock(y))$$

where $\texttt{tick}$ is a string.

## 2.2 Description of Syntax

1. A term is a character, string, number, expression or **tuple**. A tuple is a brace, '{', followed by a sequence of zero or more terms followed by brace , '}'. We define a function *length* on tuples, where $length.t$ is the number of elements in $t$. The $i$-th element of a tuple is $t[i]$, for $0 \leq i$ . If $i \geq length.t$ then $t[i]$ is an arbitrary value.

   A list is either the empty tuple $\{\ \}$, or a 2-tuple $\{a, b\}$ where $b$ is a list. For convenience we use the notation $[a, b, c, \ldots]$ to represent the list consisting of elements $a$, $b$, $c$, $\ldots$, since this notation is more succinct than the tuple representation $\{a, \{b, \{c, \{....\}\}\}\}$. Likewise, we use the notation $[a, b, c, \ldots | x]$ to represent the list consisting of elements $a$, $b$, $c$, $\ldots$, followed by list $x$. We shall use $hd.x$ and $tl.x$ to refer to the head and tail of a list. (From the definitions, $hd.x = x[0]$ and $tl.x = x[1]$. )

2. We extend the syntax of the conjunctive form to allow quantification. The syntax using quantification is:

$$(\forall i \in S :: b_i)$$

where $S$ is a list and $b_i$ is a block. For example,

$$(\forall i \in S :: x_i \stackrel{\text{def}}{=} y_i)$$

where $S$ is the list $[0, 1, 2]$ is the same as the conjunctive form:

$$(x_0 \stackrel{\text{def}}{=} y_0) \wedge (x_1 \stackrel{\text{def}}{=} y_1) \wedge (x_2 \stackrel{\text{def}}{=} y_2)$$

3. A guard is a boolean expression. The syntax of boolean expressions is not given here. We extend the syntax of the disjunctive form to allow quantification. The syntax using quantification is:

$$(\exists\, i \in S : g_i : b_i)$$

where $S$ is a list, $g_i$ is a guard and $b_i$ is a block. For example,

$$(\exists\, i \in S : i < 2 : x_i \stackrel{\mathrm{def}}{=} y_i)$$

where $S$ is the list $[0, 1, 2]$ is the same as the disjunctive form:

$$(0 < 2 \ \& \ x_0 \stackrel{\mathrm{def}}{=} y_0) \vee (1 < 2 \ \& \ x_1 \stackrel{\mathrm{def}}{=} y_1) \vee (2 < 2 \ \& \ x_2 \stackrel{\mathrm{def}}{=} y_2)$$

# 3 Semantics

## 3.1 The Program Derivation Problem in PCN

Given

1. a set $S$ of variables partitioned into a set of *input* variables and a set of *output* variables,

2. a predicate $q$ on variables in $S$, and

3. a predicate $R$ on input variables

derive a block $b$, such that input variables do not appear on the left-hand sides of directed equations, and the following safety and progress conditions are satisfied:

**Safety:**
$$(\mathcal{SAFETY}.b \ \wedge \ R) \Rightarrow q$$

**Progress:**

$$(R \wedge pre \wedge \mathcal{PROGRESS}.b) \Rightarrow (post \wedge \mathcal{P})$$

where *pre*, *post*, $\mathcal{PROGRESS}$ and $\mathcal{P}$ are defined next. Read $\mathcal{P}$ as "proper."

For a block $b$ we define $\mathcal{PROGRESS}.b$ as the predicate with syntax identical to that of $b$ except that the disjunctive form:

$$(g_0 \& b_0) \lor \ldots \lor (g_n \& b_n)$$

is replaced by

$$(\exists i :: g_i \land \mathcal{G}.g_i) \Rightarrow ((g_0 \land b_0) \lor \ldots \lor (g_n \land b_n))$$

where $\mathcal{G}$, read as "ground" is defined later. Also, in $\mathcal{PROGRESS}.b$ we treat $x \stackrel{\text{def}}{=} e$ is an uninterpreted predicate for which we shall give proof rules; we shall not, however, define $x \stackrel{\text{def}}{=} e$: All we care about are the rules that we can employ to manipulate formulae containing $\stackrel{\text{def}}{=}$.

We define $\mathcal{P}$ as :

$$\mathcal{P} = (\forall x, f, g : (x \stackrel{\text{def}}{=} f) \land (x \stackrel{\text{def}}{=} g) : f, g \text{ are identical})$$

We introduce a binary relation $\stackrel{\text{def}}{=}*$, (read as "reduces to") between variables and expressions. It is defined as follows:

$x \stackrel{\text{def}}{=}* e$, where $x$ is a variable and $e$ is an expression, if and only if,

1. $x \stackrel{\text{def}}{=} e$, or

2. $x \stackrel{\text{def}}{=} f$, and for some variable $y$ named in $f$, $y \stackrel{\text{def}}{=}* w$, and substituting $w$ for $y$ in $f$ and evaluating gives $e$.

For example:

$$((x \stackrel{\text{def}}{=} y + z) \land (y \stackrel{\text{def}}{=} z + 1) \land (z \stackrel{\text{def}}{=} 1)) \Rightarrow (x \stackrel{\text{def}}{=}* 3)$$

We define a *ground value* as a number, string of characters, *true*, *false*, or a tuple in which all elements are ground values.

From the definition of $\mathcal{P}$:

$$\mathcal{P} \Rightarrow (\forall x, \text{ and ground values } v, w : (x \stackrel{\text{def}}{=}* v) \land (x \stackrel{\text{def}}{=}* w) : v = w)$$

We define $\mathcal{G}$, read as " ground," to be a function from expressions to the boolean constants *true*, *false* as follows:

$\mathcal{G}.e$ holds if and only if:

1. for all instances of length.$x$ in $e$, there exists a tuple $v$ such that $t \stackrel{\text{def}}{=}^* v$, and

2. for all other instances of variable $x$ in $e$ (i.e., all instances in which $x$ is not an argument of length) there exists a ground value $v$ such that $t \stackrel{\text{def}}{=}^* v$.

Predicates *pre* and *post* specify the groundedness conditions on input and output variables, respectively. For a program with input $n$ and output $x$, *pre* is either $\mathcal{G}.n$ or $\mathcal{G}.(\text{length}.n)$; likewise, *post* is either $\mathcal{G}.x$ or $\mathcal{G}.(\text{length}.x)$.

# 4 Examples of Program Derivation

## 4.1 List of integers from n to 1

Let $n$ be an integer. Define predicate $h(n, x)$ as follows:

$h(n, x)$ holds if and only if:

1. $x$ is the list consisting of all integers from $n$ to 1, in decreasing order, if $n > 0$, and

2. $x$ is the empty list if $n \leq 0$.

Next, we define $h$ in the predicate calculus as:

$$h(n, x) =$$
$$((n > 0) \ \land \ (\exists xs :: (x = [n|xs]) \land h(n - 1, xs)))$$
$$\lor$$
$$((n \leq 0) \ \land \ (x = [\,]))$$

### Program Specification

Given that $n$ is an input integer and $x$ is an output list, develop a program $H(n, x)$ that establishes $h(n, x)$, i.e.,

**Safety:**
$$(\mathcal{SAFETY}.H(n, x) \ \land \ \text{int}(n)) \Rightarrow h(n, x)$$

where $\text{int}(n)$ holds if and only if $n$ reduces to an integer.

Progress

$$(\mathcal{PROGRESS}.H(n,x) \wedge \text{int}(n) \wedge \mathcal{G}.n) \Rightarrow \mathcal{G}.x$$

## Program Derivation

We manipulate the definition of $h$ to obtain program $H$. In deriving a block in disjunctive form, we use the heuristic that variables in guards are either inputs or are defined in blocks composed in parallel with the block that we are deriving. Using this heuristic, we group terms that name only $n$ into guards, and terms that name $x$ or $xs$ into the blocks associated with guards to get the program:

$$H(n,x) =$$
$$((n > 0) \ \& \ (\exists xs :: (x \stackrel{\text{def}}{=} [n|xs]) \wedge (H(n-1,xs))))$$
$$\vee$$
$$((n \leq 0) \ \& \ x \stackrel{\text{def}}{=} [\,])$$

Here the block defining $H$ is a disjunctive form $(g_0 \& b_0) \vee (g_1 \& b_1)$, where $b_0$ is the conjunctive form $(\exists xs :: (b_{0,0}) \wedge (b_{0,1}))$, and where $b_{0,0}$ is the directed equation $x \stackrel{\text{def}}{=} [n|xs]$, and $b_{0,1}$ is a program name with its arguments: $H(n-1,xs)$. Also, $g_0$ is $n > 0$, $g_1$ is $n \leq 0$, and $b_1$ is $x \stackrel{\text{def}}{=} [\,]$ .

## Program Proof

The proof of safety is trivial because $\mathcal{SAFETY}.H$ is the same as $h$.

From the definition of $\mathcal{PROGRESS}$:

$$\mathcal{PROGRESS}.H(n,x) =$$

$$((n > 0) \wedge \mathcal{G}.(n > 0)) \vee ((n \leq 0) \wedge \mathcal{G}.(n \leq 0))$$
$$\Rightarrow$$
$$((n > 0) \ \wedge \ (\exists xs :: (x \stackrel{\text{def}}{=} [n|xs]) \wedge (H(n-1,xs))))$$
$$\vee$$
$$((n \leq 0) \ \wedge \ x \stackrel{\text{def}}{=} [\,])$$

The proof of progress is by induction on $n$.
*Base Case:* $n \leq 0$ Prove that:

$$\mathcal{G}.n \wedge (n \leq 0) \wedge \mathcal{PROGRESS}.H(n, x) \;\Rightarrow\; \mathcal{G}.x$$

*Proof of Base Case:* From the definition of $\mathcal{PROGRESS}$:

$$\mathcal{G}.n \wedge (n \leq 0) \wedge \mathcal{PROGRESS}.H(n, x) \;\Rightarrow\; x \stackrel{\text{def}}{=} [\,]$$

From the definition of $\mathcal{G}$ and since $[\,]$ is a ground value:

$$x \stackrel{\text{def}}{=} [\,] \;\Rightarrow\; \mathcal{G}.x$$

The proof of the base case follows from the last two formulae.
*Induction step:* Assume:

$$\mathcal{G}.n \wedge (n \leq m) \wedge \mathcal{PROGRESS}.H(n, x) \;\Rightarrow\; \mathcal{G}.x$$

for some $m$ where $m \geq 0$, and prove

$$\mathcal{G}.n \wedge (n \leq m + 1) \wedge \mathcal{PROGRESS}.H(n, x) \;\Rightarrow\; \mathcal{G}.x$$

*Proof of Induction step:*
From the definition of $\mathcal{PROGRESS}$:
$$\mathcal{G}.n \wedge (n = m + 1) \wedge \mathcal{PROGRESS}.H(n, x)$$
$$\Rightarrow$$
$$(\exists xs :: (x \stackrel{\text{def}}{=} [n|xs]) \wedge \mathcal{PROGRESS}.H(n - 1, xs))$$

From the induction hypothesis:

$$\mathcal{PROGRESS}.H(n - 1, xs)) \;\Rightarrow\; \mathcal{G}.xs$$

$$\mathcal{G}.n \wedge \mathcal{G}.xs \wedge (x \stackrel{\text{def}}{=} [n|xs]) \;\Rightarrow\; \mathcal{G}.x$$

The proof of the induction step follows from the last three formulae.

## 4.2 Example: Membership in a List

Let $x$ be a list, let $m$ be a term, and let $r$ be a boolean; define $member(x, m, r)$ as follows:

$member(x, m, r)$ holds if and only if:

$$r = m \text{ is an element of list } x$$

Next, we define $member$ in the predicate calculus.

$member(x, m, r) =$
$$
\begin{aligned}
&((x = [\,]) &&\wedge\ (r = false)) \\
&&&\vee \\
&((x \neq [\,]) \wedge (hd.x = m) &&\wedge\ (r = true)) \\
&&&\vee \\
&((x \neq [\,]) \wedge (hd.x \neq m) &&\wedge\ member(tl.x, m, r))
\end{aligned}
$$

Note: We treat $x = [\,]$ as equivalent to $length.x = 0$.

## Program Specification

Derive a program $MEMBER(x, m, r)$, where $x$ and $m$ are inputs and $r$ is an output that establishes $member(x, m, r)$, i.e.,

**Safety:**
$$\mathcal{SAFETY}.MEMBER(x, m, r) \Rightarrow member(x, m, r)$$

**Progress:**
$$\text{all\_ground}.x \wedge \mathcal{G}.m \wedge \mathcal{PROGRESS}.MEMBER(x, m, r) \Rightarrow \mathcal{G}.r$$

where all_ground is the strongest solution of:

$$\text{all\_ground}.x = ((x = [\,]) \vee (x \neq [\,] \wedge \mathcal{G}.hd.x \wedge \text{all\_ground}.tl.x))$$

## Program Derivation

Using our heuristic, we group terms in each conjunction that do not name $r$ into guards to get the program:

$$MEMBER(x, m, r) =$$
$$((x = [\,]) \qquad\qquad \&\ r \stackrel{\text{def}}{=} false)$$
$$\lor$$
$$((x \neq [\,]) \wedge (hd.x = m) \qquad \&\ r \stackrel{\text{def}}{=} true)$$
$$\lor$$
$$((x \neq [\,]) \wedge (hd.x \neq m) \qquad \&\ MEMBER(tl.x, m, r))$$

## Program Proof

The proof of safety is trivial. The proof of progress is by induction on the length of $x$. We outline the proof here.

The groundedness of $m$ and all elements of $x$ implies that precisely one of the three guards reduces to *true*, and the other two reduce to *false*. The base case follows from the first term in the disjunction. If $x$ is not empty, and the head element of $x$ is equal to $m$, then $\mathcal{G}.r$ follows from the second term in the disjunctive form. If $x$ is not empty, and the head element of $x$ is not equal to $m$, then $\mathcal{G}.r$ follows from the third term in the disjunctive form and the induction hypothesis.

## 4.3   Example: Merge

Next consider a nondeterministic program: a (possibly unfair) merge. Let $x$, $y$ and $z$ be lists. Predicate $merge(x, y, z)$ holds if and only if $z$ is an interleaving of $x$ and $y$.

We define *merge* in the predicate calculus as follows:

$$merge(x, y, z) =$$
$$((x \neq [\,]) \quad \wedge \quad (\exists v :: (z = [hd.x|v]) \wedge (merge(tl.x, y, v))))$$
$$\lor$$
$$((y \neq [\,]) \quad \wedge \quad (\exists w :: (z = [hd.y|w]) \wedge (merge(x, tl.y, w))))$$
$$\lor$$
$$((x = [\,]) \quad \wedge \quad z = y)$$
$$\lor$$
$$((y = [\,]) \quad \wedge \quad z = x)$$

## Program Specification

Derive a program $MERGE(x, y, z)$ with inputs $x$ and $y$ and output $z$, where $MERGE(x, y, z)$ establishes $merge(x, y, z)$. The progress property we are required to prove is that if $x$ is nonempty then $z$ is nonempty. By symmetry, if $y$ is nonempty then $z$ is nonempty. We can then use induction and this progress property to prove other progress properties.

**Safety:**

$$\mathcal{SAFETY}.MERGE(x, y, z) \Rightarrow merge(x, y, z)$$

**Progress:**

$$(\mathcal{G}.(x \neq [\,]) \wedge (x \neq [\,]) \wedge \mathcal{PROGRESS}.MERGE(x, y, z)) \Rightarrow (\mathcal{G}.(z \neq [\,]) \wedge (z \neq [\,]))$$

## Program Derivation

Employing our usual heuristic gives us the program:

$$MERGE(x, y, z) =$$
$$((x \neq [\,]) \quad \& \quad (\exists v :: (z \stackrel{\mathrm{def}}{=} [hd.x|v]) \wedge (MERGE(tl.x, y, v))))$$
$$\vee$$
$$((y \neq [\,]) \quad \& \quad (\exists w :: (z \stackrel{\mathrm{def}}{=} [hd.y|w]) \wedge (MERGE(x, tl.y, w))))$$
$$\vee$$
$$((x = [\,]) \quad \& \quad (z \stackrel{\mathrm{def}}{=} y))$$
$$\vee$$
$$((y = [\,]) \quad \& \quad (z \stackrel{\mathrm{def}}{=} x))$$

## Program Proof

As usual, the proof of safety and progress are straightforward.

## 4.4   Example: All Points Shortest Path

Consider a weighted directed graph with vertices indexed $i$ where $0 \leq i < n$. Let $w[i][j]$ be the length of edge from vertex $i$ to vertex $j$. Let $d[i][j]$ be the length of the shortest path from vertex $i$ to vertex $j$. We define $d[i][j]$ in the predicate calculus by using variables $c[i][j][k]$ defined to be the length of the shortest path from vertex $i$ to vertex $j$ with intermediate vertices (if any) indexed less than $k$, for $0 \leq k \leq n$ [7]. We assume that $d$, $w$ and $c$ are tuples, appropriately indexed. The definition of $d$, and $c$ in the predicate calculus is:

$$
\begin{aligned}
sp(w, d) = \\
(\forall i, j \in [0 \ldots n-1] :: \\
(c[i][j][0] = w[i][j]) \wedge (d[i][j] = c[i][j][n]) \wedge \\
(\forall k \in [1 \ldots n] :: \\
c[i][j][k] = \min(c[i][j][k-1], (c[i][k][k-1] + c[k][j][k-1])))))
\end{aligned}
$$

## Program Specification

Derive a program $SP(w, d)$ with input $w$ and output $d$ that establishes $sp(w, d)$, i.e.,

**Safety :**

$$\mathcal{SAFETY}.SP(w, d) \Rightarrow sp(w, d)$$

**Progress:**

$$\mathcal{G}.w \wedge \mathcal{PROGRESS}.SP(w, d) \Rightarrow \mathcal{G}.d$$

## Program Derivation

The derivation of the program is trivial.

$$
\begin{aligned}
SP(w, d) = \\
(\forall i, j \in [0 \ldots n-1] :: \\
(c[i][j][0] \stackrel{\text{def}}{=} w[i][j]) \wedge (d[i][j] \stackrel{\text{def}}{=} c[i][j][n]) \wedge \\
(\forall k \in [1 \ldots n] :: \\
c[i][j][k] \stackrel{\text{def}}{=} \min(c[i][j][k-1], (c[i][k][k-1] + c[k][j][k-1])))))
\end{aligned}
$$

## Program Proof

The proof of safety is vacuous. The proof of progress is as follows. We prove by induction on $k$ that for all $k$ where $0 \leq k \leq n$:

$$\mathcal{G}.w \wedge \mathcal{PROGRESS}.SP(w,d) \Rightarrow (\forall i, j :: \mathcal{G}.c[i][j][k])$$

Then from the program we prove:

$$\mathcal{G}.w \wedge \mathcal{PROGRESS}.SP(w,d) \wedge (\forall i, j :: \mathcal{G}.c[i][j][n]) \Rightarrow (\forall i, j :: \mathcal{G}.d[i][j])$$

# 5 Termination

Our proof obligation to show termination of a program is to prove that all non-input variables named in the program are defined, i.e., for all non-input variables $x$ that appear in the program there exists some $e$ such that $x \stackrel{\text{def}}{=} e$ holds.

We leave it to the reader to prove that all the examples discussed in this paper (except Hoare's *clock*) terminate.

For reasons of efficiency in execution, we extend the conjunctive form to allow the form:

$$\mathcal{PREFIX}.(b) \wedge d$$

which is exactly the same as

$$b \wedge d$$

except that $\mathcal{PREFIX}.(b) \wedge d$ terminates some finite number of steps after $d$ terminates; therefore, our proof obligation to show termination of $\mathcal{PREFIX}.(b) \wedge d$ is merely to prove that $d$ terminates.

**Example**  Our goal is to establish the predicate $first\_primes(n, y)$ which holds if and only if $y$ is the list consisting of the first $n$ primes. We are given that $n$ is input, and $\mathcal{G}.n$, and we are required to establish $\mathcal{G}.y$, for output $y$. Let $all\_primes(x)$ hold if and only if $x$ is the (infinite) list of all primes. Let $select(n, x, y)$ hold if $y$ is a list consisting of the first $n$ elements of list $x$. (We use the same names for predicates and corresponding programs, with uppercase for programs and lowercase for predicates.) Then we can define program $FIRST\_PRIMES(n, y)$ as follows:

$$FIRST\_PRIMES(n, y) = (\mathcal{PREFIX}.ALL\_PRIMES(x)) \wedge SELECT(n, x, y)$$

Program $FIRST\_PRIMES(n, y)$ terminates in a finite number of steps after $SELECT(n, x, y)$ terminates.

## 6 Sequential Composition and Mutable Variables

Mutable variables are the variables used in imperative programming: they are declared, they have arbitrary initial values, and their values can be changed arbitrarily many times by executing assignments.

Our proof obligation regarding mutables is to demonstrate that mutables are not shared by blocks within a conjunctive form; this demonstration can be carried out syntactically.

The syntax for manipulating mutable variables using sequential composition is not important for our purposes. We prove programs that use sequential composition in the following way. We replace the sequentially composed block

$$P_1(x, y_1); \ \ldots \ ; P_k(x, y_k)$$

where $x$ is mutable and $y_i$ are not mutable, by the conjunctive form,

$$Q_1(z_0, z_1, y_1) \wedge \ldots \wedge Q_k(z_{k-1}, z_k, y_k)$$

where $z_i$ is the value of $x_i$ immediately before the execution of $P_i$, and $z$ is not mutable; we have the added proof obligation of showing that an input variable of $Q_i$ is an input variable of the entire block $Q_1(z_0, z_1, y_1) \wedge \ldots \wedge Q_k(z_{k-1}, z_k, y_k)$, or is an output variable of a block $Q_j$ where $j < i$.

## 7 Fairness

Since programs that are composed in PCN can be UNITY programs we adopt UNITY theory where fairness is needed. In the example of $MERGE(x, y, z)$ given earlier, the predicate $\mathcal{SAFETY}.MERGE(x, y, z)$ is defined recursively, and since we define recursive equations in terms of their weakest solutions, $\mathcal{SAFETY}.MERGE(x, y, z)$ can hold even if $z$ is an infinite list consisting of only elements from $x$, and none from $y$. Therefore, $MERGE(x, y, z)$ can be unfair.

We assume that we are given from UNITY theory, a way of defining predicates of programs that employ fairness; for instance, we assume that we are given a predicate $\mathcal{SAFETY}.FAIR\_MERGE(x, y, z)$ which holds if and only if $z$ is a fair interleaving of $x$ and $y$, and we do not define this predicate in PCN. Concepts such as fair interleaving are outside the basic PCN theory but we assume that we are given predicates that define such concepts. We are permitted to manipulate such predicates in PCN blocks. For instance, we can have a PCN block in the conjunctive form in which some blocks $b$ are written in UNITY, and for which we are given predicates $\mathcal{SAFETY}.b$ and $\mathcal{PROGRESS}.b$.

A UNITY-like program is written with the same syntax as a disjunctive form, with '[ ]' in place of $\vee$.

$$
\begin{array}{lll}
UNITY\text{-}program & :: & program\text{-}name(variable \prec, variable \succ) \\
& & = \\
& & UNITY\text{-}form \\
\\
UNITY\text{-}form & :: & (guard\ \&\ block)\ []\ (guard\ \&\ block) \\
& & \prec []\ (guard\ \&\ block) \succ
\end{array}
$$

Fairness is as in UNITY: in an infinite number of recursive executions of the UNITY-program, each guard is evaluated infinitely often. If, when a guard is evaluated, the guard reduces to a ground value and the guard holds, then the corresponding block is executed.

# 8 Concluding Remarks

## 8.1 Completeness

The proof rules given are not complete. One can construct programs for which one can give operational proofs of progress, but for which the proof rules given here are inadequate. Work needs to be carried out to enhance the theory to make it complete, or to identify the class of programs for which the given proof rules are adequate.

## 8.2 Operational Arguments

Programmers use parallel computers for speed. Today, we do not know how to represent the efficiency of a program executing on a parallel computer as a pair of numbers such as: the program requires $2n + 3$ steps, and $n^2 - 1$ bits of memory for an input of size $n$. Therefore, the operational model and operational arguments are extremely important in understanding performance, which after all, is the primary reason parallel computers are used [10]. This note does not describe the operational model; the interested reader is referred to [4]

## 8.3 Logic Programs

A PCN program is a normal form predicate to which we apply a small number of rules to prove progress. A PCN program is not a set of Horn clauses, and PCN does not employ unification. In addition, PCN can use mutables, sequential composition, and UNITY. Therefore, PCN programs are very different from logic programs though PCN borrows many ideas from concurrent logic programming [11, 6].

## 8.4 Functional Programming

Though we did not mention functions, they can be introduced easily as: '$f(x)$ is $y$ where $p(x, y)$.' We must, of course, prove that for each value of $x$ there is only one value of $y$ that satisfies $p(x, y)$, and in our program, $x$ must be the input variable and $y$ the output variable. Functional programs can be written in an entirely functional manner, while relational programs are written using the predicate calculus.

## 8.5 Determinism

A program is deterministic if

1. in all disjunctive forms, precisely one guard holds, and

2. prefixes are not used, and

3. UNITY is not used.

Of course, a program may be deterministic even otherwise, but then a proof is required. (Note that the execution is usually nondeterministic — in the sense that processes may be scheduled in different orders and messages may be sent at different points in the computation — even if the same output is obtained for a given input in all executions.)

PCN is designed to allow the development of programs that are obviously deterministic, or that are nondeterministic and do not use the fairness constructs of UNITY, or that do rely on UNITY. Programmers can choose the simplest form of reasoning appropriate for a given program.

## 8.6   Implementation

The syntax in the current implementation of PCN is somewhat different from that given here, but the differences are unimportant for the purposes of this paper. Also, at this time, we have not implemented UNITY programs. PCN runs on a variety of architectures thanks primarily to Stephen Taylor and Ian Foster. Furhtermore, a PCN programming environment has been built, primarily by Carl Kesselman and his colleagues at Aerospace Corporation [1].

# 9   Acknowledgment

I would like to thank Wim Feijen, Jan van de Snepscheut, Johann Lukkien, and Pieter Hofstee for their criticism.

# References


[1] Campbell, A., R.M. Chowkwanyun, C.F.Kesselman, C. A. Lee, and S. Taylor, 'A Database System to Support the Program Composition Environment', Aerospace Corp., Report No. TOR-0090(5920-05)-1, April 1990.

[2] Chandy, K. M., and J. Misra, *Parallel Program Design: A Foundation*, Addison-Wesley, Reading, Massachusetts, 1988.

[3] Chandy,K.M., and S.Taylor, 'The Composition of Concurrent Programs,' in *Proceedings Supercomputing '89*, Reno, Nevada, Nov.13-17, 1989, ACM.

[4] Chandy,K.M., and S.Taylor, 'A Primer for Program Composition Notation,' Computer Science Research Report, Caltech-CS-TR-90-10, California Institute of Technology, June 20, 1990.

[5] Dijkstra, E. W., *A Discipline of Programming*, Prentice-Hall, Englewood Cliffs, New Jersey, 1976.

[6] Foster,I., and S.Taylor, *Strand, New Concepts in Parallel Programming*, Prentice-Hall, 1989.

[7] Floyd, R. W., "Algorithm 97, Shortest Path," Communications of A.C.M., Vol. 5, pp 345, 1962.

[8] Foster,I., and S.Taylor, 'A Portable Run-Time System for PCN', Argonne National Laboratory Report No. ANL/MCS-TM-137, January 1990.

[9] Hoare, C.A.R., *Communication Sequential Processes*, Prentice-Hall International, London, U.K., 1984.

[10] Seitz, C.L., and J. Seizovic, and W. Su, 'The C Programmer's Abbreviated Guide to Multicomputer Programming', Caltech Computer Science Technical Report Caltech-CS-TR-88-1, 19 January 1988, revised 17 April 1989.

[11] Shapiro,E., 'The Family of Concurrent Logic Programming Languages,' in *ACM Computing Surveys*, Vol.21, No.3, pp412-510, September 1989.

# Lectures on Data Refinement

David Gries
Computer Science, Cornell University
Ithaca, New York 14853

## 1   Introduction

These lectures discuss the notion of *data refinement*: the replacement of one or more variables in a program by other variables, with suitable replacement of expressions and statements, to yield an equivalent program.

A data refinement can be of the "abstract to concrete" variety. Examples of this kind of refinement are the replacement of a set variable by a heap, hash table, or other data structure; the replacement of a sequence variable by some form of linked list; and the replacement of a sparse-matrix variable by a data structure that makes use of that sparseness to save space and time.

A data refinement need not be from "abstract to concrete". For example, the replacement of a pair that represents a coordinate of a point in the plane by a pair that gives the polar coordinates of that point is not from "abstract to concrete"; neither is the replacement of a boolean variable by one that contains its complement.

Data refinement could be one of the major techniques to be used in reusing program parts. The module of Modula 2 and the package of Ada are often cited as examples of what one can do in this regard. For example, once one writes a package that describes the implementation of a set variable as a heap, one should use this package over and over again. Unfortunately, the interface between the user and the package —essentially types, procedures, and functions— is too restricted, so that the decision to use such a package means too much rewriting of a program to conform to the interface. Also, conversions of representation are not handled automatically. Hence, the algorithm as presented to people begins to be far away from its implementation.

In this article, we describe a new mechanism, the *transform*, which goes further along the path of solving this problem of disparity between algorithm and program.

We begin by viewing data refinement as a coordinate transformation, because it lends itself to easier exposition, without the need for too much formality. However, the coordinate transformation is not as general as it could be. After looking at data refinement as a coordinate transformation, we will look at the more formal approach., which yields a calculus for the refinement of programs. We will also look at the development of a algorithms using data refinement.

Sections 2, 3, and 4 were lifted from reference [11] (and then modified), written by the author and Dennis Volpano. Other parts of this paper lean heavily on the references, especially [3], [14], [8], [16], and [17], . The earliest substantial work on data refinement was done by Back [2]; this work has not received the recognition it deserves because, I guess, it was ahead of its time.

It is assumed that the reader already knows the theory of weakest preconditions and has familiarity with the development of programs using that theory. See [4], [7], or any of a number of newer books on the subject.

## Table of contents



## 2   Algorithms versus programs

We desire a programming language that is closer to the notation in which scientists express their algorithms in articles and text books, because as long as the language of presentation is far removed from the programming language, too much time is wasted translating between them. For example, I like to write the conventional Huffman encoding algorithm for building a binary tree with certain properties from a given bag of trees as shown in Fig. 1, and I want this presentation to be my program as well. Contrast Fig. 1 with the presentation of the algorithm in conventional texts (e.g. [1, 99]; because there is no formal presentation language, the algorithm is usually given in an ineffective mixture of English and mathematics, which then has to be translated into a language like Pascal, including all the details about the implementations of the various bags and trees in terms of heaps, linked lists, and the like.

---

**do** $\#S \neq 1 \rightarrow$   **var** $x : tree$; $Choosemin(S, x)$; $S := S - \{x\}$;
                   **var** $y : tree$; $Choosemin(S, y)$; $S := S - \{y\}$;
                   $S := S \cup \{tree(x.v + y.v, x, y)\}$
**od**

$S$ initially contains a nonempty bag of nonempty (binary) trees; $\#S$ is the size of $S$. A tree either is empty or is a triple $(v, l, r)$ where $v$ is an integer and $l$ and $r$ are trees. $Choosemin(S, x)$ stores in $x$ a tree of $S$ with minimum $v$ field.

Figure 1: Huffman encoding in a higher-level notation

---

Allowing mathematical notation as in Fig. 1 may not be difficult. What is difficult is to describe the implementation of the "abstract" variables of the algorithm in a simple and unobtrusive fashion. In my dream programming language, it should be a trivial task to indicate that bag $S$ is to be implemented using a heap or hash table (using earlier-written descriptions of these implementations), and the fact that $S$ is to be so implemented should not cause me to change the presentation of the algorithm at all. As much as possible, Fig. 1 should remain the program, no matter how $S$ or its trees are implemented.

Once implementations have been selected for bags and trees, the program in Fig. 1 should undergo a transformation into an equivalent program expressed in terms of operations on the chosen implementations. Such transformations are, in general, *coordinate transformations*.

Two new language constructs will be introduced here: (1) the transform for describing coordinate transformations, and (2) the transformation directive for directing the use of transforms. These constructs allow us to move a step closer to the goal of merging algorithm presentation and program, while maintaining efficiency of implementation.

The material discussed here is exploratory, in that I do not have a fully-implemented language with which to experiment, and I do not give a complete description of the constructs, but only of their more important features. To set the context, assume that we are dealing with an imperative language, in the Algol 60 tradition but with more mathematical notations allowed and with more abstract data types like the set, bag, and sequence. (One can even assume the existence of a data-type definition facility, with which the programmer can introduce (the syntax of) their own data types, which will be implemented using the constructs described here.)

# 3    The coordinate transformation

The term coordinate transformation was introduced by Dijkstra [4, pg.64]. *Coordinate* and *transformation* are defined in Webster's Third New International as

- **coordinate:** any one of a set of variables or parameters used in specifying the state of a substance.

- **transformation:** the changing of an expression, formula, or statement into a different form without altering its substance or intent.

A coordinate transformation can be a transformation like a data refinement, an implementation of an abstract data type, or even a transformation unrelated to programming, like transforming a bound variable of a logical formula.


## Data refinement

Consider program segment (1). It has two *external* variables $x$ and $b$, in terms of which the program is specified. Given array $b$, the sum of $b[1..10]$ is stored in $x$. An *internal* variable $i$ is used to help perform the task.

$$(1) \qquad \begin{array}{l} [\![ \; \mathbf{var}\ i : int := 0; \\ \quad x := 0; \\ \quad \mathbf{do}\ i < 10 \rightarrow i := i + 1;\ x := x + b.i\ \mathbf{od} \\ ]\!] \\ \{x = (\Sigma k : 1 \le k \le 10 : b.k)\} \end{array}$$

(The notation $[\![\ \mathbf{var}\ i : int := 0;\ S]\!]$ denotes an Algol-like block with a local variable $i$; the scope of $i$ is its initialization and statement $S$.) Let us replace internal variable $i$ by another internal variable $j$, whose value is defined by predicate $C$, which we call the *coupling invariant*:

$$(2) \qquad C :\ i = 10 - j\ .$$

Replacing $i$ by $j$ and simplifying the loop guard yields the equivalent segment

$$(3) \qquad \begin{array}{l} [\![ \; \mathbf{var}\ j : int := 10; \\ \quad x := 0; \\ \quad \mathbf{do}\ 0 < j \rightarrow j := j - 1;\ x := x + b.(10 - j)\ \mathbf{od} \\ ]\!] \\ \{x = (\Sigma k : 1 \le k \le 10 : b.k)\} \end{array}$$

The initialization $i := 0$ has been replaced by $j := 10$ and the assignment $i := i + 1$ by $j := j - 1$. Therefore, $C$ holds at equivalent places in the two programs. So any reference to $i$ in the first program is replaced by $10 - j$ in the second program. Both programs sum the elements of $b[1..10]$, accumulating them in the same order. Here, the program segment was transformed without changing its intent, which is to store the sum of $b[1..10]$ in $x$, by replacing coordinate $i$ by coordinate $j$. We say that $i$ is represented by $j$.

A more detailed view of the previous coordinate transformation will clarify the use of coupling invariant $C$. First, introduce internal variable $j$ into program (1) and insert operations so that $C$ is always true:

(4)      $[\![$ **var** $i, j : int := 0, 10;$
       $x := 0;$
       **do** $i < 10 \rightarrow i, j := i + 1, i - 1;\ x := x + b.i$ **od**
     $]\!]$
     $\{x = (\Sigma k : 1 \leq k \leq 10 : b.k)\}$     .

In (4), $i = 10 - j$ always holds. Therefore, $i$ can be replaced by $10 - j$ and $b.i$ by $b[10 - j]$, yielding

(5)      $[\![$ **var** $i, j : int := 0, 10;$
       $x := 0;$
       **do** $10 - j < 10 \rightarrow i, j := i + 1, i - 1;\ x := x + b.(10 - j)$ **od**
     $]\!]$
     $\{x = (\Sigma k : 1 \leq k \leq 10 : b.k)\}$     .

Programs (4) and (5) are equivalent —they have the same specification in terms of the external variables— but in (4) $j$ is used only in assignments to itself and can therefore be eliminated to yield program (1), while in (5) $i$ is used only in assignments to itself and can be eliminated to yield program (3). Hence, programs (1) and (3) are also equivalent.

As another example of a coordinate transformation, due to Jan van de Snepscheut, we transform a traditional algorithm for computing the greatest common divisor $gcd(a, b)$ of two positive natural numbers $a$ and $b$ (without commenting on the reason for doing so). The traditional algorithm (6) stores $gcd(a, b)$ in variable $z$, where $a, b, z$ are the external variables and $x, y$ the internal variables.

(6)    $\{a > 0 \wedge b > 0\}$
$\llbracket$ **var** $x, y : nat := a, b;$
   $\{invariant\ P : x > 0 \wedge y > 0 \wedge gcd(x,y) = gcd(a,b)\}$
   $\{bound\ function : x + y\}$
   **do** $x > y \rightarrow x := x - y$
     $[]\ y > x \rightarrow y := y - x$
   **od**;
   $\{P \wedge x = y;\ \text{hence}\ x = y = gcd(a,b)\}$
   $z := x$
$\rrbracket$

We replace $x, y$ by two variables $u, v$ with coupling invariant

$$C : u = min(x,y) \wedge v = max(x,y)$$

to yield algorithm (7); upon termination of its loop, $u = v = max(x,y) = min(x,y)$, so both $u$ and $v$ contain $gcd(a,b)$.

(7)    $\llbracket$ **var** $u, v : nat := min(a,b), max(a,b);$
   **do** $u \neq v \rightarrow u, v := min(v - u, u), max(v - u, u)$ **od**;
   $z := v$
$\rrbracket$

This coordinate transformation illustrates that more than one variable can be transformed, and into more than one variable. It also illustrates that the coordinate transformation can involve replacements of more than expressions and assignments. In this example, we have replaced one loop by another.

We summarize the steps involved in making a coordinate transformation:

1. Write a coupling invariant $C$ that states the relation between the original variable $i$ (say) and its representation $j$ (say).

2. Replace in the program the declaration of $i$ by a declaration of $j$, where each declaration has an initialization and the two initializations together establish coupling invariant $C$.

3. Replace each statement that changes $i$ by a statement that changes $j$ such that their simultaneous execution leaves $C$ invariantly true.

4. Replace remaining subexpressions involving $i$ by equal expressions involving $j$.

These steps suggest how the coupling invariant is used to prove that the transformation yields a program that satisfies the same specification as the original program. The proof method is discussed later.

## Implementations of abstract data types

Neither program (1) nor program (3) is more abstract or concrete than the other. Many useful coordinate transformations have this flavor. But some transformations do appear to make an abstract program more concrete. For example, consider the program segment

(8)     **var** $s : set(nat)$; $j : nat$; $\cdots j \in s \cdots$

Replacing variable $s$ by a boolean array $b[0..63]$, using the coupling invariant $x \in s \equiv b.x$, would transform program (8) into

(9)     **var** $b : $ **array** $0..63$ **of** $bool$; $j : nat$; $\cdots \leq j \leq 63 \wedge b.j \cdots$

Program (9) seems more concrete than (8) because we think of set $s$ being "implemented" in terms of a lower-level array $b$. Note that only a restricted implementation of type $set(nat)$ is given, in that the set may contain only integers in $0..63$.

The abstract-to-concrete transformation is more widely known, since it corresponds to the traditional notion of implementing an abstract data type in some fashion. A coordinate transformation may be of this variety, but it need not be, as the transformation from (1) to (3) shows. [1]

## Transforming bound variables

We describe another use of the coordinate transformation, the replacement of a bound variable of a quantification, simply to illustrate that the coordinate transformation is a rather general concept and not related purely to programming. Using the relation $j = i + 5$, we can replace the expression

$$(\forall i : 0 \leq i < n : b.i)$$

by

$$(\forall j : 0 \leq j - 5 < n : b.(j - 5)) \quad .$$

As before, the transformation can be broken down into a series of steps:

$$(\forall i : 0 \leq i < n : b.i)$$
$=$     ⟨Introduce dummy $j$ with coupling invariant $j = i + 5$⟩
$$(\forall i, j : 0 \leq i < n \ \wedge \ j = i + 5 : b.i)$$
$=$     ⟨Arithmetic, substitution of equals for equals⟩

---

[1] For this reason, and at the urging of Netty van Gasteren and Wim Feijen, the term *coupling* invariant was adopted instead of *representation* invariant, which is used in [8].

$$\begin{aligned}
& \quad (\forall i, j : 0 \leq j - 5 < n \ \wedge \ i = j - 5 : b.(j-5)) \\
= & \quad \langle \text{Eliminate dummy } i \rangle \\
& \quad (\forall j : 0 \leq j - 5 < n : b.(j-5))
\end{aligned}$$

The bound variables of summations like $\Sigma_{i=0}^{n}b.i$ can likewise be replaced. Knuth [13, pp. 26-35] describes many such coordinate transformations.

# 4   A new language construct, the transform

As shown in the last section, the coordinate transformation is a rather general technique, which is applicable in many situations. Further, the same coordinate transformation may be applicable in many contexts, especially a coordinate transformation that replaces abstract variables by concrete ones. Therefore, we introduce a language construct, called the *transform*, for describing a coordinate transformation.

Coordinate transformations applied to a program are defined by the programmer by annotating variable declarations with *transformation directives*. For example, replacing **var** $i$ by **var** $i/M$ indicates that variable $i$ is to be transformed using transform $M$. Thus, if transform $M$ describes the coordinate transformation carried out on program (1) above, then program (1) annotated with a directive as in

$$\begin{aligned}
&[\![\ \textbf{var } i/M : int := 0; \\
&\ \ x := 0; \\
&\ \ \textbf{do } i < 10 \rightarrow i := i + 1;\ x := x + b.i\ \textbf{od} \\
&]\!] \\
&\{x = (\Sigma(k : 1 \leq k \leq 10 : b.k)\}
\end{aligned}$$

denotes program (3). Likewise, if $Bvector(k)$ is a transform that describes the transformation of a variable of type $set(nat)$ into a variable of type **array** $0..k-1$ **of** $bool$, then program (8) annotated with a directive as in

$$\textbf{var } s/Bvector(64) : set(nat); j : nat; \ \cdots j \in s \cdots$$

denotes program (9).

For many coordinate transformations, especially those of the abstract-to-concrete variety, the abstract program should be unencumbered as much as possible by information concerning the implementation of its high-level variables. To this end, one can specify transformation directives outside a program to indicate how variables in it are to be transformed. This is achieved by creating a separate file of transform directives. For example, to indicate that variable $s$ of the program

$$\cdots \textbf{var } s : set(nat);\ j : nat;\ \cdots j \in s \cdots$$

is to be implemented using transform *Bvector* , place the following directive in the separate file that is used when constructing the final program:

> **Use** *Bvector*(64) **on** *s*  .

Placing transformation directives outside the program allows the textual form of the abstract program to remain untouched, so that it can still be understood completely in its abstract form and yet allows us to direct its efficient implementation. Giving directives outside a program is necessary when several variables are to be transformed by a single transform (e.g. consider the transformation of *gcd* algorithm (6)).

It is possible to specify that different variables of the same type be implemented differently. For example, within a program, one set-valued variable may be implemented by a bit vector, another by a hash table, and yet another by a heap, depending on the operations the program requires on sets. Further, conversions of programmer-defined representations can be inserted automatically by the system during the transformation proces. This is a significant advance over previous mechanisms for implementing types.

## 4.1   The transform

A transform describes a coordinate transformation. It has the form

> **transform** *name*[(*parameters*)];
> *declarations* **repr** *declarations*
> {*coupling-invariant*}
> *transform-rules*
> **end**

where *parameters* is an optional list of parameters (separated by commas) ("[···]" signifies an optional phrase) and *declarations* is a list of variable declarations (separated by semicolons). The variables appearing to the left of **repr** are the *abstract* variables of the transform and those to the right are the *concrete* variables that represent the abstract ones. This terminology is used only to distinguish variables to the left of **repr**, which are to be replaced, from those on the right, which are the replacements.

The *transform-rules* prescribe the transformation of expressions and statements involving the abstract variables into expressions and statements involving the concrete variables. These expressions and statements are not just function and procedure calls, as in existing languages. A transform rule may have one of the following two forms:

> *stmtpattern* **into** *stmt*
>
> *exppattern* **into** *exp*

The first form prescribes a statement replacement, the second an expression replacement. The statement or expression to the right of **into** is the replacement for the statement or expression described by the pattern on the left.

The coordinate transformation applied to program (1) to produce (3) is described by the following transform.

```
transform M;
    var i : int    repr      var j : int
    {coupling invariant C : i = 10 − j}
 [] i := 0          into      j := 10
 [] i := i + 1      into      j := j − 1
 [] i              into      10 − j
 [] i < 10          into      0 < j
end
```

The second line of $M$ contains declarations for abstract variable $i$, the one being replaced, and concrete variable $j$, the replacement. The third line contains the coupling invariant, as a comment. It is used by the writer of the transform, as outlined earlier, in proving that the transform is correct. The first two transform rules prescribe statement replacements; the next two, expression replacements. For a pattern that describes a statement in terms of $i$, its replacement describes an equivalent one in terms of concrete variable $j$, which represents $i$, where by equivalent we mean that together they maintain the coupling invariant. For a pattern that describes an expression, the pattern and its replacement have the same type and should have equal values, one in terms of $i$ and one in terms of its representation $j$.

The line "$i < 10$ **into** $0 < j$" of $M$ is superfluous. Without it, the guard $i < 10$ of (1) would be changed to $10 − j < 10$; with it, it may be changed to $0 < j$. Its presence shows that several replacements may be possible; which one is used does not matter from a logical point of view, as long as it remains possible to eliminate all occurrences of the variable(s) being replaced. If the line were omitted, it would be reasonable for code optimization to change $10 − j < 0$ to $10 < j$.

Variable $i$ in transform $M$ is a dummy, which matches any variable $v$ (say) whose declaration is annotated with $/M$. In order for this to work properly, for each variable so annotated a distinct set of concrete variables is generated. For example, the program

```
[ var i/M, k/M : int := 0, 0;
  x := 0;
  do k < 10 →   k := k + 1; x := x + b.k od;
  i := i + 1
]
```

might be transformed into the following, where $ji$ and $jk$ are fresh variable names generated during the transformation process.

```
[ var ji, jk : int := 10, 10;
  x := 0;
```

$$\textbf{do } 10 - jk < 10 \rightarrow \quad jk := jk - 1; \ x := x + b.(10 - jk) \textbf{ od};$$
$$ji := ji - 1$$
$$\textbf{]}$$

The first two patterns of transform $M$ are rather restrictive, in that they match only assignments of the form $v := 0$ and $v := v + 1$ where $v$ is a variable of type *int* implemented by $M$. More general patterns can be written using the phrase $\langle \textbf{exp-}id : T \rangle$, which matches any expression of type $T$. In the corresponding replacement, $id$ refers to the matched expression. A more general transform $\hat{M}$ is defined below. The first transform rule prescribes the replacement of any assignment $i := e$ by an assignment $j := 10 - e$, for any expression $e$.

> **transform** $\hat{M}$;
>   **var** $i : int$        **repr**     **var** $j : int$
>   $\{coupling\ invariant\ C : i = 10 - j\}$
>   $[]\ \ i := \langle \textbf{exp-}e : int \rangle$    **into**     $j := j - 1$
>   $[]\ \ i$                       **into**     $10 - j$
> **end**

Transform *Bvector*, defined in (10), describes the coordinate transformation applied to program (8) to produce (9). It is viewed as a transform that implements an abstract data type.

(10)     $\{$ Implement a set of integers in $0..k - 1$ by a boolean array of size $k$ and an integer. Operation $s := \{\}$ requires $k$ integer assignments; all other operations take constant time. $\}$

> **transform** *Bvector*$(k : nat)$;
> **var** $s : set(nat)$           **repr**    $b : \textbf{array } 0..k - 1 \textbf{ of } bool$
> $\{coupling\ invariant : (\forall e : 0 \leq e < k : e \in s \equiv b.e)\}$
>   $[]\ \ s := \{\}$              **into**    $(\textbf{for } i : 0 \leq i < k : b.i := false)$
>   $[]\ \ \{e \in 0..k - 1\}$
>       $s := s \cup \{\langle \textbf{exp-}e : nat \rangle\}$   **into**   $b.e := true$
>   $[]\ \ \{e \in 0..k - 1\}$
>       $s := s - \{\langle \textbf{exp-}e : nat \rangle\}$   **into**   $b.e := false$
>   $[]\ \ \langle \textbf{exp-}e : nat \rangle \in s$        **into**    $0 \leq e < k \wedge b.e$
>   $[]\ \ \#s$                  **into**     $= \{$ number of true $b.i\}$
>                                   $(\#i : 0 \leq i < k : b.i)$
> **end**

Note the preconditions on two of the transform rules to indicate the conditions under which correctness of the rules is guaranteed. A precondition in a transform rule introduces additional proof obligations on the user of a transform, in that the precondition should be true before any statement or expression that is transformed using the transform rule.

A key point is that a transform can be a partial implementation of a type; it need not implement all operations of a type. For example, *Bvector* is partial because it does not implement intersection $\cap$, and it does not implement $\cup$ in all its generality, but only in its use in inserting a single value into a set. A program to be transformed using *Bvector* must restrict its use of set operations to those operations described by the patterns in *Bvector* . Many transforms describe partial implementations of types.

The user of an abstract-to-concrete transform, like *Bvector* , is generally interested only in seeing the abstract variable and the patterns of the transform, and not the concrete variables and the replacements, much like the user of a procedure is interested only in its heading and not its body. It should be possible to present the user with only this view. Also, the user will in general not have to see the program that results from the transformation.

## Definitions of representations within transforms

We now introduce an issue of particular importance, the definitions of the representations of expressions that contain variables to be transformed. An example of what we can deal with is the following. Consider

$$x := (y * x) + (x * z + 6);$$
$$\textbf{if } x + y + z > 32 \textbf{ then } \cdots$$

where all variables have type *int* . Suppose we want a transform that transforms an integer variable, like $x$ , $y$ , or $z$ , into a linked list of some form because its contents gets too large for the conventional implementation of integers. Using only **into** transform rules, it is impossible to write such a transform, because there is no way to describe the implementation of an arbitrary integer expression. What is required is the ability to define, in general, the representation of expressions from the representations of their subexpressions.

It is this additional ability to define representations of expressions that helps distinguish the transform from constructs in other languages that are used to implement abstract data types.

To illustrate this, consider a transform $BN$ that represents a boolean $b$ (say) by a natural number $j$ , where the coupling invariant is

$$(11) \quad CBN(b,j) : b \equiv j > 0$$

(We choose this example, rather than work with the variable-precision example, because the implementations of operations are far simpler.) Consider boolean variables $c, d, e$ that are represented using $BN$ by three natural-number variables $cj, dj, ej$ , respectively. Using double brackets $[\![ \ ]\!]$ around an expression to denote a function that defines the representation of that expression, we have

(12)    (a)    $[\![\,c\,]\!] = cj$   ,
       (b)    $[\![\,d\,]\!] = dj$   ,
       (c)    $[\![\,e\,]\!] = ej$   .

The representation of variable $c$ is $cj$, and the replacement for $c$ is the equal expression $cj > 0$. A representation of the expression $c \vee d$ would be $cj + dj$, since these two expressions together satisfy the coupling invariant $CBN(c \vee d, cj + dj)$. Hence, the replacement for $c \vee d$ is the equal expression $cj + dj > 0$.

We now define the $BN$-representation for all boolean expressions involving $c, d, e$ and operations $\wedge$, $\vee$, and $\neg$. The definitions of representations for constants is

(13)    (a)    $[\![\,false\,]\!] = 0$   ,
       (b)    $[\![\,true\,]\!] = 1$   .

Definitions of the representations of expressions containing operators are written recursively, as follows:

(14)    (a)    $[\![\,B1 \vee B2\,]\!] = [\![\,B1\,]\!] + [\![\,B2\,]\!]$   ,
       (b)    $[\![\,B1 \wedge B2\,]\!] = [\![\,B1\,]\!] * [\![\,B2\,]\!]$   ,
       (c)    $[\![\,\neg B1\,]\!] = \text{if } [\![\,B1\,]\!] > 0 \text{ then } 0 \text{ else } 1$   .

Proving that a definition $[\![\,LHS\,]\!] = RHS$ is correct involves proving that the coupling invariant $CBN(LHS, RHS)$ is valid, under the assumption that the operands and their representations satisfy their coupling invariants.

Given definitions (12), (13), and (14), the representation of any expression involving $c, d, e$, and the constants $true$ and $false$ can be constructed. For example, possible representations of three expressions to the left below are given to the right:

$$[\![\,c \vee d \vee e\,]\!] \qquad cj + dj + ej$$
$$[\![\,(c \vee d) \wedge d\,]\!] \qquad (cj + dj) * dj$$
$$[\![\,c \vee \neg d\,]\!] \qquad cj + (\text{if } dj > 0 \text{ then } 0 \text{ else } 1)$$

With an additional transformation rule

$$b := b1 \textbf{ into } j := j1$$

(where both $b$ and $b1$ have known $BN$-representations), any assignment statement involving $c, d, e$, and the constants $true$ and $false$ can be transformed by first building the representation of the expression of the assignment and then using this additional rule. For example, the assignment $e := c \vee d \vee e$ would be transformed into

$$ej := cj + dj + ej \quad .$$

Note carefully the difference between the representation of an expression and the transformation of an expression. With coupling invariant $CBN$ described above in (11), the representation of variable $c$ is $cj$, and $CBN(c, cj)$ holds, but the transformation of $c$ is $cj > 0$.

In the notation of the transform, the transform rule that defines a representation has the form

$$[/N]\ exppattern \ \textbf{repr}\ exp$$

where $N$ is the name of the transform for which the representation is being defined; $/N$ may be omitted if the transform rule occurs within the definition of $N$. For example, the complete transform $BN$ is given below.

(15)     **transform** $BN$;

| | | | | |
|---|---|---|---|---|
| | **var** $b : bool$ | | **repr** | $j : nat$ |
| | \{*coupling invariant* $CBN(b, j) : b \equiv j > 0$\} | | | |
| [] | (1) | $false : bool$ | **repr** | $0$ |
| [] | (2) | $true : bool$ | **repr** | $1$ |
| [] | (3) | $b1 \lor b2$ | **repr** | $j1 + j2$ |
| [] | (4) | $b1 \land b2$ | **repr** | $j1 * j2$ |
| [] | (5) | $\neg b$ | **repr** | **if** $j = 0$ **then** $1$ **else** $0$ |
| [] | (6) | $\langle \text{exp-}x \rangle : bool$ | **repr** | **if** $x$ **then** $1$ **else** $0$ |
| [] | (7) | $b$ | **into** | $j > 0$ |
| [] | (8) | $b := b1$ | **into** | $j := j1$ |

    **end**

The transform rules on lines (1)-(6) are representation definitions. Each pattern in these rules describes an expression whose representation is being defined. The expression has type $bool$, the type of abstract variable $b$, and the corresponding representation on the right side has type $nat$, the type of concrete variable $j$. For each of these representation rules $LHS$ **repr** $RHS$, coupling invariant $CBN(LHS, RHS)$ holds.

An occurrence of abstract variable $b$ within a pattern matches any variable or expression whose representation is known to be $BN$. Concrete variable $j$ can then be used on the right side of **repr** to denote its representation. A representation definition may refer to more than one operand whose representation is known to be $BN$ using the following convention. If $b$ is the abstract variable, then $bi$ for any digit $i$ matches any variable or expression whose representation is known to be $BN$; in the replacement, the concrete variable followed by $i$ denotes the corresponding representation. Examples of this notation appear in lines 3, 4, and 8 of transform $BN$. This notation is allowed only when none of the abstract or concrete variables ends in a digit.

The transform rules on lines (1) and (2) of $BN$ define the representations of constants, whereas the rule on line (6) defines a representation of an expression whose representation is not known. It is used only when necessary, to facilitate the use of some other transform rule (see conversions of representation, discussed later).

## 4.2   The transformation process

We present a method of transforming a program according to transformation directives in order to give the reader a feeling for the problems involved. The transformation process begins by generating concrete variables for the variables to be transformed, replacing the declarations of the variables that are to be transformed by declarations of the concrete variables, and marking in some fashion all occurrences of the original variables as variables whose representations are known. For example, using an underbar to denote an expression whose $BN$-representation has been constructed, the program segment

$$(16) \qquad \textbf{var } c/BN, d/BN, e : bool;$$
$$\cdots$$
$$d := false;$$
$$\textbf{if } (c \lor d) \land (c \lor e) \textbf{ then } \cdots$$

would be transformed by the first step into

$$(17) \qquad \textbf{var } cj, dj : nat; e : bool;$$
$$\cdots$$
$$d := false;$$
$$\textbf{if } (\underline{c} \lor \underline{d}) \land (\underline{c} \lor e) \textbf{ then } \cdots$$

(note that $e$ is not being transformed), where we have

$$(18) \qquad \text{(a) } [\![ c ]\!] = cj \quad,$$
$$\qquad \qquad \text{(b) } [\![ d ]\!] = dj \quad.$$

Thereafter, representations are constructed for expressions that contain subexpressions whose representations are known until a state is reached in which a transform rule can be used to introduce the representation.

We illustrate this process by transforming assignment $\underline{d} := false$ of (17). The first step determines the representation of $false$, but does not involve a transformation. The second step provides the transformation.

$$\underline{d} := false$$
$$\rightarrow \quad \langle \text{use pattern (1) of } BN \rangle$$
$$\underline{d} := \underline{false} \quad \text{where } [\![ false ]\!] = 0$$
$$\rightarrow \quad \langle \text{use pattern (8) of } BN \rangle$$
$$dj := 0$$

We illustrate the process again on a more complicated example, transforming expression $(\underline{c} \lor \underline{d}) \land (\underline{c} \lor e)$ of (17). The expression is annotated with type $bool$ in order to denote the type that is required in this context.

$$((\underline{c} \vee \underline{d}) \wedge (\underline{c} \vee e)) : bool$$
$$\rightarrow \quad \langle \text{Use line (3) of } BN \text{, with } b1, b2 := c, d \,\rangle$$
$$((\underline{c \vee d}) \wedge (\underline{c} \vee e)) : bool, \quad \text{where } [\![\, c \vee d \,]\!] = cj + dj$$
$$\rightarrow \quad \langle \text{Use line (6) of } BN \text{ to convert } e \,\rangle$$
$$((\underline{c \vee d}) \wedge (\underline{c} \vee \underline{e})) : bool, \quad \text{where } [\![\, e \,]\!] = (\textbf{if } e \textbf{ then } 1 \textbf{ else } 0)$$
$$\rightarrow \quad \langle \text{Use line (3) of } BN \text{, with } b1, b2 := c, e \,\rangle$$
$$((\underline{c \vee d}) \wedge (\underline{c \vee e})) : bool,$$
$$\text{where } [\![\, c \vee e \,]\!] = cj + (\textbf{if } e \textbf{ then } 1 \textbf{ else } 0)$$
$$\rightarrow \quad \langle \text{Use line (4) of } BN \text{, with } b1, b2 := c \vee d, c \vee e \,\rangle$$
$$((\underline{c \vee d \wedge (c \vee e)}) : bool,$$
$$\text{where } [\![\, (c \vee d) \wedge (c \vee e) \,]\!] = (cj + dj) * (cj + (\textbf{if } e \textbf{ then } 1 \textbf{ else } 0))$$
$$\rightarrow \quad \langle \text{Use line (7) of } BN \text{ to transform into a } bool \,\rangle$$
$$(cj + dj) * (cj + (\textbf{if } e \textbf{ then } 1 \textbf{ else } 0)) > 0$$

Note how most of the work involved constructing the representation of the expression. The only transformation occurred on the last line, which is actually a conversion of representation.


## 4.3  Conversions of representation

Any necessary conversions of representation are inserted during the transformation process, provided they have been described suitably. Note that these are conversions of representation only, and not of values from one type to another. For example, an expression $x + y + z$ can remain in this abstract form no matter how integer variables $x$, $y$ and $z$ are implemented (perhaps using variable-precision arithmetic), with automatic conversions of representation being inserted as necessary.

In some cases, a conversion of representation is introduced automatically in order to be able to continue to transform the program. This is illustrated in the previous example, where the reference to $e$ was transformed into **if** $e$ **then** 1 **else** 0, using line (6) of transform $BN$. There may be many ways to perform conversions so that the transformation process can continue. It doesn't matter which way is chosen, as long as the transformation process terminates successfully. However, it may be useful to allow the programmer to decide, interactively with the help of the system, which to use.


## 4.4  Proof obligations for coordinate transformations

Proof obligations may be partitioned into those of the transform user and those of the transform author. The user of a transform must prove only that the precondition of a pattern that matches some part of their program is true at that point in the program where the part appears. For example, note the preconditions on two of the transform rules in transform $Bvector$ to indicate the conditions under which the correctness of the rules is guaranteed. This kind of proof obligation is required when the transform implements partial operations, and the program must follow the imposed restrictions.

The proof obligations of the transform author are classified according to the kind of transform rule:

**Expression replacements p into r.** Pattern $p$ and replacement $r$ must have the same type. Pattern $p$ describes a class of expressions in terms of variables and expressions whose representations are known; $r$ describes equal expressions in terms of those representations. The correctness of the pair is shown by proving that $p = r$, under the assumption that (0) the precondition of $p$ holds and (1) the abstract variables referenced in $p$ and their representation satisfy the coupling invariant. For example, we show that line (7) of $BN$ is correct:

$$
\begin{array}{ll}
& b \\
= & \langle \text{Coupling invariant } CBN(b, j) \text{ holds} \rangle \\
& j > 0
\end{array}
$$

**Statement replacements p into r.** Pattern $p$ describes a class of statements involving variables and expressions whose representations are known; $r$ describes equivalent statements in terms of those representations. The correctness of the pair is proved by showing that their simultaneous execution maintains the coupling invariant in any state in which the precondition of the pattern is true. (The correctness concern will be more complicated in cases that other variables that are not being transformed are assigned and in cases where $p$ is nondeterministic; see Sect. 5.6.) For example, we show that line (8) of $BN$ is correct:

$$
\begin{array}{ll}
& wp(``b, j := b1, j1", b \equiv j > 0) \\
= & \langle \text{Definition of } wp \rangle \\
& b1 \equiv j1 > 0 \\
= & \langle \text{Coupling invariant } CBN(b1, j1) \text{ holds} \rangle \\
& true
\end{array}
$$

**Representation definitions /M p repr r**, where $M$ is a transform formula, $p$ is a pattern whose type is that of the abstract variable of $M$, and $r$ is a replacement whose type is that of the representation of the abstract variable. It defines the representation of a class of expressions of the form $[\![ p ]\!] = r$. For example, the pattern on line (3) of $BN$ defines $[\![ b1 \lor b2 ]\!] = j1 + j2$. To prove that a representation definition $CBN(p, r)$ holds. Here is the proof that this definition is correct:

$$
\begin{array}{ll}
& b1 \lor b2 \\
= & \langle \text{Coupling invariant } CBN(b1, j1) \text{ holds} \rangle \\
& j1 > 0 \lor b2 \\
= & \langle \text{Coupling invariant } CBN(b2, j2) \text{ holds} \rangle \\
& j1 > 0 \lor j2 > 0
\end{array}
$$

$$= \quad \langle\, j1 \text{ and } j2 \text{ are natural numbers; predicate calculus} \rangle$$
$$j1 + j2 > 0$$

## 4.5  Discussion of the transform

The main advantage of using the transform is that an algorithm and its encoding in the programming language can be closer together, in fact are usually the same except for the addition of transformation directives. Thus, there is less need for "pseudo code", or for a "design language" that is different from the programming language.

This advantage came from (0) allowing different variables to be implemented differently, in a simple fashion; (1) allowing for more general operations to be implemented, and not just procedure and function calls; (2) introducing the notion of automatically building the representation of an expression from the representations of its subexpressions; and (3) allowing the automatic insertion of conversion of representations of variables and expressions during the transformation process.

The transform supports the use of modern programming methodology based on developing proof and program hand-in-hand. In fact, it encourages a method of programming (and program presentation) in which one first develops a correct program and then determines how to implement its variables to make it efficient. With this paradigm, we see the transform as providing for more frequent use of program parts (the program part being the transform). For example, we could have a library of transforms, part of an (electronic) software engineer's handbook, with 50 different transforms, each giving a different implementation of the set and each written in terms of the high-level mathematical operations on sets. Included would be implementations of a set as a bit vector, an array, a heap, several different kinds of hash tables, and different versions of balanced trees, 2-3 trees, and red-black trees. Relatively few people use such implementations currently, mostly because they are not easily accessible. However, the library of transforms would be accessible to all simply through the use of transformation directives.

Also, some algorithms are most advantageously presented in terms of a high-level algorithm followed by an intricate, specialized, data refinement that can be expressed in a transform. A good example appears in [8], where an algorithm dealing with a set variable and sequence variable is presented and then the two variables are transformed using one transform. The coordinate transformation is such that a sort operation on the sequence disappears completely, and the structure of the program is radically changed.

A number of algorithms have been advantageously written and presented in terms of coordinate transformations, (e.g. [6] and [8]), so the concept of coordinate transformation has proved its usefulness. We will investigate the use of coordinate transformations in presenting programs in a later section.

Whether the transform itself, as presented here, will turn out to be a useful programming concept may only be ascertained by implementing it and experimenting with it. After all, a prime motivation for its development has been its use in larger programs, and for that, one needs an implementation. To this end, a programming language, *Polya* , is being designed at Cornell, which includes the transform as well as a data type definition facility. We are building a prototype implementation in order to investigate the feasibility of the ideas presented here.

The transform evolved out of similar ideas presented in [8], which introduced the notion of a coupling invariant for describing the relationship between abstract and concrete variables. The earlier view of using a function from a concrete domain to an abstract domain, introduced in [12], proved unsuitable as soon as *restricted* implementations were allowed instead of full implementations of data types. That is, in some cases a function from concrete to abstract need not exist, in that several abstract values are represented by a single concrete one.

# 5 A calculus for data refinement

We now introduce a calculus for data refinement —a set of rules for manipulating programs by replacing some variables by others. This calculus substantiates and generalizes the work on coordinates transformations discussed in the previous sections.

The first substantial work in this area of formal refinement was by Back [2]. The articles more often referenced are by Morris [17] and Morgan and Gardiner [16], which convey the same message but are technically different in places. The basic material in this section is culled from [17]and [16].

We present only an outline of the material on data refinement. For example, most of the proofs of theorems concerning data refinement are omitted.

## 5.1 Procedural refinement

By a procedural refinement, we mean the replacement of one program statement by another without changing the correctness of the program. Thus, we say that statement $S$ is refined by statement $T$, and write $S \subseteq T$, if for all predicates $R$,

$$(19) \quad wp(S, R) \Rightarrow wp(T, R) \quad \text{for all states.}$$

For example, we have $i := i + 2 \subseteq i := i + 1; i := i + 1$. As another example, we have

$$\textbf{if } true \rightarrow x := 2 \ [] \ true \rightarrow x := 3 \textbf{ fi} \subseteq x := 2$$

It has been proved that operator $\subseteq$ is *monotonic*, by which we mean that if $S \subseteq T$, and if $S$ is a substatement of a program, then replacing $S$ by $T$ in the program yields a refinement of that program. (This property holds for all the convential sequential statements but not for parallel ones.) Formally, this is stated as follows. Let $P$ be a program, and assume that it contains an occurrence of a name $X$ at a place where a statement may occur. Then

$$S \subseteq T \;\Rightarrow\; P[X := S] \subseteq P[X := T] \quad,$$

where $P[X := S]$ denotes the textual replacement within $P$ of identifier $X$ by $S$.

This property of monotonicity of refinement substantiates and formalizes informal techniques of refinement that people have been using for years. It is at the heart of what Wirth called *step-wise reinement*.


## 5.2  The statements to be used

We shall later refer to or use the statments of Dijkstra's guarded-command notation: the assignment, sequencing, the conditional statement, and the loop. We assume the reader knows the weakest-precondition definitions of these statements. In addition, we will be using a few other statements, which we now introduce.

For $P$ a predicate, $\{P\}$ placed in a program is typically a comment, indicating what is expected to be true at that place. We use $\{P\}$ not as a comment but as an *assertion statement*, which is defined in terms of weakest preconditions as follows:

$$(20) \quad wp(\{P\}, R) \;\equiv\; P \wedge R \quad.$$

Thus, execution of $\{P\}$ is defined only in states in which $P$ is true, and its execution is equivalent to execution of *skip*. In fact, one can easily show that $\{P\}$ is procedurally refined by *skip*: $\{P\} \subseteq skip$, which means that replacing $\{P\}$ by *skip* does not affect the correctness of a program. Often, we omit the semicolon between an assertion statement and a following statement, writing $\{P\}; S$ as $\{P\} S$.

The next statement we introduce is the generalized assignment. Let $v$ be a list of (distinct) variables and $Q$ a predicate. Then, execution of the statement $v\!:\!-Q$ assigns values to $v$ to truthify $Q$. For example, execution of $b, c\!:\!-(b = c)$ assigns values to $b$ and $c$ in order to truthify $b = c$. Three procedural refinements of $b, c\!:\!-(b = c)$ are $b := a$, $a := b$, and $a, b := 6, 6$.

The weakest-precondition definition of the generalized assignment is

$$(21) \quad wp(\text{``}v\!:\!-Q\text{''}, R) \;\equiv\; (\forall v : Q : R) \quad.$$

If no set of values satisfies $Q$, then $(\forall v : Q : R)$ is *true*, so execution of $v\!:\!-Q$ establishes $R$ in *all* initial states. Thus, $v\!:\!-Q$ is a miraculous statement. In this case, there will be no way to refine $v\!:\!-Q$ into a statement that can be implemented. The programming language allows miracles, but the part of the programming language that can be implemented does not.

The assertion statement and the generalized assignment statement together form a specification statement. For example, $\{P\}\ v{:}{-}Q$ specifies a program that in a state satisfying $P$ stores in $v$ to truthify $Q$. Thus, it is equivalent to the Hoare triple $\{P\}\ S\ \{Q\}$ where $S$ is restricted to assignments to $v$. In this way, the assertion and generalized assignment statements can be used to incorporate specifications and programming-language statements. [2]

The introduction of a specification statement has been advocated by Hoare and others for some time, even though it is not implementable, for it places specification and program on the same level, allowing us to move freely between them in our calculations. Just to give an idea how a calculation can be performed, we calculate one refinement of the specification $\{a > 0\}\ a,b{:}{-}(a = b)$. Again, we use $R[x := y]$ to denote textual substitution for the variables in list $x$ by the corresponding expressions in list $y$.

$$
\begin{aligned}
& wp(\text{``}\{a > 0\};\ a,b{:}{-}(a = b)\text{''},\ R) \\
= \quad & \langle\text{Definition of semicolon and generalized assignment}\rangle \\
& wp(\{a > 0\}, (\forall a, b : a = b : R)) \\
= \quad & \langle\text{Definition of assertion statement}\rangle \\
& a > 0\ \wedge\ (\forall a, b : a = b : R) \\
= \quad & \langle\text{One-point rule}\rangle \\
& a > 0\ \wedge\ R[b := a] \\
\Rightarrow \quad & \langle\text{Predicate calculus}\rangle \\
& R[b := a] \\
= \quad & \langle\text{Definition of assignment}\rangle \\
& wp(\text{``}b := a\text{''},\ R)
\end{aligned}
$$

So $\{a > 0\}\ a,b{:}{-}(a = b)\ \subseteq\ b := a$, and the specification can be replaced by the assignment.

Finally, we define the block, which introduces local variables with initial values. Let $v$ be a list of variables, $P$ a predicate, and $S$ a statement. Then the statement

$$\lbrack\!\lbrack\ \mathbf{var}\ v{:}{-}P;\ S\rbrack$$

introduces local variables $v$, whose scope is $P$ and $S$. Executing a block consists of assigning values to $v$ that truthify $P$ and executing $S$. Formally, the block is defined as

$$(22) \qquad wp(\lbrack\!\lbrack\ \mathbf{var}\ v{:}{-}P;\ S\rbrack,\ R) \equiv (\forall v : P : wp(S, R))$$

where it is assumed that $R$ does not contain free occurrences of the variables of $v$.

----

[2] Morgan and Gardiner [16] do not use the assertion statement but instead introduce a specification statement $v : [P, Q]$, which has the same meaning as $\{P\};\ v{:}{-}Q$.

## 5.3  The definition of data refinement

The purpose of this section is to describe conditions under which a block in which $a$ is declared as local is procedurally refined by a block in which $c$ is declared as local, where $a$ and $c$ are (disjoint) lists of variables:

$$[\![ \ \mathbf{var}\ a{:}{-}Pa;\ Sa ] \ \subseteq\ [\![ \ \mathbf{var}\ c{:}{-}Pc;\ Sc ]$$

For example, if $a$ is a set variable, then $c$ could be a list of variables that describe a heap or a hash-table implementation of $a$. Variable $a$, predicate $Pa$, and statement $Sa$ are called the *abstract* variable, predicate and statement; similarly, $c$, $Pc$, and $Sc$ are called *concrete*.

Thus, this refinement is the generalization of the coordinate transformation discussed in earlier sections.

The conditions under which the above refinement holds involve a *coupling invariant $C$* that couples $a$ and $c$, just as in the earlier sections dealing with coordinate transformations. Also needed is a new relation, called data-refinement, which gives the necessary conditions on $Sa$ and $Sc$. Note that $Sa$ does not refer to $c$ and $Sc$ does not refer to $a$. Let $Ra$ be a predicate that does not contain $c$ as free. We say that $Sa$ is *data-refined* by $Sc$ (with regard to $a$, $c$, and $C$), written $Sa \sqsubseteq Sc$, if the following holds: [3]

$$(23) \quad (\exists a : C : wp(Sa, Ra)) \ \Rightarrow\ wp(Sc, (\exists a : C : Ra)) \quad \text{(for all } Ra\text{)}$$

Some understanding of this definition can be given in terms of Fig. 2. The top of the diagram represents the abstract state and execution of $Sa$ in it; the bottom represents the concrete state and execution of $Sc$ in it.

Both the antecedent and the consequent of (23) have $c$ as free, but not $a$, since $a$ is bound by the existential quantifiers. Consider a state that satisfies the LHS of (23). The LHS indicates that in this state, there is an abstract value $a$ that is coupled to $c$ such that execution of $Sa$ will terminate in a state satisfying $Ra$.

Consider a state that satisfies the RHS of (23). The RHS indicates that in this state, execution of $Sc$ terminates in a state in which there exists an abstract value that is coupled to $c$ for which $Ra$ holds.

Definition (23) of data refinement can be simplified if coupling invariant $C$ is *functional*, which means that for each value of $c$ there exists at most one value for $a$ for which $C$ holds. If $C$ is functional, then $C$ can be written in the form

$$(24) \quad C \ \equiv\ Cc \wedge a = e$$

where $Cc$ and $e$ do not contain free occurrences of $a$. With $C$ in this form, we manipulate (23) as follows.

---

[3] Properly, we should write $Sa \sqsubseteq_{a,c,C} Sc$ in order to indicate the abstract variables, concrete variables, and coupling invariant. We omit the subscripts when they are evident from the context.

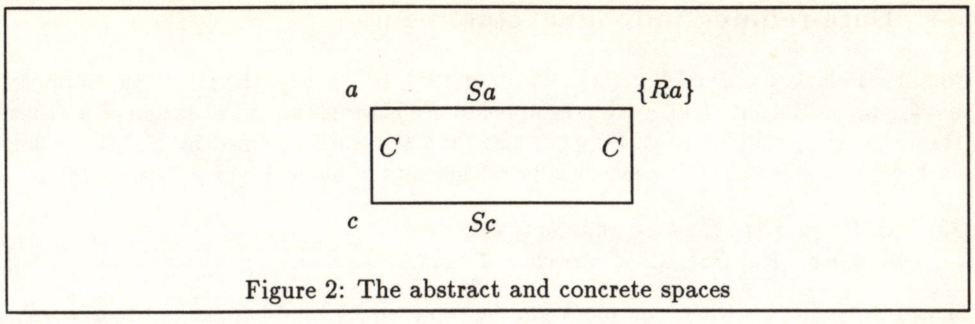

Figure 2: The abstract and concrete spaces

$$
\begin{aligned}
& (\exists a : C : wp(Sa, Ra)) \;\Rightarrow\; wp(Sc, (\exists a : C : Ra)) \\
=\;\; & \langle (24),\ \text{twice} \rangle \\
& (\exists a : Cc \wedge a = e : wp(Sa, Ra)) \;\Rightarrow\; wp(Sc, (\exists a : Cc \wedge a = e : Ra)) \\
=\;\; & \langle \text{Trading, twice} \rangle \\
& (\exists a : a = e : Cc \wedge wp(Sa, Ra)) \;\Rightarrow\; wp(Sc, (\exists a : a = e : Cc \wedge Ra)) \\
=\;\; & \langle \text{One-point rule, twice; } a \text{ is not free in } Cc \rangle \\
& Cc \wedge wp(Sa, Ra)[a := e] \;\Rightarrow\; wp(Sc, Cc \wedge Ra[a := e])
\end{aligned}
$$

This rule is simpler. We will see similar simplifications later on when functional coupling invariants are used. But let us give an example to show that functional coupling invariants do not always exist. This happens mainly when partial or restricted implementations are used.

Suppose we are implementing an "undo" facility. A special button exists on the keyboard, which can be used to undo previous keystrokes in a first-in-last-out discipline. For example, hitting the undo button 4 times undoes the last 4 keystrokes. Our implementation of the undo facility uses a stack $S$ to maintain information for undoing keystrokes. Whenever a key is hit, information is stored on $S$. When the undo button is pressed, the information at the top of $S$ is removed and used to undo a keystroke.

$S$ is unbounded and memory is finite, so want to replace $S$ by an array $b$ that contains the top 100 elements of $S$. This is a restricted implementation, and it means that the undo button can be used to undo at most the last 100 keystrokes. This is a reasonable restriction. But now the concrete variable $b$ is coupled to an infinite number of abstract stacks $S$ —all those with the same top values as are in $b$.

Whenever the concrete variable implements a restricted form of the abstract variables, a non-functional coupling invariant may result. Furthermore, almost all our implementations of types are restricted in some fashion —the only one that isn't is *bool*.

## 5.4  Data-refining individual statements

Consider a block $[\![$ **var** $a{:}{-}Pa;\ Sa]\!]$ . Theorem (25), below, says the following. Suppose that for any initialization of $c$ (that establishes $Pc$ ) there is an initialization of $a$ (that establishes $Pa$ ) coupled to it. Suppose also that $Sa$ is data-refined by $Sc$ . Then the block $[\![$ **var** $a{:}{-}Pa;\ Sa]\!]$ is procedurally refined by the block $[\![$ **var** $c{:}{-}Pc;\ Sc]\!]$ :

(25)   If $Pc \Rightarrow (\exists a : C : Pa)$ and $Sa \sqsubseteq Sc$ , then
       $[\![$ **var** $a{:}{-}Pa;\ Sa]\!] \ \subseteq\ [\![$ **var** $c{:}{-}Pc;\ Sc]\!]$ .

Thus, (25) indicates when one block can be replaced by another, or, in other words, when a set of variables $a$ can be replaced by a set of variables $c$ .

In case coupling invariant $C$ is functional, $C \equiv Cc \wedge a = e$ , (25) reduces to

(26)   If $Pc \Rightarrow Cc \wedge Pa[a := e]$ and $Sa \sqsubseteq Sc$ , then
       $[\![$ **var** $a{:}{-}Pa;\ Sa]\!] \ \subseteq\ [\![$ **var** $c{:}{-}Pc;\ Sc]\!]$ .

In both (25) and (26), a key to the validity of the replacement of $a$ by $c$ is that statement $Sa$ should be data-refined by its replacement $Sc$ , i.e. $Sa \sqsubseteq Sc$ must hold. However, definition (23) of data refinement is rather complicated. By investigating separately the statements of the language, we will see that the refinement of each can be found more simply. In fact, we will be able to *calculate* data refinements. This means that in many cases, once the concrete variables $c$ to replace $a$ are chosen and coupling invariant $C$ has been formulated, the refinements of each of the various parts of $a$ can be calculated (and then further refined); there is no need to guess them.

We now show how the various types of statements $a$ can be refined. For some of the types, we give two refinements: the general one and then the refinement in case $C$ is functional.

**Case 0:  $Sa$ does not refer to abstract variables.** If $Sa$ does not refer to $a$ , then $Sa \sqsubseteq Sa$ .

**Case 1: Assertion statement.** For an assertion statement $\{P\}$ , we have

(27)   $\{P\} \sqsubseteq \{(\exists a : C : P)\}$

and, if $C$ is functional, $C = Cc \wedge a = c$ ,

(28)   $\{P\} \sqsubseteq \{Cc \wedge P[a := e]\}$

**Case 2: The specification statement.** Let $x$ be a (possibly empty) list of variables that are distinct from the abstract and concrete variables. Then

(29)   $\{P\}\, a, x{:}{-}Q \sqsubseteq \{(\exists a : C : P)\}\, c, x{:}{-}(\exists a : C : Q)$

Further, we can show that the concrete specification shown above is the best possible, in the following sense. Any concrete program $Sc$ is a data refinement of $\{P\}\, a, x{:}{-}Q$ iff it is a procedural refinement of $\{(\exists a : C : P)\}\, c, x{:}{-}(\exists a : C : Q)$ :

(30)  For all concrete programs $Sc$, $\{P\}\, a, x{:}{-}Q \sqsubseteq Sc$
if and only if $\{(\exists a : C : P)\}\, c, x{:}{-}(\exists a : C : Q) \subseteq Sc$.

In case $C$ is functional, $C \equiv Cc \wedge a = e$, (29) and (30) reduce to

(31)  $\{P\}\, a, x{:}{-}Q \sqsubseteq \{Cc \wedge P[a := e]\}\, c, x{:}{-}Cc \wedge Q[a := e]$ .

(32)  For all concrete programs $Sc$, $\{P\}\, c, x{:}{-}Q \sqsubseteq Sc$
if and only if $\{Cc \wedge P[a := e]\}\, c, x{:}{-}Cc \wedge Q[a := e] \subseteq Sc$.

**Case 3: Simple assignment to common variables with a functional coupling invariant.** Let $x$ be variables that are in neither $a$ nor $c$ and let $C$ be functional, $C \equiv Cc \wedge a = e$. Then we have

(33)  $\{P\}\, x := f \sqsubseteq \{Cc \wedge P[a := e]\}\, x := f[a := e]$

**Case 4: Sequential composition.** Data refinement distributes through sequential composition:

(34)  If $Sa \sqsubseteq Sc$ and $Ta \sqsubseteq Tc$, then $Sa; Ta \sqsubseteq Sc; Tc$.

**Case 5: Blocks.** Local blocks can be carried over to the concrete specification:

(35)  $[\![\ \mathbf{var}\ v{:}{-}P;\ S]\!] \sqsubseteq [\![\ \mathbf{var}\ v{:}{-}(\exists a : C : P);\ T]\!]$  follows from $S \sqsubseteq T$

**Case 6: Conditional statement.** We consider data-refining a conditional statement $\mathbf{if}\ (\square P_i \to S_i)\ \mathbf{fi}$. Write

$$Q_i \ \equiv\ (\forall a : C : P_i)$$

$Q_i$ will be the replacement for guard $P_i$. Suppose $S_i \sqsubseteq T_i$, for each $i$. Suppose further that $(\exists a : C : (\exists i :: P_i)) \Rightarrow (\exists i :: Q_i)$ holds. Then

(36)  $\mathbf{if}\ (\square P_i \to S_i)\ \mathbf{fi} \sqsubseteq \mathbf{if}\ (\square Q_i \to T_i)\ \mathbf{fi}$

Condition $(\exists a : C : (\exists i :: P_i)) \Rightarrow (\exists i :: Q_i)$ ensures that at least one guard of the concrete conditional statement is true, so that the concrete conditional statement does not abort.

If $C$ is functional, $C \equiv Cc \wedge a = e$, $Q_i$ reduces to $Cc \wedge P_i[a := e]$, so (36) reduces to

(37)  $\mathbf{if}\ (\square P_i \to S_i)\ \mathbf{fi} \sqsubseteq \mathbf{if}\ (\square P_i[a := e] \to T_i)\ \mathbf{fi}$

**Case 7: Loop.** Refining a loop is similar to refining a conditional statement. Consider the loop $\{I\}\ \mathbf{do}\ P \to S\ \{I\}\ \mathbf{od}$, where the appearances of assertion $I$ indicate that $I$ is to be a loop invariant. Define

$$Q \ \equiv\ (\forall a : C \wedge I : P)\ ,$$
$$\hat{Q} \ \equiv\ (\forall a : C \wedge I : \neg P)\ .$$

Then $\{I\}\,\mathbf{do}\,P \to S\,\{I\}\,\mathbf{od}\ \sqsubseteq\ \mathbf{do}\,Q \to T\,\mathbf{od}$ follows from $S \sqsubseteq T$ and $(\exists a : C \wedge I : Q \vee \hat{Q})$.

If $C$ is functional, $C \equiv Cc \wedge a = e$, this reduces to

$$\mathbf{do}\,P \to S\,\mathbf{od}\ \sqsubseteq\ \mathbf{do}\,Q[a := e] \to T\,\mathbf{od}\ \ \text{follows from } S \sqsubseteq T$$

In writing these theorems concerning the data-refinement of statements, I have made substantial use of the material in [16] and [17]. These two papers tackle data refinement by calculation in a similar fashion, but there are enough differences between them to make a comparison difficult.

## 5.5  Using the data-refinement rules

In Sect. 4.1, we introduced the transform $Bvector(k : nat)$, which we repeat here with $a$ and $c$ instead of $s$ and $b$ for the abtract and concrete variables:

$(38)$  $\quad$ **transform** $Bvector(k : nat)$;
$\qquad$ **var** $a : set(nat)$ $\qquad\qquad$ **repr** $\quad$ $c :$ **array** $0..k-1$ **of** $bool$
$\qquad$ $\{coupling\ invariant : (\forall i : 0 \leq i < k : i \in a \equiv c.i)\}$
$\qquad$ $[]$ $\ a := \{\,\}$ $\qquad\qquad\qquad$ **into** $\quad$ (**for** $i : 0 \leq i < k : c.i := false$)
$\qquad$ $[]$ $\ \{x \in 0..k-1\}$
$\qquad\qquad$ $a := a \cup \{\langle \text{exp-}x : nat \rangle\}$ **into** $\quad$ $c.x := true$
$\qquad$ $[]$ $\ \{x \in 0..k-1\}$
$\qquad\qquad$ $a := a - \{\langle \text{exp-}x : nat \rangle\}$ **into** $\quad$ $c.x := false$
$\qquad$ $[]$ $\ \langle \text{exp-}x : nat \rangle \in a$ $\qquad\quad$ **into** $\quad$ $0 \leq x < k \wedge c.x$
$\qquad$ $[]$ $\ \#a$ $\qquad\qquad\qquad\qquad$ **into** $\quad$ $= \{\,\text{number of true } c.i\,\}$
$\qquad\qquad\qquad\qquad\qquad\qquad\qquad\quad$ $(\#i : 0 \leq i < k : c.i)$

$\qquad$ **end**

Here, we want to prove, using the rules of data refinement just introduced, that a transformation effected using this transform is indeed a procedural refinement. Such a transformation replaces a block of the form

$$[\!\![\ \mathbf{var}\ a : set(nat){:}{-}(a = \{\,\}); \ Sa ]\!\!]$$

by a block of the form

$$[\!\![\ \mathbf{var}\ c\ \mathbf{array}\ 0..k-1\ \mathbf{of}\ bool{:}{-}(c = false);\ Sc ]\!\!]\ \ ,$$

where $c = false$ is shorthand for $(\forall i : 0 \leq i < k : \neg c.i)$. Let the coupling invariant, expressed in functional form, be $true \wedge a = e$ where $e = \{i \mid 0 \leq i < k \wedge c.i\}$.

Assume that all references to $a$ within $Sa$ are either statements or expressions as given in transform $Bvector$.

Let us first consider the translation of expressions. A glance at the rules in the previous section for refining statements (when the coupling invariant is functional) reveals that many of them have an expression of the form $P[a := e]$. On the other hand, transform $Bvector$ calls for replacing not the variable $a$ but expressions of the form $\#a$ and $x \in a$. However, we can prove the following lemma.

**Lemma.** Let $P$ be any expression whose only references to $a$ are in subexpressions of the form $\#a$ and $x \in a$, and let $\overline{P}$ denote that expression with occurrences of $\#a$ and $x \in a$ replaced by $(\#i : 0 \leq i < k : c.i)$ and $c.x$, respectively. Then, in any state in which the coupling invariant $a = e$ holds and in which $0 \leq x < k$ holds, $P[a := e] = \overline{P}$.

Formally, the proof of the lemma has to be by induction on the structure of the expression; we leave the proof to the reader.

With this little lemma, we can prove that the transformation given by $Bvector$ does indeed result in a correct data refinement. Since the coupling invariant is functional, by (26), we must prove

$$c = false \;\Rightarrow\; true \wedge (a = \{\})[a := e] \quad \text{and} \quad Sa \sqsubseteq Sc$$

It is a simple matter to prove that the first condition holds (using the lemma), and we concentrate on proving $Sa \sqsubseteq Sc$. We do this by induction on the structure of $Sa$ and on the understanding that its components that reference $s$ are as given in the transform. This means that each type of statement that could be $Sc$ must be investigated. We prove three cases and leave the rest to the reader.

Suppose $Sa$ does not refer to abstract variables. Then the transformation has no effect and $Sa$ and $Sc$ are the same, and, $Sa \sqsubseteq Sa$ holds.

Suppose $Sa$ is an assertion statement $\{P\}$. Then we know that $Sa \sqsubseteq true \wedge P[a := e]$. We have,

$$
\begin{array}{ll}
& \{P\} \\
\sqsubseteq & \quad \langle 27 \rangle \\
& \{true \wedge P[a := e]\} \\
= & \quad \langle \text{Predicate calculus} \rangle \\
& \{P[a := e]\} \\
= & \quad \langle \text{The Lemma} \rangle \\
& \{\overline{P}\}
\end{array}
$$

so $\{P\} \sqsubseteq \{\overline{P}\}$ and $\{P\}$ can be replaced by $\{\overline{P}\}$.

Suppose $Sa$ is a statement $\{x \in 0..k-1\}\ a := a - \{x\}$ for some expression $x$. We must prove that $Sa \sqsubseteq c.x := false$. None of the rules given in the previous subsection can be used in this case, but we can prove $Sa \sqsubseteq c.x := false$ using the Gries/Prins rule (43). Thus, we have to prove

$$\{x \in 0..k-1\} \wedge a = e \;\Rightarrow\; wp(c.x := false, \neg wp(a := a - \{x\}, \neg a = e))$$

This simple proof is left to the reader. Note that the refinement $c.x := \mathit{false}$ will be further transformed if $x$ contains $a$, but the lemma discussed above can be used to show that this transformation yields a data refinement.

## 5.6  Three equivalent data-refinement definitions

We have been working with definition (23) of data refinement, which we repeat here. Let $Ra$ denote a predicate that contains no concrete variable (only abstract variables and other variables). Then $Sa$ is data-refined to $Sc$ (using abstract variable $a$, concrete variable $c$, and coupling invariant $C$), written $Sa \sqsubseteq Sc$, if

$$(39) \quad (\exists a : C : wp(Sa, Ra)) \;\Rightarrow\; wp(Sc, (\exists a : C : Ra)) \quad \text{(for all } Ra\text{)}$$

There are two other equivalent predicates that we want to discuss here. First, in the earlier section on the coordinate transformation, we indicated that $\{P\}\, Sa$ could be replaced by $Sc$ if

$$(40) \quad P \wedge C \wedge wp(Sa, \mathit{true}) \;\Rightarrow\; wp(Sc, wp(Sa, C))$$

—i.e. if under certain conditions execution of $Sc$ followed by $Sa$ maintained the coupling invariant. This rule is too weak if $Sa$ is nondeterministic. For example, suppose $a$ is a set, suppose $Sa$ appends either 2 or 3 to $a$, and suppose $Sc$ appends 2 to the representation of $a$. $Sc$ is certainly a suitable refinement of $Sa$; it is simply less deterministic. But (40) is too weak to prove $Sa \sqsubseteq Sc$.

One needs a rule that formalizes the statement

(41)  Execution of $Sc$ terminates in a state in which it is possible for execution of $Sa$ to truthify $C$.

To formalize it, we first have to introduce a double negative and write it as

(42)  Execution of $Sc$ terminates in a state in which it is not guaranteed for execution of $Sa$ to truthify $\neg C$.

Adding the necessary antecedent, statement (42) can be formalized as

$$(43) \quad P \wedge C \wedge wp(Sa, \mathit{true}) \;\Rightarrow\; wp(Sc, \neg wp(Sa, \neg C))$$

Further, if $Sa$ assigns to a variable $x$ that is not an abstract variable, then $Sc$ should assign the same value to $x$. We can introduce this requirement into (43) by writing both $Sa$ and $Sc$ as a function of $x$, as follows. We have $Sa \sqsubseteq Sc$ exactly when

$$(44) \quad P \wedge C \wedge wp(Sa.x, \mathit{true}) \wedge x = z \;\Rightarrow\; wp(Sc.z, \neg wp(Sa.x, \neg(x = z \wedge C)))$$

Rule (44), introduced by Gries and Prins in [8], was proved equivalent to the Morris-Morgan-Gardiner rule (39) by Chen and Udding in [3]. Chen and Udding also derived a third equivalent rule. Let $Ra$ and $Rc$ denote predicates that do not contain concrete and abstract variables, respectively. Their rule is

$$(45) \quad [C \Rightarrow Ra = Rc] \Rightarrow [C \wedge wp(Sa, Ra) \Rightarrow wp(Sc, Rc)]$$

where the brackets around the antecedent and consequent denote implicit quantification over all states. Thus, this rule can be interpreted as: validity of $C \Rightarrow Ra = Rc$ implies validity of $C \wedge wp(Sa, Ra) \Rightarrow wp(Sc, Rc)$. See [5] for a full discussion of this notation.

Let's give names to the three equivalent forms of definition of data-refinement: call (39) the Morris/Morgan rule, (44) or its restricted form (44) the Gries/Prins rule, and (45) the Chen/Udding rule.

Morris/Morgan refers to the general predicate $Ra$, Chen/Udding to $Ra$ and $Rc$, and Gries/Prins to neither. Morris/Morgan has an existential quantifier in both its antecedent and its consequent, Chen/Udding only in its consequent, and Gries/Prins in neither. Instead of quantification, Gries/Prins has the double negation, and it is also complicated by the need to distinguish the common variables that are assigned in $Sa$ and $Sc$. Finally, Morris/Morgan and Chen/Udding simplify if the coupling invariant is functional, while Gries/Prins does not. These rules seem truly different, and it may seem odd that they are equivalent.

Which rule should one use in proving that $Sa \sqsubseteq Sc$? That depends on the circumstances. For proving properties of data refinement, for example, proving all the theorems listed previously for the refinement of the various types of statements, Morris/Morgan appears easier to use. For proving correctness of data refinements of particular programs, the case is not so clear. Let us look at an example, taken from [14] —other examples can also be found in [14].

Suppose an integer $a$ is to be replaced by a boolean $c$ using the coupling invariant

$$C : c \equiv odd.a \quad .$$

Statement $a := 0$ is to be replaced by $c := false$, so we must prove $a := 0 \sqsubseteq c := false$. We give the proof using all three rules.

**Morris/Morgan proof.** In this proof, we manipulate (39) (instantiated with $Sa$, $Sc$, and $C$) in order to show that it is implied by $true$. We have

$$(\exists a : C : wp(a := 0, Ra)) \Rightarrow wp(c := false, (\exists a : C : Ra))$$
$$= \quad \langle \text{Def of } wp \text{ for assignment, Definition of } C \rangle$$
$$(\exists a : c = odd.a : Ra[a := 0]) \Rightarrow (\exists a : false = odd.a : Ra[c := false])$$
$$= \quad \langle a \text{ not free in } Ra[a := 0] \, ; \, c \text{ not free in } Ra \rangle$$
$$(\exists a : c = odd.a : true) \wedge Ra[a := 0]) \Rightarrow (\exists a : false = odd.a : Ra)$$
$$= \quad \langle (\exists a : c = odd.a : true) \text{ is true; pred.calc.} \rangle$$

$$Ra[a := 0]) \;\Rightarrow\; (\exists a : false = odd.a : Ra)$$
$$\Leftarrow \quad \langle \text{Trading; Instantiation, } a = 0 \rangle$$
$$Ra[a := 0]) \;\Rightarrow\; false = odd.0 \wedge Ra[a := 0]$$
$$= \quad \langle \text{Predicate calculus} \rangle$$
$$true$$

**Gries/Prins proof.** In this case, $P$ and $wp(Sa, true)$ are both $true$, so the antecedent reduces to $C$. We set aside the antecedent and manipulate the consequent:

$$wp(``c := false", \neg wp(a := 0, \neg c = odd.a))$$
$$= \quad \langle \text{Definition of } wp \text{ of assignment} \rangle$$
$$wp(c := false, \neg\neg(c = odd.0))$$
$$= \quad \langle \text{Cancel double negation} \rangle$$
$$wp(c := false, c = odd.0)$$
$$= \quad \langle \text{Definition of } wp \text{ of assignment} \rangle$$
$$false = odd.0$$
$$= \quad \langle \text{Predicate calculus} \rangle$$
$$true$$

**Chen/Udding proof.** We set aside the antecedent, $[C \;\Rightarrow\; Ra = Rc]$, and manipulate the consequent. Note that instantiation $a, c := 0, false$ in the antecedent yields $false = odd.0 \Rightarrow (Ra = Rc)[a, c := 0, false]$. Since its antecedent is true, $a$ is not free in $Rc$, and $c$ is not free in $Ra$, this predicate reduces to $Ra[a := 0] = Rc[c := false]$. We manipulate the consequent:

$$C \wedge wp(a := 0, Ra) \;\Rightarrow\; wp(c := false, Rc)$$
$$= \quad \langle \text{Definition of } wp \text{ for assignment, twice} \rangle$$
$$C \wedge Ra[a := 0] \;\Rightarrow\; Rc[c := false]$$
$$= \quad \langle \text{Shuffling} \rangle$$
$$C \;\Rightarrow\; (Ra[a := 0] \;\Rightarrow\; Rc[c := false])$$
$$= \quad \langle \text{Instantiated antecedent, } Ra[a := 0] = Rc[c := false] \rangle$$
$$C \;\Rightarrow\; true$$
$$= \quad \langle \text{Predicate calculus} \rangle$$
$$true$$

In this one example, the Gries/Prins formulation seems easier to use. The Morris/Morgan formulation has to eliminate the existential quantifiers and deal with the general term $Ra$. Notice how the Chen/Udding formulation uses the instantiated antecedent of the rule to eliminate $Ra$ and $Rc$ in the consequent. This method is usually used to eliminate $Ra$ and $Rc$.

# 6  Using data refinement to present programs

As we have mentioned, one major use of data refinement and coordinate transformation is the implementation of abstract data types. In this setting, data refinement is a vehicle for the reuse of program parts, the program part being the transform (module, or (package) that describes an implementation.

There is a second important use of data refinement, the development and presentation of algorithms. In the presentation of many algorithms that deal with complicated data structures, the presentation is often clouded by myriads of implementation details. *Indexitis*, the overuse of subscripted variables and subscripts upon subscripts, is often a symptom of the problem. In such cases, it may be better to first develop (or present) the algorithm at a higher level —e.g. in terms of bags, sets, sequences, trees, or graphs— can allow the algorithmic aspects of the problem to be stated simply and clearly. Then, a data refinement or coordinate transformation can be used to show how the high-level concepts are expressed in terms of the lower-level ones, arrays, linked lists and the like.

Time and space do not permit a detailed explanation here. The reader is referred to some of the papers that describe algorithms in this fashion: [6], [8], [9], and [10].

# References


[1] Aho, A.V., J.E. Hopcroft and J.D. Ullman. *Date Structures and Algorithms*. Addison-Wesley, Reading, 1985, 96-102.

[2] Back, R.-J. On the correctness of refinement steps in program development. Report A-1978-4, Computer Science Department, University of Helsinki, 1978.

[3] Chen, W., and J.T. Udding. Towards a calculus of data refinement. LNCS 375, *Mathematics of Program Construction*, Springer Verlag, New York, 1989.

[4] Dijkstra, E.W. *A Discipline of Programming*. Prentice Hall, Englewood Cliffs, New Jersey, 1976.

[5] Dijkstra, E.W., and C.S. Scholten. *Predicate Calculus and Program Semantics*. Springer Verlag, New York, 1989.

[6] Feijen, W.H.J., A.J.M. van Gasteren, and D. Gries. In-situ inversion of a cyclic permutation. *IPL 24*, 1 (January 1987), 11-14.

[7] Gries, D. *The Science of Programming*. Springer Verlag, New York, 1981.

[8] Gries, D., and J. Prins. A new notion of encapsulation. *Proc. SIGPLAN 85 Symposium on Language Issues in Programming Environments*. Seattle, Washington, June 1985, 131-139.

[9] Gries, D., and J. Prins. McLaren's masterpiece. *Science of Computer Programming 8* (1987), 139-145.

[10] Gries, D., and J. van de Snepscheut. Inorder traversal of a binary tree and its inversion. Tech. Rpt. 87-876, Computer Science Department, Cornell University, November 1987.

[11] Gries, D., and D. Volpano. The Transform —a new language construct. *Structured Programming 11* (January 1990), 1-10.

[12] Hoare, C.A.R. Proof of correctness of data representations. *Acta Informatica 1* (1972), 271-281.

[13] Knuth, D.E. *The Art of Programming, Vol I, Fundamental Algorithms*. Addison-Wesley, Reading, 1963.

[14] Lutz, E. *IPL*.

[15] Morgan, C. Data refinement by miracles. *IPL 26* (1987), 243-246.

[16] Morgan, C., and P.H.B. Gardiner. Data refinement by calculation. *Acta Informatica 27*, (1990), 481-503.

[17] Morris, J.M. The laws of data refinement. *Acta Informatica 26* (1989), 287-308.

# Refinement algebra proves correctness of compilation

C.A.R. Hoare and He Jifeng

Oxford University Computing Laboratory, 11 Keble Road, Oxford, OX1 3QD, UK.

**Abstract:** A compiler is specified by a description of how each construct of the source language is translated into a sequence of object code instructions. The meaning of the object code can be defined by an interpreter written in the source language itself. A proof that the compiler is correct must show that interpretation of the object code is at least as good (for any relevant purpose) as the corresponding source program. The proof is conducted using standard techniques of data refinement. All the-calculations are based on algebraic laws governing the source language. The theorems are expressed in a form close to a logic program, which may be used as a compiler prototype, or as a check on the results of a particular compilation. It is suggested that this formal framework provides appropriate interfaces for compiler implementors, and hardware designers, as well as users of the language.

**Keywords:** Algebra, imperative programming, refinement, compiler correctness.

## 1 Introduction

Compilation is specified as a relation between a source program $p$ (in some high level source language HL) and the corresponding object code $c$ (in some target machine language ML). Further details of compilation are given by a symbol table $\Psi$, mapping the global identifiers of $p$ to storage locations of the target machine. This compilation relation will be abbreviated as a predicate

$$\mathcal{C} \, p \, c \, \Psi$$

The internal structure of $p$, $c$ and $\Psi$ will be elaborated as the need arises.

Improvement is a relation between a program $q$ and a program $p$ that holds whenever for any purpose the observable behaviour of $q$ is as good as or better than that of $p$; more precisely, if $q$ satisfies every specification satisfied by $p$, and maybe more. For example, in a procedural language, a program is better if it terminates more often and/or gives a more determinate result. This relation is written

$$p \sqsubseteq q$$

where $\sqsubseteq$ is necessarily transitive and reflexive (a preorder).

If $p$ and $q$ are programs operating on different data spaces, they cannot be directly compared. But if $r$ is a translation from the data space of $q$ to that of $p$, we can then compare them before and after the translation

$$r \, ; \, p \sqsubseteq q \, ; \, r$$

where the semicolon denotes sequential composition. The relation $r$ is known as a simulation or refinement; such simulations are the basis of several modern program development techniques, e.g. VDM, Z [7, 9].

To define compiler correctness precisely, we need to ascribe meanings to $p$, $c$ and $\Psi$. Let $\hat{c}$ be a formal description of the behaviour of the target machine executing the machine code $c$. Let $\hat{p}$ similarly be an abstract behavioural definition of the meaning of the program $p$. Finally, let $\hat{\Psi}$ be a transformation which assigns to each global identifier $x$ of the source program the value of the corresponding location $\Psi x$ in the store of the target computer. In order to prove that the object code $c$ is a correct translation of $p$, we need to show that

$$\hat{\Psi} \, ; \, \hat{p} \sqsubseteq \hat{c} \, ; \, \hat{\Psi}$$

We therefore define

$$\mathcal{C} p c \Psi \stackrel{def}{=} \hat{\Psi} \, ; \, \hat{p} \sqsubseteq \hat{c} \, ; \, \hat{\Psi}$$

The task of proof-oriented compiler design is to develop a mathematical theory of the predicate $\mathcal{C}$. This should include enough theorems to enable the implementor to select correct object code for each construct of the programming language. It may be that the theory allows choice between several different codes for the same source program; this gives some scope for later optimisation by selecting the most efficient alternative.

As example of the kind of theorem by which the designer specifies a compiler, consider the following possible theorems (using $<$ and $>$ as sequence brackets and $\cdot$ for catenation):

$$\mathcal{C}(x := y) \; < \mathtt{load}(\Psi y), \; \mathtt{store}(\Psi x) > \; \Psi$$
$$\mathcal{C}\mathtt{SKIP} <> \; \Psi$$
$$\mathcal{C}(p1 \, ; p2)\,(c1 \cdot c2)\,\Psi \qquad \textit{whenever } \mathcal{C}p1 \, c1 \, \Psi \textit{ and } \mathcal{C}p2 \, c2 \, \Psi$$

When these have been proved, the implementor of the compiler knows that a simple assignment can be translated into the pair of commands shown above; that a $\mathtt{SKIP}$ command translates to an empty code sequence; and that sequential composition can be translated by concatenating the translations of its two components.

The form of these theorems is extraordinarily similar to that of a recursively defined Boolean function, which could be used to check that a particular compilation of a safety critical program has been successful. It is also quite similar to a logic program, which could be useful if a prototype of the compiler is needed, perhaps for bootstrapping purposes. It is also similar to a conventional procedural compiler, structured according to the principle of recursive descent. Finally, it is quite similar to an attribute grammar for the language; and this permits the use of standard techniques for splitting a compiler into an appropriate number of passes. Thus the collection of theorems serves as an appropriate interface between the designers of a compiler and its implementors.

# 2  Interfaces

The correctness condition has three free variables ($p$, $\Psi$, and $c$), and is comprehensible only in a formal framework in which the following definitions are available

1. the semantics of the language HL, which defines the relation between a program $p$ and its behaviour $\hat{p}$.

2. the architecture of the machine, which defines the relationship between machine code $c$ in the language ML and its behaviour $\hat{c}$.

3. the improvement relation $\sqsubseteq$ and the transformation $\hat{\Psi}$.

Our first task is to choose the most appropriate framework within which to formalise the definition of the details of HL, ML and $\sqsubseteq$. In this selection, the most important criterion is that each formal definition must be suitable as the interface to another group of engineers responsible for its implementation and/or use. Furthermore each definition should be self-sufficient for the needs of each class of user, and protect them from the details and complexities of the other interfaces. This means that there will be a significant intellectual gap between the interfaces; and so closing each gap will be a significant mathematical achievement, not a mere recoding exercise. All responsibility for mutual consistency of the interfaces should be taken by the designers of the high level language and its compiler. Of course, in the actual implementation of the compiler there may be many internal interfaces, for example between successive compilation passes and the loader. Such interfaces are of no concern to any user of the compiler, and will not be treated in this paper.

## 2.1  The High Level Language

In the design of an implementation of a programming language, the first and absolute requirement is a perfect comprehension of its meaning. If the implementation is to be supported by mathematical proof, the meaning must be expressed by some mathematical definition which forms the basis of the reasoning. A wide variety of formalisms have been proposed for this purpose, and there is difficulty in choosing between them.

The user of a programming language also needs a sufficient understanding of it to discharge an obligation to deliver programs which meet their specification. The specification of a program can be expressed as a mathematical predicate, describing observable characteristics of the behaviour of the program when executed. For example, in the case of a conventional language the specification states the desired relationship between the initial and the final values of the global variables of the program, together with any interactions (input or output) in which it engages during execution. From the users' point of view, the best formalisation of the meaning of a programming language is one that relates programs to the specifications which they satisfy.

The specification-oriented semantics of the programming language formally defines the relation between $p$ and $\hat{p}$ in the introduction: $\hat{p}$ is just the strongest behavioural specification satisfied by $p$. However in mathematics it is permissible (and indeed universal practice) to use notations directly to stand for what they mean. We can profit enormously by this convention if we just allow the program text itself to mean its own strongest specification; thus we define ^ on programs as the identity function:

$$\hat{p} \stackrel{def}{=} p$$

## 2.2 Machine Language

The whole purpose of designing a compiler is that programmers should not write in machine code. It is the responsibility of the compiler to ensure that execution of the target code should have the same (or better) behaviour than that ascribed to the source code. So it is important that the applications programmer should never need to see the specification of the machine code.

The true user of the formal specification of the target machine code is the hardware engineer, who is responsible for design of the logic of the computer itself, and for certifying that it meets its specification. The formalism that most simplifies this task will be highly operational, including details of the structure of the registers, storage, and input/output devices, and of how they are updated by each instruction. Experience shows that a direct and widely understood way of describing these things is by the notations of a simple procedural programming language.

Apart from familiarity and operational bias, there is another advantage of procedural notations, that of modularity. The definition can start with a simple subset of the components of the machine state, and the subset of instructions which refer to them. New components can later be added, as required to deal with the further instructions which update them. All the earlier simpler program fragments leave the new components unchanged; they remain correct, and their proofs are still valid, and do not have to be changed when moving from the simpler instruction set to the more complex.

But a decisive advantage of defining the target machine code in a high level language can be obtained if the language used for writing the interpreter is actually the source language itself, whose semantics are already known. So the relation between the object code $c$ and its behaviour $\hat{c}$ is that $\hat{c}$ is a program written in a high level language. Its initial state is known to contain the object code $c$ produced by the compiler, and its task is to interpret that code, engaging in the input and output instructions and producing eventually the final machine state. It is the task of the hardware designer to implement the interpreter directly by the microcode and data paths of a computer.

## 2.3 The Improvement Relation

The improvement relation is now one that must be defined between two programs $p$ and $\hat{c}$; both of them are expressed in the same high level programming language, which has a known specification-oriented semantics. The only possible definition of improvement must be in terms of specifications; the better program must satisfy every specification satisfied by the worse program, and maybe more. Thus whenever the worse program $p$ is known to be adequate for a particular task, the better one $q$ will be too. The usual notation for this is

$$p \sqsubseteq q .$$

This kind of direct comparison is only possible between programs operating on the same global variables, and specifications expressed in terms of them. But our interpreter $\hat{c}$

operates on variables denoting registers and store of the target computer, whereas the source program $p$ operates on variable identifiers usually chosen by the programmer. In order to compare them, we need to know the relationship between the programmer's variable names and the machine locations to which they have been allocated by the compiler. This is defined by the compiler's symbol table $\Psi$, which maps each variable of the program $p$ to the address of the storage location allocated to hold its value. We therefore define $\hat{\Psi}$ as a sort of symbolic dump; it assigns to each high level variable $x$ the value of its corresponding storage location, thereby effecting a transformation from the low level machine state to the implicit state of the high level language program. Thus $\hat{\Psi}$ can be written as an assignment

$$x := M[\Psi x]; \ y := M[\Psi y]; \ldots$$

where $x, y \ldots$ are program variables and $M$ is an array representing the main store of the target machine. Now we can define the concept of improvement as one that holds between programs when their corresponding states have been transformed by $\hat{\Psi}$:

$$\hat{\Psi} ; p \sqsubseteq \hat{c} ; \hat{\Psi}$$

On the left hand of this formula, $\hat{\Psi}$ initialises the program variables from the target machine store; and on the right hand it extracts the final values of the variables after execution of the object code. It may seem strange that the compiler itself is allowed to determine $\Psi$, which plays such an important role in the proof of its own correctness. For example, is there a danger that $\hat{\Psi}$ might be the program `Abort`, which would make the compilation relation trivially true? Fortunately, the construction of $\hat{\Psi}$ avoids these and other dangers.

It is one of the goals of this paper to make a convincing case that the theorems being proved are actually relevant and conducive to the correctness of the compiler. For this, no formal proof could ever be given, since then the formalism of this proof would need to be convincingly argued instead.

## 2.4  Proof Strategy

The compilation predicate is defined as an inequation. The easiest way to prove such inequations is by algebraic transformations and simplifications. Such transformations can readily be checked or even performed by computer. But each transformation must be justified by an algebraic law which is valid for the language in which the reasoning is conducted. So these algebraic laws must themselves be proved from the specification-oriented semantics of the programming language.

Thus it seems a good idea to separate from the task of proving a particular compiler the more general task of deriving a sufficient set of algebraic laws applicable to all programs expressed in the language. These laws are of especial value in optimisation of the source or target codes. The sufficiency of such a set of algebraic laws can be established by an appropriate kind of normal form theorem. In effect, a complete set of laws can act as an algebraic specification of the meaning of the high level programming language; and one which is far more suited to the needs of compiler design than the specification-oriented semantics from which they were derived. Another advantage of algebraic laws is that of modularity and generality: each of them is valid in many different programming languages, and they often remain valid when the language is extended.

The intellectual gap between a specification-oriented semantics and an algebraic semantics is a significant one, and well worth treating as a separate task, which should ideally be carried out by the designers of the programming language. In the remainder of this paper, we will assume that the task has already been completed. In the next section we state without proof the algebraic laws which we will assume to be true of our chosen source language, and state some general theorems that can be proved from them. The following sections will show how the theorems can be used to prove a simple compiler.

# 3 Algebraic Laws

The basic laws defining Dijkstra's language [3] of guarded commands are given in [5]. Some of the more useful ones are repeated here for convenience. We take the simplifying view that all expressions always deliver a value.

Sequential composition is associative and has unit SKIP.

**Law 1**     (1) $(p\,;\,q)\,;\,r\,=\,p\,;(q\,;\,r)$
         (2) $p\,;\,\text{SKIP}\,=\,\text{SKIP}\,;\,p\,=\,p$

Conditionals are coproducts.

**Law 2**     (1)  **if** $true$ **then** $p$ **else** $q\,=\,p\,=\,$ **if** $false$ **then** $q$ **else** $p$
         (2) (**if** $b$ **then** $p$ **else** $q$)$\,;\,r\,=\,$ **if** $b$ **then** $(p;r)$ **else** $(q;r)$
         (3)  **if** $b$ **then** $p$ **else** $p\,=\,p$
         (4)  **if** $b$ **then** (**if** $b\,\vee\,c$ **then** $p$ **else** $q$) **else** $r\,=\,$ **if** $b$ **then** $p$ **else** $r$
         (5)  **if** $b$ **then** (**if** $c$ **then** $p$ **else** $q$) **else** $r\,=$
            **if** $c$ **then** (**if** $b$ **then** $p$ **else** $r$) **else** (**if** $b$ **then** $q$ **else** $r$)
         (6)  **if** $b$ **then** $p$ **else** $q\,=\,$ **if** $\neg\,b$ **then** $q$ **else** $p$

We do not deal in this paper with assignment to subscripted variables. However the following laws are still valid if $x$ and $e$ contain subscripted variables, provided the subscripts are constants, not subject to change by program.

**Law 3**     (1) If $e$ does not mention $x$, $(x := e\,;\,y := e)\,=\,(y := e\,;\,x := y)$
         (2) $(x := e\,;\,x := f(x))\,=\,(x := f(e))$
         (3) $(x := x)\,=\,\text{SKIP}$
         (4) $(x,\,y := e,\,y)\,=\,(x := e)$
         (5) $(x := e\,;\,\textbf{if}\,b(x)\,\textbf{then}\,p\,\textbf{else}\,q)\,=$
            (**if** $b(e)$ **then** $(x := e\,;\,p)$ **else** $(x := e\,;\,q))$
         (6) If $x$ and $y$ are distinct, and $e$ does not mention $y$, and $x$ does not appear
            in $f$   $(x := e\,;\,y := f)\,=\,(y := f\,;\,x := e)$

$\sqsubseteq$ is an $\omega$-complete partial order, i.e., it is reflexive, transitive and antisymmetric, and any ascending chain $\{p_i\}$ has a least upper bound $\sqcup_i p_i$ satisfying

$$\sqcup_i p_i \sqsubseteq q \qquad \textit{iff} \qquad \textit{for all } i : p_i \sqsubseteq q$$

Sequential composition is continuous, i.e., it preserves the least upper bound of ascending chains.

**Law 4** $\qquad p \,;\, \sqcup_n q_n \,;\, r \,=\, \sqcup_n (p \,;\, q_n \,;\, r)$

`Abort` is the bottom of $\sqsubseteq$ and left zero of sequential composition.

**Law 5** $\qquad$ (1) $\texttt{Abort} \sqsubseteq p$
$\qquad\qquad$ (2) $\texttt{Abort} \,;\, p = \texttt{Abort}$
$\qquad\qquad$ (3) $p \,;\, \texttt{Abort} = \texttt{Abort}, \quad$ provided that $p$ contains neither input nor output.

$\sqsubseteq$ has a greatest lower bound operator $\sqcap$, representing non-deterministic choice:

$$(r \sqsubseteq p \textit{ and } r \sqsubseteq q) \qquad \textit{iff} \qquad r \sqsubseteq (p \sqcap q)$$

From this it follows that

$$(p \sqcap q) \sqsubseteq p \textit{ and } p \sqcap q = q \sqcap p$$

Sequential composition distributes through non-deterministic choice.

**Law 6** $\qquad p \,;\, (q_1 \sqcap q_2) \,;\, r = (p \,;\, q_1 \,;\, r) \sqcap (p \,;\, q_2 \,;\, r)$

Let $D$ be a nonempty set of values. The command $x :\in D$ assigns to $x$ the value of an arbitrary member of $D$.

**Law 7** $\qquad$ (1) $x := e \,=\, x :\in \{e\}$
$\qquad\qquad$ (2) $(x := e \,;\, y :\in f^{-1}(x)) \,=\, (y :\in f^{-1}(e) \,;\, x := e)$
$\qquad\qquad\qquad$ provided that $x$ and $y$ are distinct, and $y$ is not in $e$ and $f$ is surjective.
$\qquad\qquad$ (3) $y :\in f^{-1}(fy) \sqsubseteq \texttt{SKIP}$ if $f$ is total.

The command **var** $x$ introduces a new variable $x$, and the command **end** $x$ ends the scope of $x$. Declaration and end commands obey the following laws.

**Law 8** $\qquad$ (1) If $p$ contains no occurrence of $x$, then
$\qquad\qquad\qquad p \,;\, \mathbf{var}\ x \,=\, \mathbf{var}\ x \,;\, p, \qquad\qquad p \,;\, \mathbf{end}\ x \,=\, \mathbf{end}\ x \,;\, p$
$\qquad\qquad$ (2) If $x$ does not appear in the expression $b$ then
$\qquad\qquad\qquad (\mathbf{var}\,x \,;\, \mathbf{if}\ b\ \mathbf{then}\ p\ \mathbf{else}\ q) \,=\, \mathbf{if}\ b\ \mathbf{then}\ (\mathbf{var}\,x \,;\, p)\ \mathbf{else}\ (\mathbf{var}\,x \,;\, q)$
$\qquad\qquad\qquad (\mathbf{end}\,x \,;\, \mathbf{if}\ b\ \mathbf{then}\ p\ \mathbf{else}\ q) \,=\, \mathbf{if}\ b\ \mathbf{then}\ (\mathbf{end}\,x \,;\, p)\ \mathbf{else}\ (\mathbf{end}\,x \,;\, q)$
$\qquad\qquad$ (3) Initialisation of a variable can only make a program more deterministic, and the final value of a variable is irrelevant.
$\qquad\qquad\qquad \mathbf{var}\,x \sqsubseteq \mathbf{var}\ x \,;\, x :\in D$
$\qquad\qquad\qquad \mathbf{end}\,x \,=\, x :\in D \,;\, \mathbf{end}\ x$
$\qquad\qquad$ (4) Let $U$ be the domain of the variable $x$, then
$\qquad\qquad\qquad \mathbf{end}\ x \,;\, \mathbf{var}\ x \,=\, x :\in U \sqsubseteq \texttt{SKIP}$
$\qquad\qquad\qquad \mathbf{var}\ x \,;\, \mathbf{end}\ x \,=\, \texttt{SKIP}$
$\qquad\qquad$ (5) If $x$ and $y$ are distinct, then

$$\mathbf{var}\ x\ ;\ \mathbf{var}\ y\ =\ \mathbf{var}\ x, y\ =\ \mathbf{var}\ y\ ;\ \mathbf{var}\ x$$
$$\mathbf{end}\ x\ ;\ \mathbf{end}\ y\ =\ \mathbf{end}\ x, y\ =\ \mathbf{end}\ y\ ;\ \mathbf{end}\ x$$

We define an assertion as causing abortion if false

$$\{b\}\ \overset{def}{=}\ \mathbf{if}\ b\ \mathbf{then}\ \mathtt{SKIP}\ \mathbf{else}\ \mathtt{Abort}$$

**Law 9**　　　(1) $\{b\} \sqsubseteq \mathtt{SKIP}$
　　　　　　(2) If $e$ does not contain $x$ then
$$x := e\ =\ (x := e\ ;\ \{x = e\})\quad and\quad (\{x = e\}\ ;\ x := e)\ =\ \{x = e\}$$

Least fixed points are defined for a program $P(X)$ containing the program variable $X$:

$$\mu X.\ P(X)\ \overset{def}{=}\ \textstyle\bigsqcup_i P^i(\mathtt{Abort})$$

where
$$P^0(X)\ \overset{def}{=}\ X$$
$$P^{k+1}(X)\ \overset{def}{=}\ P(P^k(X))$$

Algebraically they can be characterized by the following laws

**Law 10**　　　(1) $\mu X.\ P(X)\ =\ P(\mu X.P(X))$
　　　　　　(2) If $P(Y) \sqsubseteq Y$ then $\mu X.\ P(X) \sqsubseteq Y$

The **while** is defined

$$\mathbf{while}\ b\ \mathbf{do}\ p\ \overset{def}{=}\ \mu X.\ \mathbf{if}\ b\ \mathbf{then}\ (p\ ;\ X)\ \mathbf{else}\ \mathtt{SKIP}$$

**Law 11**　　　$\mathbf{while}\ b\ \mathbf{do}\ p\ =\ \mathbf{if}\ b\ \mathbf{then}\ (p\ ;\ \mathbf{while}\ b\ \mathbf{do}\ p)\ \mathbf{else}\ \mathtt{SKIP}$

Proof: Direct from the definition of loop and law 10(1).

The following law is surprisingly important, mainly in proving the correctness of sequential composition.

**Law 12**　　　If $b \Rightarrow c$ then $\mathbf{while}\ b\ \mathbf{do}\ p\ ;\ \mathbf{while}\ c\ \mathbf{do}\ p\ =\ \mathbf{while}\ c\ \mathbf{do}\ p$

Proof:　　　$\{LHS \sqsubseteq RHS\}$:

Let $P(X) \overset{def}{=} \mathbf{if}\ b\ \mathbf{then}\ (p\ ;\ X)\ \mathbf{else}\ \mathtt{SKIP}$ and prove by induction that

$$P^n(\mathtt{Abort})\ ;\ RHS\ \sqsubseteq\ RHS$$

From law 5(1) and 5(2) it follows that

$$P^0(\mathtt{Abort})\ ;\ RHS\ =\ \mathtt{Abort}\ \sqsubseteq\ RHS$$

By induction one has

$$P^{n+1}(\mathtt{Abort})\ ;\ RHS$$

$$= \quad \{def \ of \ P^{n+1}\}$$
$$( \ \textbf{if} \ b \ \textbf{then} \ (p; \ P^n(\text{Abort})) \ \textbf{else} \ \text{SKIP}); \ RHS$$
$$= \quad \{law \ 2(2) \ and \ 1(2)\}$$
$$\textbf{if} \ b \ \textbf{then} \ (p; \ P^n(\text{Abort}); \ RHS) \ \textbf{else} \ RHS$$
$$\sqsubseteq \quad \{induction \ hypothesis, \ monotonicity\}$$
$$\textbf{if} \ b \ \textbf{then} \ (p; \ RHS) \ \textbf{else} \ RHS$$
$$= \quad \{b \ \Rightarrow \ c \ and \ law \ 2(4)\}$$
$$\textbf{if} \ b \ \textbf{then} \ ( \ \textbf{if} \ c \ \textbf{then} \ (p; \ RHS) \ \textbf{else} \ \text{SKIP}) \ \textbf{else} \ RHS$$
$$= \quad \{law \ 11\}$$
$$\textbf{if} \ b \ \textbf{then} \ RHS \ \textbf{else} \ RHS$$
$$= \quad \{law \ 2(3)\}$$
$$RHS$$

By the continuity of sequential composition (law 4) we conclude

$$LHS = \mu X.\, P(X);\ RHS \sqsubseteq RHS$$

$\{RHS \sqsubseteq LHS\}$: $LHS$

$$= \quad \{law \ 11\}$$
$$( \ \textbf{if} \ b \ \textbf{then} \ (p; \ \textbf{while} \ b \ \textbf{do} \ p) \ \textbf{else} \ \text{SKIP}); \ RHS$$
$$= \quad \{law \ 2(2) \ and \ 1(2)\}$$
$$\textbf{if} \ b \ \textbf{then} \ (p; \ LHS) \ \textbf{else} \ RHS$$
$$= \quad \{b \ \Rightarrow \ c \ and \ law \ 2(4)\}$$
$$\textbf{if} \ b \ \textbf{then} \ (\textbf{if} \ c \ \textbf{then} \ (p; \ LHS) \ \textbf{else} \ \text{SKIP}) \ \textbf{else} \ RHS$$
$$= \quad \{law \ 11\}$$
$$\textbf{if} \ b \ \textbf{then} \ (\textbf{if} \ c \ \textbf{then} \ (p; \ LHS) \ \textbf{else} \ \text{SKIP})$$
$$\textbf{else} \ (\textbf{if} \ c \ \textbf{then} \ (p; \ RHS) \ \textbf{else} \ \text{SKIP})$$
$$\sqsupseteq \quad \{RHS \sqsupseteq LHS, \ monotonicity\}$$
$$\textbf{if} \ b \ \textbf{then} \ (\textbf{if} \ c \ \textbf{then} \ (p; \ LHS) \ \textbf{else} \ \text{SKIP})$$
$$\textbf{else} \ (\textbf{if} \ c \ \textbf{then} \ (p; \ LHS) \ \textbf{else} \ \text{SKIP})$$
$$= \quad \{law \ 2(3)\}$$
$$\textbf{if} \ c \ \textbf{then} \ (p; \ LHS) \ \textbf{else} \ \text{SKIP}$$

From law 10(2) it follows that

$$LHS \sqsupseteq \mu X.\,( \ \textbf{if} \ c \ \textbf{then} \ (p; \ X) \ \textbf{else} \ \text{SKIP}) = RHS$$

**Law 13**  \quad If $\neg b(e)$ then $(x := e; \ \textbf{while} \ b(x) \ \textbf{do} \ p) \ = \ x := e$

Proof: \quad $LHS$
$$= \quad \{law \ 11\}$$

254

$$x := e \; ; \; \textbf{if } b(x) \textbf{ then } (p \, ; \, \textbf{while } b(x) \textbf{ do } p) \textbf{ else } \text{SKIP}$$
$= \quad \{law \; 3(5) \; and \; 1(2)\}$
$$\textbf{if } b(e) \textbf{ then } (x := e \, ; \, p \, ; \, \textbf{while } b(x) \textbf{ do } p) \textbf{ else } x := e$$
$= \quad \{given \; \neg \, b(e) \; and \; law \; 2(1)\}$
$$RHS$$

The following law enables us to transform a tail-recursion to a loop

**Law 14**       If $r$ contains neither input nor output then
$$\mu X.r(r \, ; \, \textbf{if } b \textbf{ then } (p \, ; \, X) \textbf{ else } q) = (r \, ; \, \textbf{while } b \textbf{ do } (p \, ; \, r)) \, ; \, q$$

Proof: Let
$$P(X) \;\overset{def}{=}\; r \, ; \, \textbf{if } b \textbf{ then } (p \, ; \, X) \textbf{ else } q$$
$$Q(X) \;\overset{def}{=}\; \textbf{if } b \textbf{ then } (p \, ; \, r \, ; \, X) \textbf{ else } \text{SKIP}$$

and prove by induction that for all $k \geq 0$
$$P^k(\text{Abort}) = r \, ; \, Q^k(\text{Abort}) \, ; \, q$$

For $k = 0$ one has
$$P^0(\text{Abort})$$
$= \quad \{def \; of \; P^0\}$
$$\text{Abort}$$
$= \quad \{antecedent \; and \; law \; 5(2) \; and \; 5(3)\}$
$$r; \text{Abort} \, ; \, q$$
$= \quad \{def \; of \; Q^0\}$
$$r \, ; \, Q^0(\text{Abort}) \, ; \, q$$

By induction one has
$$P^{k+1}(\text{Abort})$$
$= \quad \{def \;\; P(X)\}$
$$r \, ; \, \textbf{if } b \textbf{ then } (p \, ; \, P^k(\text{Abort})) \textbf{ else } q$$
$= \quad \{induction \; hypothesis\}$
$$r \, ; \, \textbf{if } b \textbf{ then } (p \, ; \, r \, ; \, Q^k(\text{Abort}) \, ; \, q) \textbf{ else } q$$
$= \quad \{law \; 2(2)\}$
$$r \, ; \, \textbf{if } b \textbf{ then } (p \, ; \, r \, ; \, Q^k(\text{Abort})) \textbf{ else } \text{SKIP} \; ; \; q$$
$= \quad \{def \; of \; Q(X)\}$
$$r \, ; \, Q^{k+1}(\text{Abort}) \, ; \, q$$

which leads to
$$LHS$$
$= \quad \{def \;\; \mu X.P(X)\}$
$$\sqcup_k P^k(\text{Abort})$$
$= \quad \sqcup_k(r \, ; \, Q^k(\text{Abort}) \, ; \, q)$
$= \quad \{law \; 4\}$
$$r \; ; \; \sqcup_k Q^k(\text{Abort}) \; ; \; q$$
$= \quad \{def \;\; \textbf{while } b \textbf{ do } p\}$
$$RHS \qquad\qquad\qquad\qquad \square$$

# 4   The Interpreter

In our simple example, we shall treat a machine with just four components.

$m : rom \rightarrow instruction$ is the store occupied by the code of the program.

$M : ram \rightarrow word$        is the store used for variables (word-length unspecified).

$P : rom$            is the pointer to the current instruction.

$A : word$          is the only general-purpose register.

We will consider only four instructions, each defined by a fragment of code describing its effect for all $n : ram$ and $j : rom$.

$$\text{effect}(load(n)) \stackrel{def}{=} (A := M[n] \,;\, P := P + 1)$$

$$\text{effect}(store(n)) \stackrel{def}{=} (M[n] := A \,;\, P := P + 1)$$

$$\text{effect}(jump(j)) \stackrel{def}{=} P := j$$

$$\text{effect}(cond(j)) \stackrel{def}{=} \textbf{if } A \textbf{ then } P := P + 1 \textbf{ else } P := j$$

For all undefined instructions $i$, we define $\text{effect}(i) \stackrel{def}{=} \texttt{Abort}$. This places upon the compiler designer the obligation to ensure that such an instruction is never executed. More importantly, it leaves open the option of later definition of additional instructions. All earlier proofs remain valid, because the interpreter can only be improved by making it $\texttt{Abort}$ less often.

We have now defined the effect of each single machine instruction. But an interpreter is required to interpret a whole sequence of instructions held in $m$ between two specified locations, say $s$ and $f$. The interpreter $\mathcal{I}sf$ accomplishes means of a loop

$$\mathcal{I}sf \stackrel{def}{=} P := s \,;\, \textbf{while } P < f \textbf{ do } step \,;\, \{P = f\}$$

where $step$ abbreviates $\text{effect}(m[P])$, and satisfies

**Law 15**      If $e$ does not mention $P$, then $(P := e \,;\, step) = (P := e \,;\, \text{effect}(m[e]))$

The final assertion $\{P = f\}$ places upon the compiler designer an obligation tocompile code which exits normally through its end, without any wild forward jumps. This is essential to prove correctness of sequential composition. A hardware implementation of the interpreter is permitted to ignore the assertions; that can only make it better than the one which is known to be correct.

Elementary facts about the interpreter can be derived by symbolic execution. For example

**Lemma 1**      (1) $\mathcal{I}ss = P := s$

                       (2) If $m[f - 1] = jump(s)$ then $\mathcal{I}(f - 1)f = \mathcal{I}sf$

                       (3) If $s < f$ and $m[s] = jump(f)$ then $\mathcal{I}sf = P := f$

                       (4) If $m[s] = load(l)$ and $m[s + 1] = store(n)$

                            then $\mathcal{I}s(s + 2) = A := M[l] \,;\, M[n] := M[l] \,;\, P := s + 2$

                       (5) If $j - s > 1$ and $m[s] = load(l)$ and $m[s + 1] = cond(j)$ then

                            $(P := s \,;\, \textbf{while } P < s + 2 \textbf{ do } step) =$

                            $(A := M[l] \,;\, \textbf{if } M[l] \textbf{ then } P := s + 2 \textbf{ else } P := j)$

Proof:    (1)      $\mathcal{I}ss$

$$= \{def\ of\ \mathcal{I}\}$$
$$P := s\,;\ \textbf{while}\ P < s\ \textbf{do}\ step\,;\ \{P = s\}$$
$$= \{law\ 13\}$$
$$P := s\,;\ \{P = s\}$$
$$= \{law\ 9(2)\}$$
$$P := s$$

$$(2)\qquad \mathcal{I}(f-1)f$$
$$= \{def\ of\ \mathcal{I}\ and\ law\ 11\}$$
$$P := f-1\,;\ \textbf{if}\ P < f\ \textbf{then}\ (step\,;\ \textbf{while}\ P < f\ \textbf{do}\ step)$$
$$\textbf{else}\ \text{SKIP}\,;\ \{P = f\}$$
$$= \{law\ 3(5)\ and\ 1(2)\}$$
$$\textbf{if}\ f-1 < f\ \textbf{then}\ (P := f-1\,;\ step\,;\ \textbf{while}\ P < f\ \textbf{do}\ step)$$
$$\textbf{else}\ P := f-1\,;\ \{P = f\}$$
$$= \{law\ 2(1)\}$$
$$P := f-1\,;\ step\,;\ \textbf{while}\ P < f\ \textbf{do}\ step\,;\ \{P = f\}$$
$$= \{law\ 15\ and\ def\ of\ jump\}$$
$$P := f-1\,;\ P := s\,;\ \textbf{while}\ P < f\ \textbf{do}\ step\,;\ \{P = f\}$$
$$= \{law\ 3(2)\}$$
$$P := s\,;\ \textbf{while}\ P < f\ \textbf{do}\ step\,;\ \{P = f\}$$
$$= \{def\ of\ \mathcal{I}\}$$
$$\mathcal{I}sf$$

The remaining conclusions can be established in a similar way.

**Lemma 2** $\qquad \mathcal{I}ss\,;\ \mathcal{I}sf\ =\ \mathcal{I}sf\ =\ \mathcal{I}sf\,;\ \mathcal{I}ff$

Proof:
$$\mathcal{I}ss\,;\ \mathcal{I}sf$$
$$= \{lemma\ 1(1)\ and\ def\ of\ \mathcal{I}\}$$
$$P := s\,;\ P := s\,;\ \textbf{while}\ P < f\ \textbf{do}\ step\,;\ \{P = f\}$$
$$= \{law\ 3(2)\ and\ def\ of\ \mathcal{I}\}$$
$$\mathcal{I}sf$$
$$= \{def\ of\ \mathcal{I}\ and\ law\ 9(2)\}$$
$$P := s\,;\ \textbf{while}\ P < f\ \textbf{do}\ step\,;\ \{P = f\}\,;\ P := f$$
$$= \{def\ of\ \mathcal{I}\ and\ lemma\ 1(1)\}$$
$$\mathcal{I}sf\,;\ \mathcal{I}ff$$

In the remainder of this paper we will use $m[s : f)$ to stand for the machine code held in $m$ between $s$ and $f$. We now design fragments of machine code corresponding to each control construct of HL. In each case, we need to prove that $\mathcal{I}$ distributes inward to its components.

**Lemma 3**     If $j \leq f$ then $\mathcal{I}sf \sqsupseteq \mathcal{I}sj ; \mathcal{I}jf$

Proof:     $\mathcal{I}sj ; \mathcal{I}jf$

$= \quad \{def\ of\ \mathcal{I}\}$

$\quad\quad P := s ;\ \textbf{while}\ P < j\ \textbf{do}\ step ; \{P = j\}$

$\quad\quad P := j ;\ \textbf{while}\ P < f\ \textbf{do}\ step ; \{P = f\}$

$= \quad \{law\ 9(2)\}$

$\quad\quad P := s ;\ \textbf{while}\ P < j\ \textbf{do}\ step ; \{P = j\}$

$\quad\quad \textbf{while}\ P < f\ \textbf{do}\ step ; \{P = f\}$

$\sqsubseteq \quad \{law\ 9(1)\ and\ 1(2)\}$

$\quad\quad P := s ;\ \textbf{while}\ P < j\ \textbf{do}\ step ;$

$\quad\quad \textbf{while}\ P < f\ \textbf{do}\ step ; \{P = f\}$

$= \quad \{antecedent\ and\ law\ 12\}$

$\quad\quad P := s ;\ \ \textbf{while}\ P < f\ \textbf{do}\ step ; \{P = f\}$

$= \quad \{def\ of\ \mathcal{I}\}$

$\quad\quad \mathcal{I}sf$

**Corollary**     If $j \leq f$ then

$\quad\quad \mathcal{I}sf = P := s ;\ \textbf{while}\ P < j\ \textbf{do}\ step ;\ \textbf{while}\ P < f\ \textbf{do}\ step ; \{P = f\}$

**Lemma 4**     If $s + 2 < j \leq f$

and $m[s] = load(n)$ and $m[s + 1] = cond(j)$ and $m[j - 1] = jump(f)$

then   $\mathcal{I}sf \sqsupseteq A := M[n] ;\ \textbf{if}\ M[n]\ \textbf{then}\ (\mathcal{I}(s{+}2)(j{-}1) ; \mathcal{I}ff)\ \textbf{else}\ \mathcal{I}jf$

Proof:     $\mathcal{I}sf$

$= \quad \{antecedent\ and\ corollary\ of\ lemma\ 3\}$

$\quad\quad P := s ;\ \textbf{while}\ P < s + 2\ \textbf{do}\ step ;$

$\quad\quad \textbf{while}\ P < f\ \textbf{do}\ step ; \{P = f\}$

$= \quad \{lemma\ 1(5)\}$

$\quad\quad A := M[n] ;\ \textbf{if}\,M[n]\ \textbf{then}\ P := s + 2\ \textbf{else}\ P := j ;$

$\quad\quad \textbf{while}\ P < f\ \textbf{do}\ step ;\ \{P = f\}$

$= \quad \{law\ 2(2)\ and\ def\ of\ \mathcal{I}\}$

$\quad\quad A := M[n] ;\ \textbf{if}\ M[n]\ \textbf{then}\ \mathcal{I}(s + 2)f\ \textbf{else}\ \mathcal{I}jf$

$\sqsupseteq \quad \{antecedent\ and\ lemma\ 3\}$

$\quad\quad A := M[n] ;\ \textbf{if}\ M[n]\ \textbf{then}\ (\mathcal{I}(s + 2)(j - 1) ; \mathcal{I}(j - 1)f)\ \textbf{else}\ \mathcal{I}jf$

$= \quad \{antecedent\ and\ lemma\ 1(3)\}$

$\quad\quad A := M[n] ;\ \textbf{if}\ M[n]\ \textbf{then}\ (\mathcal{I}(s + 2)(j - 1) ; P := f)\ \textbf{else}\ \mathcal{I}jf$

$= \quad \{lemma\ 1(1)\}$

$\quad\quad A := M[n] ;\ \textbf{if}\ M[n]\ \textbf{then}\ (\mathcal{I}(s + 2)(j - 1) ; \mathcal{I}ff)\ \textbf{else}\ \mathcal{I}jf$

**Lemma 5**     If $s + 1 < f - 1$
and $m[s] = load(n)$ and $m[s+1] = cond(f)$ and $m[f-1] = jump(s)$
then $\mathcal{I}sf \sqsupseteq A := M[n]$; **while** $M[n]$ **do** $(\mathcal{I}(s+2)(f-1)$; $A := M[n])$; $\mathcal{I}ff$

Proof:

$\quad\quad\quad \mathcal{I}sf$

$= \quad \{antecedent\ and\ corollary\ of\ lemma\ 3\}$

$\quad\quad P := s$; **while** $P < s + 2$ **do** $step$;

$\quad\quad$ **while** $P < f$ **do** $step$; $\{P = f\}$

$= \quad \{antecedent\ and\ lemma\ 1(5)\}$

$\quad\quad A := M[n]$; **if** $M[n]$ **then** $P := s + 2$ **else** $P := f$;

$\quad\quad$ **while** $P < f$ **do** $step$ ; $\{P = f\}$

$= \quad \{law\ 2(2)\ and\ def\ of\ \mathcal{I}\}$

$\quad\quad A := M[n]$; **if** $M[n]$ **then** $\mathcal{I}(s+2)f$ **else** $\mathcal{I}ff$

$\sqsupseteq \quad \{antecedent\ and\ lemma\ 3\}$

$\quad\quad A := M[n]$; **if** $M[n]$ **then** $\mathcal{I}(s+2)(f-1)$; $\mathcal{I}(f-1)f)$ **else** $\mathcal{I}ff$

$= \quad \{antecedent\ and\ lemma\ 1(2)\}$

$\quad\quad A := M[n]$; **if** $M[n]$ **then** $\mathcal{I}(s+2)(f-1)$; $\mathcal{I}sf)$ **else** $\mathcal{I}ff$

which leads to $\quad \mathcal{I}sf$

$\sqsupseteq \quad \{law\ 10(2)\}$

$\quad\quad \mu X.(A := M[n]$; **if** $M[n]$ **then** $\mathcal{I}(s+2)(f-1)$; $X)$ **else** $\mathcal{I}ff)$

$= \quad \{law\ 14\}$

$\quad\quad A := M[n]$; **while** $M[n]$ **do** $(\mathcal{I}(s+2)(f-1)$; $A := M[n])$; $\mathcal{I}ff$

## 5   The Simulation

The symbol table $\Psi$ is a finite injection

$$\Psi : identifier \mapsto ram$$

We will use $dom(\Psi)$ and $ran(\Psi)$ to denote the domain and the range of $\Psi$ respectively. $ran(\Psi) \vartriangleleft M$ is the store $M$ after removal from its domain of those locations used by the symbol table $\Psi$.

Let $PSTATE$ and $MSTATE$ stand for the program state and the machine state respectively. Define the simulation

$$\hat{\Psi} : MSTATE \rightarrow PSTATE$$

as declaration and initialisation of of all non-local variables of the program, followed by "forgetting" of the machine state.

$$\hat{\Psi} \stackrel{def}{=} \mathbf{var}\ x, y, \ldots, z;$$
$$x, y, \ldots, z := M[\Psi x], M[\Psi y], \ldots, M[\Psi z];$$
$$\mathbf{end}\ P, A, M$$

Since $\hat{\Psi}$ is a surjective mapping from the machine state to the program state, we can define its *inverse* as follows:

$$\hat{\Psi}^{-1} \;\overset{def}{=}\; \mathbf{var}\ P,\ A,\ M\ ;$$
$$M[\Psi x],\ M[\Psi y],\ \ldots,M[\Psi z]\ :=\ x,\ y,\ \ldots,\ z\ ;$$
$$\mathbf{end}\ x,\ y,\ \ldots,z$$

Algebraically, the inverse of the simulation $\hat{\Psi}$ can be characterised by the following laws:

$$\hat{\Psi}^{-1}\,;\,\hat{\Psi}\ =\ \mathtt{SKIP}_{PSTATE} \qquad\qquad \hat{\Psi}\,;\,\hat{\Psi}^{-1}\ \sqsubseteq\ \mathtt{SKIP}_{MSTATE}$$

where $\mathtt{SKIP}_T$ denotes the identity mapping on the set $T$. In the rest of this paper the subscript of $\mathtt{SKIP}$ will be dropped when it can be deduced from the context.

We define the function $[\Psi]$

$$[\Psi](p)\ \overset{def}{=}\ \hat{\Psi}\,;\,p\,;\,\hat{\Psi}^{-1}$$

**Lemma 6** $\qquad \hat{\Psi}\,;\,p\ \sqsubseteq\ q\,;\,\hat{\Psi}\ \Leftrightarrow\ [\Psi](p)\ \sqsubseteq\ q$

Proof:

$$\hat{\Psi}\,;\,p\ \sqsubseteq\ q\,;\,\hat{\Psi}\ \Rightarrow\ \{\textit{composition is monotonic}\}$$
$$(\hat{\Psi}\,;\,p\,;\,\hat{\Psi}^{-1})\ \sqsubseteq\ (q\,;\,\hat{\Psi}\,;\,\hat{\Psi}^{-1})$$
$$\Rightarrow\ \{\hat{\Psi}\,;\,\hat{\Psi}^{-1}\ \sqsubseteq\ \mathtt{SKIP}\ \textit{and law } 1(2)\}$$
$$(\hat{\Psi}\,;\,p\,;\,\hat{\Psi}^{-1})\ \sqsubseteq\ q$$
$$\Rightarrow\ \{\textit{def of } [\Psi]\}$$
$$[\Psi](p)\ \sqsubseteq\ q$$
$$\Rightarrow\ \{\textit{composition is monotonic}\}$$
$$([\Psi](p)\,;\,\hat{\Psi})\ \sqsubseteq\ (q\,;\,\hat{\Psi})$$
$$\Rightarrow\ \{\hat{\Psi}^{-1}\,;\,\hat{\Psi}\ =\ \mathtt{SKIP}\ \textit{and law } 1(2)\}$$
$$\hat{\Psi}\,;\,p\ \sqsubseteq\ q\,;\,\hat{\Psi}$$

From the lemma we conclude that $[\Psi](p)$ is the *weakest specification* of a correct implementation of $p$ under the given symbol table $\Psi$.

**Lemma 7** $\qquad [\Psi](\mathtt{Abort})\ =\ \mathtt{Abort}$

Proof: Since $\hat{\Psi}$ and $\hat{\Psi}^{-1}$ contain no input or output, from law 5 one has
$$\hat{\Psi}\,;\,\mathtt{Abort}\ =\ \mathtt{Abort}\,;\,\hat{\Psi}^{-1}\ =\ \mathtt{Abort}$$

**Lemma 8** $\qquad [\Psi](p\,;\,q)\ =\ [\Psi](p)\,;\,[\Psi](q)$

Proof:

$$[\Psi](p\,;\,q)$$
$$=\ \{\textit{def of } [\Psi]\}$$
$$\hat{\Psi}\,;\,(p\,;\,q)\,;\,\hat{\Psi}^{-1}$$

$$= \{\hat{\Psi}^{-1}; \hat{\Psi} = \text{SKIP } and \ law \ 1(2)\}$$
$$\hat{\Psi}; (p; \hat{\Psi}^{-1}; \hat{\Psi}; q); \hat{\Psi}^{-1}$$
$$= \{law \ 1(1) \ and \ def \ of \ [\Psi]\}$$
$$[\Psi](p); [\Psi](q)$$

**Lemma 9**    $[\Psi](\text{SKIP}) = \hat{\Psi}; \hat{\Psi}^{-1}$

Proof: Direct from law 1(2).

**Lemma 10**    If $e$ does not mention program variables $x, y, \ldots z$, then
(1) $P := e \sqsupseteq [\Psi](\text{SKIP})$
(2) $A := e \sqsupseteq [\Psi](\text{SKIP})$

Proof:    (1)    $P := e$

$\sqsupseteq$  $\{law \ 1(2) \ and \ \text{SKIP} \sqsupseteq \hat{\Psi}; \hat{\Psi}^{-1}\}$
$P := e; \hat{\Psi}; \hat{\Psi}^{-1}$

$=$  $\{def \ of \ \hat{\Psi}\}$
$P := e; \mathbf{var}\ x, y, \ldots, z;$
$x, y, \ldots, z := M[\Psi x], M[\Psi y], \ldots, M[\Psi z];$
$\mathbf{end}\ P, A, M; \hat{\Psi}^{-1}$

$=$  $\{antecedent \ and \ law \ 8(1)\}$
$\mathbf{var}\ x, y, \ldots, z; P := e$
$x, y, \ldots, z := M[\Psi x], M[\Psi y], \ldots, M[\Psi z];$
$\mathbf{end}\ P, A, M; \hat{\Psi}^{-1}$

$=$  $\{antecedent \ and \ law \ 3(6)\}$
$\mathbf{var}\ x, y, \ldots, z;$
$x, y, \ldots, z := M[\Psi x], M[\Psi y], \ldots, M[\Psi z]; P := e$
$\mathbf{end}\ P, A, M; \hat{\Psi}^{-1}$

$=$  $\{law \ 8(3)\}$
$\mathbf{var}\ x, y, \ldots, z;$
$x, y, \ldots, z := M[\Psi x], M[\Psi y], \ldots, M[\Psi z];$
$\mathbf{end}\ P, A, M; \hat{\Psi}^{-1}$

$=$  $\{def \ of \ \hat{\Psi} \ and \ lemma \ 9\}$
$[\Psi](\text{SKIP})$

(2). Similar to (1).

**Lemma 11**    $[\Psi](p \sqcap q) = [\Psi](p) \sqcap [\Psi](q)$

Proof: Direct from distributivity of sequential composition (law 6).

Now we wish to extend the definition of $[\Psi]$ to expressions:

$$[\Psi](e) \;\overset{def}{=}\; e[M[\Psi x]/x,\; M[\Psi y]/y,\; \ldots,\; M[\Psi z]/z]$$

where $e[v/x]$ denotes the expression formed by replacing each free occurrence of $x$ in $e$ by $v$.

**Lemma 12**    $[\Psi](\textbf{if } b \textbf{ then } p \textbf{ else } q) = \textbf{ if } [\Psi](b) \textbf{ then } [\Psi](p) \textbf{ else } [\Psi](q)$

Proof:

$$[\Psi](\textbf{if } b \textbf{ then } p \textbf{ else } q)$$

$= \{def\ of\ [\Psi]\}$

$\hat{\Psi}\;;\; \textbf{if } b \textbf{ then } p \textbf{ else } q\;;\; \hat{\Psi}^{-1}$

$= \{def\ of\ \hat{\Psi}\ and\ law\ 2(2)\}$

$\textbf{var } x,\, y,\, \ldots,\, z\,;$

$x,\, y,\, \ldots,\, z\; :=\; M[\Psi x],\, M[\Psi y],\, \ldots,\, M[\Psi z]\,;$

$\textbf{end } P,\, A,\, M\,;$

$\textbf{if } b \textbf{ then } (p\,;\, \hat{\Psi}^{-1}) \textbf{ else } (q\,;\, \hat{\Psi}^{-1})$

$= \{law\ 8(2)\}$

$\textbf{var } x,\, y,\, \ldots,\, z\,;$

$x,\, y,\, \ldots,\, z\; :=\; M[\Psi x],\, M[\Psi y],\, \ldots,\, M[\Psi z]\,;$

$\textbf{if } b$

$\textbf{then } (\textbf{end } P,\, A,\, M\,;\, p\,;\, \hat{\Psi}^{-1})$

$\textbf{else } (\textbf{end } P,\, A,\, M\,;\, q\,;\, \hat{\Psi}^{-1})$

$= \{law\ 3(5)\ and\ 8(2)\}$

$\textbf{if } b[M[\Psi x]/x,\, M[\Psi y]/y,\, \ldots M[\Psi z]/z]$

$\textbf{then } (\hat{\Psi}\,;\, p\,;\, \hat{\Psi}^{-1})$

$\textbf{else } (\hat{\Psi}\,;\, q\,;\, \hat{\Psi}^{-1})$

$= \{def\ [\Psi]\}$

$\textbf{if } [\Psi](b) \textbf{ then } [\Psi](p) \textbf{ else } [\Psi](q)$

**Lemma 13**    $[\Psi](\textbf{while } b \textbf{ do } p) = \mu Y.(\textbf{if } [\Psi](b) \textbf{ then } ([\Psi](p)\,;\, Y) \textbf{ else } [\Psi](\text{SKIP}))$

Proof:    Let $Y \overset{def}{=} \textbf{while } b \textbf{ do } p$.

$\{RHS \sqsubseteq LHS\}$:    From law 11 we have

$Y\quad =\quad \textbf{if } b \textbf{ then } (p\,;\, Y) \textbf{ else } \text{SKIP}$

$\quad\;\;\Rightarrow\quad \{lemma\ 12\ and\ 8\}$

$$[\Psi](Y) = \textbf{if } [\Psi](b) \textbf{ then } ([\Psi](p)\,;\,[\Psi](Y)) \textbf{ else } [\Psi](\text{SKIP})$$

$\Rightarrow \quad \{law\ 10(2)\}$

$$RHS \sqsubseteq [\Psi](Y)$$

$\Rightarrow \quad \{def\ of\ Y\}$

$$RHS \sqsubseteq LHS$$

$\{LHS \sqsubseteq RHS\}$: First we have

$$[\Psi](p)\,;\,q$$

$= \quad \{def\ of\ [\Psi]\}$

$$\hat{\Psi}\,;\,p\,;\,\hat{\Psi}^{-1}\,;\,q$$

$\sqsupseteq \quad \{law\ 1(2)\ and\ \hat{\Psi}\,;\,\hat{\Psi}^{-1} \sqsubseteq \text{SKIP}\}$

$$\hat{\Psi}\,;\,p\,;\,\hat{\Psi}^{-1}\,;\,q\,;\,\hat{\Psi}\,;\,\hat{\Psi}^{-1}$$

$= \quad \{def\ of\ [\Psi]\}$

$$[\Psi](p\,;\,\hat{\Psi}^{-1}\,;\,q\,;\,\hat{\Psi}) \qquad\qquad (*)$$

Then from law 11 it follows that

$$RHS = \textbf{if } [\Psi](b) \textbf{ then } ([\Psi](p)\,;\,RHS) \textbf{ else } [\Psi](\text{SKIP})$$

$\Rightarrow \quad \{(*)\ and\ monotonicity\}$

$$RHS \sqsupseteq \textbf{if } [\Psi](b) \textbf{ then } [\Psi](p\,;\,\hat{\Psi}^{-1}\,;\,RHS\,;\,\hat{\Psi}) \textbf{ else } [\Psi](\text{SKIP})$$

$\Rightarrow \quad \{lemma\ 12\}$

$$RHS \sqsupseteq [\Psi](\textbf{if } b \textbf{ then } (p\,;\,\hat{\Psi}^{-1}\,;\,RHS\,;\,\hat{\Psi}) \textbf{ else } \text{SKIP})$$

$\Rightarrow \quad \{monotonicity\}$

$$(\hat{\Psi}^{-1}\,;\,RHS\,;\,\hat{\Psi}) \sqsupseteq$$
$$\hat{\Psi}^{-1}\,;\,[\Psi](-\textbf{if } b \textbf{ then } (p\,;\,\hat{\Psi}^{-1}\,;\,RHS\,;\,\hat{\Psi}) \textbf{ else } \text{SKIP})\,;\,\hat{\Psi}$$

$\Rightarrow \quad \{\hat{\Psi}^{-1}\,;\,\hat{\Psi} = \text{SKIP}\ and\ law\ 1(2)\}$

$$(\hat{\Psi}^{-1}\,;\,RHS\,;\,\hat{\Psi}) \sqsupseteq \textbf{if } b \textbf{ then } (p\,;\,\hat{\Psi}^{-1}\,;\,RHS\,;\,\hat{\Psi}) \textbf{ else } \text{SKIP}$$

$\Rightarrow \quad \{law\ 10(2)\}$

$$(\hat{\Psi}^{-1}\,;\,RHS\,;\,\hat{\Psi}) \sqsupseteq Y$$

$\Rightarrow \quad \{monotonicity\}$

$$(\hat{\Psi}\,;\,\hat{\Psi}^{-1}\,;\,RHS\,;\,\hat{\Psi}\,;\,\hat{\Psi}^{-1}) \sqsupseteq (\hat{\Psi}\,;\,Y\,;\,\hat{\Psi}^{-1})$$

$\Rightarrow \quad \{\hat{\Psi}\,;\,\hat{\Psi}^{-1} \sqsubseteq \text{SKIP}\ and\ law\ 1(2)\}$

$$RHS \sqsupseteq [\Psi](Y)$$

$\Rightarrow \quad \{def\ of\ Y\}$

$$RHS \sqsupseteq LHS$$

**Corollary**      If $b$ is a variable

then $[\Psi](\textbf{while } b \textbf{ do } p) = \textbf{while } M[\Psi b] \textbf{ do } [\Psi](p)\,;\,[\Psi](\text{SKIP})$

Proof:      $[\Psi](\textbf{while } b \textbf{ do } p)$

$\qquad\quad = \quad \{lemma\ 13\}$

$$\mu Y.\,(\textbf{if}\,[\Psi](b)\,\textbf{then}\,([\Psi](p)\,;\,Y)\,\textbf{else}\,[\Psi](\texttt{SKIP}))$$
$$=\quad \{antecedent\ and\ [\Psi](e)\}$$
$$\mu Y.\,(\textbf{if}\,M[\Psi b]\,\textbf{then}\,([\Psi](p)\,;\,Y)\,\textbf{else}\,[\Psi](\texttt{SKIP}))$$
$$=\quad \{law\ 14\}$$
$$\textbf{while}\,M[\Psi b]\,\textbf{do}\,[\Psi](p)\,;\,[\Psi](\texttt{SKIP})$$

**Lemma 14**  If $e$ does not mention variables $P$, $A$ and $M$
then $[\Psi](x := e)\ =\ M[\Psi x] := [\Psi](e)\,;\,[\Psi](\texttt{SKIP})$

Proof:  $[\Psi](x := e)$

$=\ \{def\ of\ [\Psi]\}$

**var** $x,\,y,\,\ldots,\,z$;

$x,\,y,\,\ldots,z\ :=\ M[\Psi x],\,M[\Psi y],\,\ldots,\,M[\Psi z]$;

**end** $P,\,A,\,M$ ; $x := e$ ; $\hat{\Psi}^{-1}$

$=\ \{law\ 8(1)\}$

**var** $x,\,y,\,\ldots,\,z$;

$x,\,y,\,\ldots,z\ :=\ M[\Psi x],\,M[\Psi y],\,\ldots,\,M[\Psi z]$; $x := e$;

**end** $P,\,A,\,M$ ; $\hat{\Psi}^{-1}$

$=\ \{law\ 3(2)\ and\ def\ of\ [\Psi](e)\}$

**var** $x,\,y,\,\ldots,\,z$;

$ex,\,y,\,\ldots,z\ :=\ [\Psi](e),\,M[\Psi y],\,\ldots,\,M[\Psi z]$;

**end** $P,\,A,\,M$ ; $\hat{\Psi}^{-1}$

$=\ \{law\ 8(3)\ and\ 7(1)\}$

**var** $x,\,y,\,\ldots,\,z$;

$x,\,y,\,\ldots,z\ :=\ [\Psi](e),\,M[\Psi y],\,\ldots,\,M[\Psi z]$;

$M[\Psi x] := x$; **end** $P,\,A,\,M$ ; $\hat{\Psi}^{-1}$

$=\ \{law\ 3(1),\ which\ applies\ because\ \Psi x\ is\ a\ constant\ subscript\ of\ M\}$

**var** $x,\,y,\,\ldots,\,z$;

$M[\Psi x] := [\Psi](e)$; $x,\,y,\,\ldots,z\ :=\ M[\Psi x],\,M[\Psi y],\,\ldots,\,M[\Psi z]$;

**end** $P,\,A,\,M$ ; $\hat{\Psi}^{-1}$

$=\ \{law\ 8(1)\ and\ lemma\ 9\}$

$M[\Psi x] := [\Psi](e)$; $[\Psi](\texttt{SKIP})$

**Lemma 15**  If $v \in dom(\Psi)$ and $\Phi = \{v\} \trianglelefteq \Psi$
then $[\Phi](\textbf{var}\,v\,;\,p\,;\,\textbf{end}\,v) \sqsubseteq [\Psi](p)$

Proof: First we want to establish the following facts:

$$\hat{\Psi}\ =\ \{def\ of\ \hat{\Psi}\}$$

$$\mathbf{var}\ v, x, \ldots, z\,;$$
$$v, x, \ldots, z := M[\Psi v], M[\Psi x], \ldots, M[\Psi z]\,;$$
$$\mathbf{end}\ P,\ A,\ M$$
$$=\quad \{law\ 8(1)\ and\ def\ of\ \hat{\Phi}\}$$
$$\mathbf{var}\ v\,;\ v := M[\Psi v]\,;\ \hat{\Phi} \qquad\qquad (*)$$

$$\hat{\Psi}^{-1} \quad = \quad \{def\ of\ \hat{\Psi}^{-1}\}$$
$$\mathbf{var}\ P,\ A,\ M\,;$$
$$M[\Psi v], M[\Psi x], \ldots, M[\Psi z] := v, x, \ldots, z\,;$$
$$\mathbf{end}\ v, x, \ldots, z$$
$$\sqsupseteq \quad \{law\ 7(1)\ and\ 8(3)\}$$
$$\mathbf{var}\ P,\ A,\ M\,;$$
$$M[\Psi x], \ldots, M[\Psi z] := x, \ldots, z\,;$$
$$\mathbf{end}\ v, x, \ldots, z$$
$$= \quad \{law\ 8(5)\ and\ def\ of\ \hat{\Phi}^{-1}\}$$
$$\hat{\Phi}^{-1}\,;\ \mathbf{end}\ v \qquad\qquad (**)$$

Thus one has
$$[\Phi](\mathbf{var}\ v\,;\ p\,;\ \mathbf{end}\ v)$$
$$= \quad \{def\ of\ [\Phi]\}$$
$$\hat{\Phi}\,;\ \mathbf{var}\ v\,;\ p\,;\ \mathbf{end}\ v\,;\ \hat{\Phi}^{-1}$$
$$= \quad \{antecedent\ and\ law\ 8(1)\}$$
$$\mathbf{var}\ v\,;\ \hat{\Phi}\,;\ p\,;\ \hat{\Phi}^{-1}\,;\ \mathbf{end}\ v$$
$$\sqsubseteq \quad \{conclusion\ (**),\ law\ 8(3)\ and\ 7(1)\}$$
$$\mathbf{var}\ v\,;\ v := M[\Psi v]\,;\ \hat{\Phi}\,;\ p\,;\ \hat{\Psi}^{-1}$$
$$= \quad \{conclusion\ (*)\}$$
$$\hat{\Psi}\,;\ p\,;\ \hat{\Psi}^{-1}$$
$$= \quad \{def\ of\ [\Psi]\}$$
$$[\Psi](p)$$

## 6   The Compiling Specification

The compiling specification $\mathcal{C}pc\Psi$ means

$$\hat{\Psi}\,;\ p \sqsubseteq \hat{c}\,;\ \hat{\Psi}$$

In this paper, the object code is given by $m[s : f)$, where $m$ is the code store, $s$ is the starting address and $f$ is the finishing address for execution. $\hat{c}$ is thus defined as $\mathcal{I}sf$. From lemma 6 it follows that the compiling specification can also be expressed by

$$\mathcal{C}pm[s : f)\Psi \stackrel{def}{=} [\Psi](p) \sqsubseteq \mathcal{I}sf$$

We can now begin to prove the series of theorems by which a compiler designer specifies to its implementor the valid code for each of the constructions of the source language.

**Theorem 1** {SKIP}     $\mathcal{C}\,\mathrm{SKIP}\,m[s:s)\Psi$

Proof: From lemma 1(1) and lemma 10(1) we have     $\mathcal{I}ss = P := s \sqsupseteq [\Psi](\mathrm{SKIP})$

There is no need to specify an unique object code for each source program construction. For example, lemma 1(3) also permits us to prove

**Theorem 1a**

$$\text{If } s < f \text{ and } m[s] = jump(f) \text{ then } \mathcal{C}\,\mathrm{SKIP}\,m[s:f)\Psi.$$

Sequential composition is the subject of a more complicated theorem; it specifies that a sequential composition can be compiled by recursively compiling its operands.

**Theorem 2** {composition}     If $s \leq j \leq f$ and $\mathcal{C}pm[s:j)\Psi$ and $\mathcal{C}qm[j:f)\Psi$
then $\mathcal{C}(p;q)m[s:f)\Psi$

Proof: The conclusion follows from the fact that

$$\mathcal{I}sf$$
$$\sqsupseteq \quad \{antecedent \text{ and } lemma\ 3\}$$
$$\mathcal{I}sj\,;\,\mathcal{I}jf$$
$$\sqsupseteq \quad \{antecedent\}$$
$$[\Psi](p)\,;\,[\Psi](q)$$
$$= \quad \{lemma\ 8\}$$
$$[\Psi](p\,;\,q)$$

**Theorem 3** {conditional}     If $b$ is a variable and $s + 2 < j \leq f$
and $m[s] = load(\Psi b)$ and $m[s+1] = cond(j)$ and $m[j-1] = jump(f)$
and $\mathcal{C}pm[s+2:j-1)\Psi$ and $\mathcal{C}qm[j:f)\Psi$
then $\mathcal{C}(\textbf{if } b \textbf{ then } p \textbf{ else } q)m[s:f)\Psi$

Proof:     $\mathcal{I}sf$
$$\sqsupseteq \quad \{antecedent \text{ and } lemma\ 4\}$$
$$A := M[\Psi b]\,;\, \textbf{if } M[\Psi b] \textbf{ then } (\mathcal{I}(s+2)(j-1)\,;\,\mathcal{I}ff) \textbf{ else } \mathcal{I}jf$$
$$= \quad \{lemma\ 1(1)\}$$
$$A := M[\Psi b]\,;\, \textbf{if } M[\Psi b] \textbf{ then } (\mathcal{I}(s+2)(j-1)\,;\, P := f) \textbf{ else } \mathcal{I}jf$$
$$\sqsupseteq \quad \{lemma\ 10 \text{ and } antecedent\}$$
$$[\Psi](\mathrm{SKIP})\,;\, \textbf{if } M[\Psi b] \textbf{ then } ([\Psi](p)\,;\, [\Psi](\mathrm{SKIP})) \textbf{ else } [\Psi](q)$$
$$= \quad \{lemma\ 8,\ law\ 1(2) \text{ and } def\ of\ [\Psi](e)\}$$

$$[\Psi](\mathrm{SKIP})\,;\ \mathbf{if}\ [\Psi](b)\ \mathbf{then}\ [\Psi](p)\ \mathbf{else}\ [\Psi](q)$$

$$=\quad \{lemma\ 12\}$$

$$[\Psi](\mathrm{SKIP})\,;\ [\Psi](\mathbf{if}\ b\ \mathbf{then}\ p\ \mathbf{else}\ q)$$

$$=\quad \{lemma\ 8\ and\ law\ 1(2)\}$$

$$[\Psi](\mathbf{if}\ b\ \mathbf{then}\ p\ \mathbf{else}\ q)$$

**Theorem 4** {loop}    If $b$ is a variable and $s+1 < f-1$
and $m[s] = load(\Psi(b))$ and $m[s+1] = cond(f)$
and $m[f-1] = jump(s)$
and $\mathcal{C}pm[s+2:f-1)\Psi$
then $\mathcal{C}(\mathbf{while}\ b\ \mathbf{do}\ p)m[s:f)\Psi$

Proof:    $\mathcal{I}sf$

$\sqsupseteq\quad \{antecedent\ and\ lemma\ 5\}$

$A := M[\Psi b]\,;$

$\mathbf{while}\ M[\Psi b]\ \mathbf{do}\ (\mathcal{I}(s+2)(f-1)\,;\ A := M[\Psi b])\,;\ \mathcal{I}ff$

$\sqsupseteq\quad \{antecedent\ and\ lemma\ 10\ and\ 1(1)\}$

$[\Psi](\mathrm{SKIP})\,;\ \mathbf{while}\ M[\Psi b]\ \mathbf{do}\ ([\Psi](p)\,;\ [\Psi](\mathrm{SKIP}))\,;\ [\Psi](\mathrm{SKIP})$

$=\quad \{lemma\ 8\ and\ law\ 1(2)\}$

$[\Psi](\mathrm{SKIP})\,;\ \mathbf{while}\ M[\Psi b]\ \mathbf{do}\ [\Psi](p)\,;\ [\Psi](\mathrm{SKIP})$

$=\quad \{corollary\ of\ lemma\ 13\}$

$[\Psi](\mathrm{SKIP})\,;\ [\Psi](\mathbf{while}\ b\ \mathbf{do}\ p)$

$=\quad \{lemma\ 8\ and\ law\ 1(2)\}$

$[\Psi](\mathbf{while}\ b\ \mathbf{do}\ p)$

**Theorem 5** {non-determinism}    If $\mathcal{C}pm[s:f)\Psi$ then $\mathcal{C}(p \sqcap q)m[s:f)\Psi$

Proof:    $\mathcal{I}sf$

$\sqsupseteq\quad \{antecedent\}$

$[\Psi](p)$

$\sqsupseteq\quad \{def\ of\ \sqcap\ \}$

$[\Psi](p) \sqcap [\Psi](q)$

$=\quad \{lemma\ 11\}$

$[\Psi](p \sqcap q)$

Since $p \sqcap q = q \sqcap p$, we have immediately

**Theorem 5a**    If $\mathcal{C}qm[s:f)\Psi$ then $\mathcal{C}(p \sqcap q)m[s:f)\Psi$

This gives the implementor freedom to implement $p \sqcap q$ either as $p$ or as $q$. This freedom may be used to improve the quality of the compiler by selecting whichever gives the more efficient object code, at least in the cases where this is statically checkable.

If the source program would abort, the compiler is allowed to generate any object code whatsoever.

**Theorem 6** {Abort}     $\mathcal{C}\mathsf{Abort}m[s:f)\Psi$

Proof: Direct from lemma 7 and law 5(1).

The theorem for declaration applies only when the declared variable is distinct from all global variables. (In the absence of recursion, this can be achieved by preliminary substitution of a fresh name for the local variable). In any case, the body of the block is compiled with a symbol table $\Psi$, an injection formed by extending the domain of $\Phi$ to include the declared variable $v$. Thus $\Phi$ equals $\{v\} \lhd \Psi$, i.e., the result of removing $v$ from the domain of $\Psi$.

**Theorem 7** {declaration}     If $v \in dom(\Psi)$ and $\Phi = \{v\} \lhd \Psi$ and $\mathcal{C}pm[s:f)\Psi$
$$\text{then } \mathcal{C}(\mathbf{var}\ v\,;\,p\,;\,\mathbf{end}\ v)m[s:f)\Phi$$

Proof:
$$\mathcal{I}sf$$
$$\sqsupseteq \quad \{antecedent\}$$
$$[\Psi](p)$$
$$\sqsupseteq \quad \{antecedent\ and\ lemma\ 15\}$$
$$[\Phi](\mathbf{var}\ v\,;\,p\,;\,\mathbf{end}\ v)$$

**Theorem 8** {assignment}     If $m[s] = load(\Psi y)$ and $m[s+1] = store(\Psi x)$
$$\text{then } \mathcal{C}(x := y)m[s:s+2)\Psi$$

Proof: Direct from lemma 14 and lemma 1(4).

# 7　Conclusion

This paper has illustrated a structure and methodology for proof of correctness of a compiler design; the proofs are primarily algebraic transformations to establish refinement relations. But the example programming language is very small, and there is at present no evidence that the methods will generalise to more powerful source languages or more complex target machines.

One hopeful aspect of the method is that each feature of the language is introduced by a separate theorem. So new features should be easy to add, as well as new, more efficient, transformations of the existing features.

On the other hand, it is clear that major new features such as recursion or concurrency (or worse still, both of them) will require significant additional complexity, for example

1. inclusion of new components such as stacks into the machine state;

2. formulation of invariant assertions (e.g. non-interference and stack integrity) and their insertion into the interpreter.

3. proliferation of compiler tables to deal with scopes and types.

Ideally, these additions should preserve the validity of all the proofs about the simpler aspects of the language.

Jonathan Bowen has noticed that all the theorems which specify the code of the compiler have the form of Horn clauses, and can therefore be readily translated and executed as logic programs in PROLOG. He has used this principle to write a compiler from a subset of occam [6] to the machine code of a transputer [1,2].

The style of our proofs is almost exclusively algebraic, and their correctness can in principle be checked by an algebraic term rewriting system. Augusto Sampaio [8] is currently performing these checks with the aid of OBJ3 [4]. The labour of the task can be much reduced by the careful structuring of algebraic theories, which has been attempted in this paper.

# Acknowledgement

The research reported here is a contribution to an international project ProCoS (Provably correct systems), supported by the European Community under ESPRIT II Basic Research Actions. It owes much to the inspiration and assistance of our colleagues on that project, and to detailed scrutiny by Carroll Morgan, Oege de Moor and Augusto Sampaio.

# References

[1] J. Bowen, He Jifeng and P. Pandya. An Approach to Verifiable Compiling Specification and Prototyping. Lecture Notes in Computer Science 456, Springer-Verlag, 45–60, 1990.

[2] J. Bowen. From Programs to Object Code and back again using Logic Programming. ProCoS Document OU JB 7/1, 1990.

[3] E. W. Dijkstra. Guarded commands, non-determinacy, and formal derivation of programs. Comm. ACM 18 (8): 453–7, 1975.

[4] J. Goguen and T. Winkler. Introducing OBJ3. SRI International Technical Report SRI-CSL-88-9, 1988.

[5] C.A.R. Hoare and others. Laws of Programming. Comm. ACM 30 (8): 672–86, 1987. [Reprinted in this volume].

[6] INMOS Limited. occam 2 Reference Manual. Prentice Hall International, 1988.

[7] C.B. Jones. Systematic Software Development Using VDM. Prentice Hall International, 1986.

[8] A. Sampaio. A comparative study of theorem provers: proving correctness of compiling specifications. Oxford University Computing Laboratory Technical Report PRG-TR-20-90, 1990.

[9] J.M. Spivey. Understanding Z – A Specification Language and its Formal Semantics. Cambridge Tracts in Computer Science 3. Cambridge University Press, 1988.

# Chapter 4

# Distributed Systems

Distributed Systems in general show a high degree of combinatorial complexity. Therefore for distributed systems the classical principles of levels of abstractions for specifications and refinements are especially needed. Only if it is possible to provide an appropriate formal foundation for modelling distributed systems including specification and refinement calculi, one can hope that we will be able to design and develop distributed systems in a reliable way.

J.A. Bergstra

M. Broy

E.-R. Olderog

# Process Algebra with Signals and Conditions


J.C.M. Baeten

*Department of Software Technology, CWI,*
*P.O. Box 4079, 1009 AB Amsterdam, The Netherlands*

*Programming Research Group, University of Amsterdam,*
*P.O. Box 41882, 1009 DB Amsterdam, The Netherlands*


J.A. Bergstra

*Programming Research Group, University of Amsterdam,*
*P.O. Box 41882, 1009 DB Amsterdam, The Netherlands*

*Department of Philosophy, Utrecht University,*
*Heidelberglaan 2, 3584 CS Utrecht, The Netherlands*



Several new operators are introduced on top of the algebra of communicating processes (ACP) from [BK 84] in order to incorporate stable signals in process algebra. Semantically this involves assigning labels to nodes of process graphs in addition to the actions that serve as labels of edges. The labels of nodes are called signals. In combination with the operators of BPA, two signal insertion operators allow to describe each finite tree labeled with actions and signals. In the context of parallel processes there is a new feature connected with signals: the signal observation mechanism. This mechanism is organised on basis of a signal observation function in very much the same way as the communication mechanism of ACP is organised around a communication function. Also, we discuss conditionals on signals and processes, and make much use of them in our examples.



*1980 Mathematics Subject Classification (1985 revision):* 68Q10, 68Q55, 68Q45, 68Q40.
*1987 CR Categories:* F.1.2, F.3.2, D.1.3, D.3.1, D.4.1.
*Key words & Phrases:* process algebra, ACP, signals, observation, condition, guard, guarded command.
*Note:* This research was sponsored in part by ESPRIT under contract 432, METEOR as well as under contract 3006, CONCUR. This paper is a revised and extended version of [BE 88a].


## 1. INTRODUCTION

This paper is concerned primarily with concrete process algebra in the sense of [BB 88]. Concrete process algebra is that part of process algebra that does not involve Milner's silent action $\tau$ [MI 80, MI 89] or the empty action $\varepsilon$ due to Vrancken [VR 86]. This paper is a revised and substantially extended version of [BE 88a]. An extended abstract of [BE 88a] was published as [BE 88b].

Process algebra is a well-established part of computer science ([AB 84], [BW 90], [HE 88], [HO 85], [MI 80, 89]). We discuss process algebra in the setting of ACP (see [BK 84], [BW 90]). The new feature that we hope to contribute to process algebra with this paper is the presence of explicit signals of a persistent nature. In addition we discuss the use of conditional expressions in process algebra. The original set-up of process algebra inherited from Milner's CCS views processes semantically as trees of actions. These actions are best thought of as atomic actions because otherwise the intuition behind the axioms becomes rather obscure. A mechanism of particular importance, that has not yet been analyzed in the setting of ACP, is the presence of visible aspects of the state of a process. Usually in process algebra the state of a process can only be understood (or observed) via the actions that can be performed from that state. In the set-up that will be presented here some aspects of the system state are visible not so much through the actions that will follow but much more directly as signals that persist in time for some extended duration. Typically, one might think of some light signal on a control panel. Several signals may be visible simultaneously and one is easily led to a boolean algebra of signals based on an empty signal and a composition operator on signals. The main step is to introduce two operators called signal insertion operators. In terms of a graph model of processes these signal insertion operators are able to place a signal as a label to a node in the graph. Given finite sets A and AS, using the root signal insertion operator and the terminal signal insertion operator it is possible to describe every finite tree with edges labeled by atomic actions from A and nodes labeled by compositions of atomic signals from AS. Without problem the operators can be used to give guarded recursive definitions of infinite processes with signals.

There is a need for motivation for this work. It seems that every extension of the operator set of process algebra makes the subject less focussed and comprehensible in the eyes of mathematically oriented observers. Our point of view, however, is that whenever it is difficult to model process phenomena in terms of existing operators of process algebra there is a reason to investigate extensions of the formalism. An essential restriction is that existing axioms remain valid in a each new setting. In other words one may extend the set of operators and axioms of ACP but may not modify it grossly. This does not imply that we expect that all phenomena can eventually be modeled in process algebra. On the contrary it is quite likely that process algebra will come to real limits sooner or later. The reason for this restriction is that it should be quite clear when a departure of process algebra is needed in order to find an adequate modeling of some sequential or concurrent mechanism. Respecting these restrictions, it is excluded that process algebra becomes a moving target altogether which claims universality but in fact shows lack of commitment to its basic principles. Now these basic principles are exactly the operators and axioms of BPA, PA and ACP (see e.g. [BK 84]) or the exposition below) but not the semantic models that have been defined for these axiom systems. Therefore we feel that the approach of this paper is methodologically adequate: all axioms of ACP are preserved, more operators and axioms (and in fact processes) are added that nevertheless all satisfy the axioms of ACP. Moreover the recursive definitions of the new operators all have the spirit of term rewriting that underlies the design of ACP.

The term rewriting analysis of ACP with signals has been carried out in BROUWER [BR 90]. He uses this analysis to obtain an *implementation* of ACP with signals. Moreover he has designed syntax that allows to incorporate signals in the process specification language PSF of MAUW & VELTINK [MV 90].

Of course it remains to be seen that the approach to the problem of incorporating stable and nonatomic signals in the framework of process algebra as presented in this paper is a useful one. As weak points we see the following ones:

(i) There is very little structure on the signals in terms of data structuring. Of course such a structure can be found by simply requiring that the signals come from some sort in an abstract data type, but that is certainly not a connection between process structure and data structure.

(ii) In cases where a signal is present before as well as after an action this model presents in an unavoidable way an atomic interruption of the signal during the execution of the action. This interruption is not always adequate in view of the intuitions behind the mechanism.

(iii) It is not clear that both the root signal insertion operation and the terminal signal insertion operator are needed. Having two primitive operators for the simple concept of putting a label on a node seems quite overdone at first sight. Nevertheless the presence of both operators gives rise to a flexible and efficient algebra. It turns out that in the presence of an empty step terminal signal insertion is not needed.

Besides signals, this paper introduces a conditional expression and relates it with the guarded command of [BBMV 89]. These constructs are unproblematic and could (should) have been introduced in the ACP setting much earlier indeed. In designing the syntax for the extra operators we will make use of algebraic abstract syntax. Thus as much as possible the syntax will be no more than an algebraic signature and diagrams will be added to illustrate these signatures.

## 2. ADDING SIGNALS TO BASIC PROCESS ALGEBRA

### 2.1 BPA AND ACS

In order to make the paper somewhat self contained all axioms of process algebra from [BK 84] are repeated. The reader is supposed to have some familiarity with these axioms, however, as there is no explanation of the intuitions behind the original operators of process algebra. Let A be a finite set. The elements of A will be called atomic actions. Every atomic action is an element of P, the sort of processes. There are also two binary operators on P, viz. + (alternative composition) and $\cdot$ (sequential composition). The core system BPA (Basic Process Algebra) over this signature has the following axioms ($x,y,z \in P$).

| | |
|---|---|
| $x + y = y + x$ | A1 |
| $(x + y) + z = x + (y + z)$ | A2 |
| $x + x = x$ | A3 |
| $(x + y) \cdot z = x \cdot z + y \cdot z$ | A4 |
| $(x \cdot y) \cdot z = x \cdot (y \cdot z)$ | A5 |

TABLE 1. BPA.

In the sequel a rather large number of new operators and axioms will be presented. The first step is to introduce the algebra ACS (algebra of composed signals). This involves the introduction of a sort **AS** of atomic signals which is in fact a parameter for the design of the algebras and which should be defined specifically for each application.

The interpretation of **AS** will be a finite set and for each element of this finite set a constant is added in **AS**. Then there is sort **CS** of composed signals. **CS** is just the power set of **AS** and it is equipped with the usual operators on sets. The axioms of ACS are as follows ($a, b \in$ **AS**, $u, v, w \in$ **CS**).

| | |
|---|---|
| $u \cup v = v \cup u$ | CS1 |
| $(u \cup v) \cup w = u \cup (v \cup w)$ | CS2 |
| $u \cup u = u$ | CS3 |
| $u \cup \varnothing = u$ | CS4 |
| $u \cap v = v \cap u$ | CS5 |
| $u \cap (v \cap w) = (u \cap v) \cap w$ | CS6 |
| $u \cap u = u$ | CS7 |
| $u \cap \varnothing = \varnothing$ | CS8 |
| $u \cup (v \cap w) = (u \cup v) \cap (u \cup w)$ | CS9 |
| $u \cap (v \cup w) = (u \cap v) \cup (u \cap w)$ | CS10 |
| $a \in \varnothing = \mathrm{false}$ | CS11 |
| $a \in \{a\} = \mathrm{true}$ | CS12 |
| $a \in \{b\} = \mathrm{false} \quad \text{if } a \neq b$ | CS13 |
| $a \in u \cup w = (a \in u) \vee (a \in w)$ | CS14 |
| $a \in u \cap w = (a \in u) \wedge (a \in w)$ | CS15 |

TABLE 2. ACS.

The signature of ACS in pictured in figure 1.

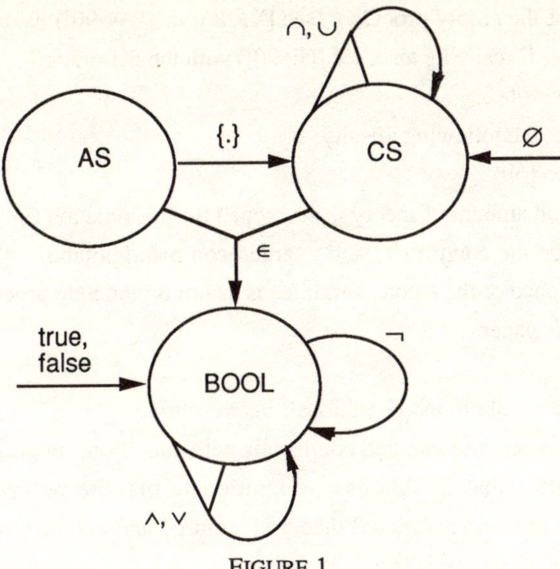


FIGURE 1.

## 2.2 CONDITIONALS

A useful feature in the coming examples is the addition of *conditionals* or *guards* to our language. It should be noted that this extension is independent of the presence of signals. We use letters $\phi,\psi$ to range over conditions COND. We first consider the ternary operator $.\triangleleft.\triangleright.: P \times COND \times P \rightarrow P$. The expression $x\triangleleft\phi\triangleright y$ should be read as **if** $\phi$ **then** x **else** y. We take this notation from [HHJ+ 87]. We think the notation with triangles is preferable over the one with key words, because the former notation does not have scoping problems and is more mathematical. We will also add this operator to our signal algebra, i.e. we will have a ternary operator $.\triangleleft.\triangleright.: CS \times COND \times CS$.

The advantage of adding the conditional expressions is easily understood in the use of parametrized process specifications. Let for instance, for $n \in \mathbb{N}$, $P(n)$ be the process $a^{n}{\cdot}b$. One obtains a uniform specification for the $P(n)$ as follows ($n \geq 0$):

$$P(n) = b \triangleleft n{=}0 \triangleright a{\cdot}P(pred(n))$$

(here **pred** is the predecessor function: $pred(0) = 0$ and $pred(n{+}1) = n$).

Besides this operator, we will use a variant of it that is a binary operator, the *guarded command*, that was introduced in [BBMV 89]. We have the notation $.{:}\rightarrow.: COND \times P \rightarrow P$. The expression $\phi{:}\rightarrow x$ is read as **if** $\phi$ **then** x. We have a similar operator on signals. There is the basic identity:

$$\phi{:}\rightarrow x = x\triangleleft\phi\triangleright\delta.$$

We see that the existence of this operator presupposes the presence of the $\delta$ constant. This is why the ternary operator is more basic and therefore preferable to the binary operator. Nevertheless, the binary operator will turn out to be very useful in examples.

In the presence of the empty process $\varepsilon$ (see [VR 86] or [BW 90]) we can even use a *unary* operator $\{.\}: \text{COND} \to P$ (called *guard*, see [GP 90]) with the definition:

$$\{\phi\} = \varepsilon \triangleleft \phi \triangleright \delta.$$

Notice that this implies the following identity:

$$\phi{:}{\to}x = \{\phi\}{\cdot}x.$$

In [GP 90] a substantial amount of theory is developed for this notation for guards. In this paper, we will concentrate on the conditional and guarded command notation. As the guard operator presupposes the existence of the $\varepsilon$ constant, it leads us out of concrete process algebra, and thus out of the scope of this paper.

Throughout this text we will discuss conditionals in two settings:

i.   conditional expressions and guarded commands with conditions ranging over the set BOOL = {true, false}. Notice that in this case, conditionals may be removed from all closed expressions. It follows that no additional theory is required and that no consistency problem is raised whatsoever. The signature is shown in figure 2.

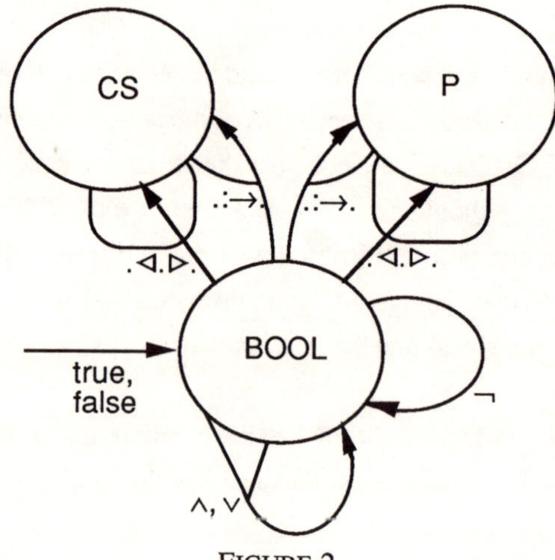


FIGURE 2.

ii.   conditional expressions and guarded commands in the more general case that the condition is taken from the free Boolean algebra $\mathbb{B}_n$ with generators $\theta_1,...,\theta_n$. In this case we will usually add a distributive law for each new operator. Only in the case of the state operator some complications arise that can be solved in a natural way however. The signature is shown in figure 3.

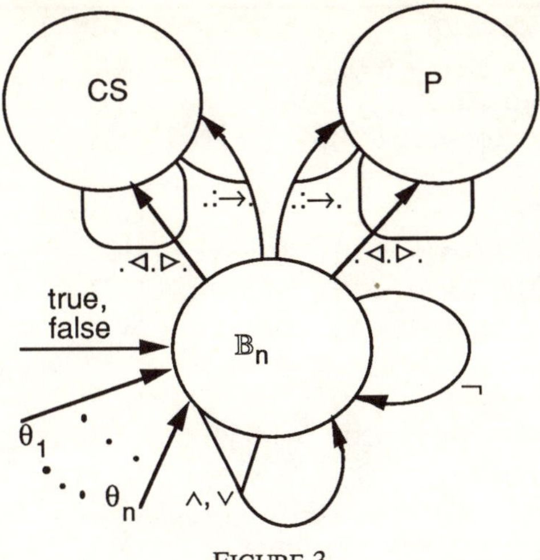


FIGURE 3.

We first consider case (i). The following axioms are obvious. $x,y \in P$, $u,v \in CS$.

| | |
|---|---|
| $x \triangleleft true \triangleright y = x$ | CO1 |
| $x \triangleleft false \triangleright y = y$ | CO2 |
| $u \triangleleft true \triangleright v = u$ | CO3 |
| $u \triangleleft false \triangleright v = v$ | CO4 |
| | |
| $true :\rightarrow x = x$ | GC1 |
| $false :\rightarrow x = \delta$ | GC2 |
| $true :\rightarrow u = u$ | GC3 |
| $false :\rightarrow u = \varnothing$ | GC4 |

TABLE 3. Conditionals and guarded command.

As we will see, the two operators are interdefinable. In the sequel, we will most often give the extra laws for the ternary operator, as this operator embodied a more classical construction, and has the advantages outlined above. Sometimes this is not possible, however. We will also use $:\rightarrow$ in case the law to be formulated would look incomprehensible for $.\triangleleft.\triangleright.$. The laws for the other operator are usually derivable using the first four axioms of the following table. This table gives extra axioms needed in case (ii), when we have conditions ranging over an arbitrary finite Boolean algebra.

As an example, we show two calculations:

i. $x \triangleleft (\phi \vee \psi) \triangleright y = (\phi \vee \psi) :\rightarrow x + \neg(\phi \vee \psi) :\rightarrow v = (\phi \vee (\neg\phi \wedge \psi)) :\rightarrow x + (\neg\phi \wedge \neg\psi) :\rightarrow y =$
$= (\phi :\rightarrow x) + (\neg\phi :\rightarrow (\psi :\rightarrow x)) + (\neg\phi :\rightarrow (\neg\psi :\rightarrow y)) = (\phi :\rightarrow x) + (\neg\phi :\rightarrow (x \triangleleft \psi \triangleright y)) =$
$x \triangleleft \phi \triangleright (x \triangleleft \psi \triangleright y)$.

ii. $x \triangleleft \neg\phi \triangleright y = \neg\phi :\rightarrow x + \neg(\neg\phi) :\rightarrow y = \phi :\rightarrow y + (\neg\phi) :\rightarrow x = y \triangleleft \phi \triangleright x$.

$$\phi:\!\rightarrow x = x \triangleleft \phi \triangleright \delta \qquad\qquad \text{CG1}$$
$$x \triangleleft \phi \triangleright y = \phi:\!\rightarrow x + (\neg\phi):\!\rightarrow y \qquad\qquad \text{CG2}$$
$$\phi:\!\rightarrow u = u \triangleleft \phi \triangleright \varnothing \qquad\qquad \text{CG3}$$
$$u \triangleleft \phi \triangleright v = \phi:\!\rightarrow u \cup (\neg\phi):\!\rightarrow v \qquad\qquad \text{CG4}$$

$$\phi:\!\rightarrow \varnothing = \varnothing \qquad\qquad \text{GC5}$$
$$\phi:\!\rightarrow (u \cup v) = (\phi:\!\rightarrow u) \cup (\phi:\!\rightarrow v) \qquad\qquad \text{GC6}$$
$$(\phi \vee \psi):\!\rightarrow u = (\phi:\!\rightarrow u) \cup (\psi:\!\rightarrow u) \qquad\qquad \text{GC7}$$
$$\phi:\!\rightarrow (\psi:\!\rightarrow u) = (\phi \wedge \psi):\!\rightarrow u \qquad\qquad \text{GC8}$$

$$\phi:\!\rightarrow \delta = \delta \qquad\qquad \text{GC9}$$
$$\phi:\!\rightarrow (x + y) = (\phi:\!\rightarrow x) + (\phi:\!\rightarrow y) \qquad\qquad \text{GC10}$$
$$(\phi \vee \psi):\!\rightarrow x = (\phi:\!\rightarrow x) + (\psi:\!\rightarrow x) \qquad\qquad \text{GC11}$$
$$\phi:\!\rightarrow (\psi:\!\rightarrow x) = (\phi \wedge \psi):\!\rightarrow x \qquad\qquad \text{GC12}$$
$$(x \cdot z) \triangleleft \phi \triangleright (y \cdot z) = (x \triangleleft \phi \triangleright y) \cdot z \qquad\qquad \text{CO5}$$

TABLE 4. Conditionals over a Boolean algebra.

LEMMA. Consider the algebra CS with operators $\cap, \cup, \{.\}, \in$ and $:\!\rightarrow$ over the Boolean algebra $\mathbb{B}_n$ with generators $\theta_1,...,\theta_n$. Let $u$ be a closed term over this algebra. Then $u$ can be written in the form

$$u = \bigcup_{i=1}^{2^n} (\phi_i :\!\rightarrow u_i),$$

where for $1 \leq i \leq 2^n$, $u_i \in$ CS and the $\phi_i$ range over all 'complete' conjunctions of literals. A *literal* is a term of the form $\theta_i$ or $\neg\theta_i$.

THEOREM. Consider the algebra BPA with operators $+, \cdot$ and $:\!\rightarrow$ over the Boolean algebra $\mathbb{B}_n$ with generators $\theta_1,...,\theta_n$. Let $t$ be a closed term over this algebra. Then $t$ can be written in the form

$$t = \sum_{i \in I} (\phi_i :\!\rightarrow a_i) \cdot t_i + \sum_{i \in J} (\phi_i :\!\rightarrow b_i),$$

where $I, J \subseteq \{1,...,2^n\}$, $a_i, b_i \in A$, $t_i$ again closed terms and the $\phi_i$ are as above. Here, we use the convention that a sum over an empty set is considered equal to $\delta$.

A consequence of this theorem is that one may view BPA with guarded commands as a variation of BPA with atomic actions $\phi_i :\!\rightarrow a_i$.

## 2.3 ROOT AND TERMINAL SIGNAL INSERTION OPERATORS

The next operators to be introduced are the signal insertion operators. $[.,.]$ is the root signal insertion operator and $\langle .,. \rangle$ is the terminal signal insertion operator. The intuition behind these operators is that both assign labels (signals) to the states of processes. Root signal insertion places a signal at the root node of a process. Terminal signal insertion places one and the same signal at each terminal node of a process. If one is interested solely in processes that show signals exclusively in nonterminal states one may as well forget about the terminal signal insertion operator. Leaving out all axioms involving terminal insertion from the coming sections one will obtain an appropriate description of root signal insertion.

Whereas in process algebra one usually confines oneself to labeling the transitions and perhaps to some labeling of the nodes that is directly related to the mechanism of transition labeling, here it is intended to have labelings of states of processes with the same status as the labelings of the state transitions by means of atomic actions. With some effort it turns out that an algebraic specification of the resulting notion of processes can be given that indeed constitutes a conservative enrichment of ACP (at least regarding identities between finite closed process expressions). The following 10 equations are added to BPA thus obtaining BPAS (BPA with signals). Remarks on models for BPAS are given in section 7.

$$
\begin{array}{ll}
[u, x] \cdot y = [u, x \cdot y] & \text{RS1} \\
[u, x] + y = [u, x + y] & \text{RS2} \\
[u, [v, x]] = [u \cup v, x] & \text{RS3} \\
[\varnothing, x] = x & \text{RS4}
\end{array}
$$

TABLE 5. Root signal insertion.

The first axiom expresses the fact that the root of a sequential product is the root of its first component. Axiom RS2 can be given in a more symmetric form as follows:

$$[u, x] + [v, y] = [u \cup v, x + y].$$

This equation depends on the fact that the roots of two processes in an alternative composition are identified. Therefore signals must be combined. The third axiom expresses the fact that there is no sequential order in the presentation of signals. Of course one might imagine that a sequential ordering on signals is introduced, but we think that the introduction of such a sequential ordering is far from obvious (it also leads to problems concerning the associativity of the parallel composition operator). The combination of the signals is taking 'both' of them whereas $x + y$ has to choose between $x$ and $y$. The equations below regard terminal signal insertion.

$$
\begin{array}{ll}
\langle x \cdot y, u \rangle = x \cdot \langle y, u \rangle & \text{TS1} \\
\langle x + y, u \rangle = \langle x, u \rangle + \langle y, u \rangle & \text{TS2} \\
\langle \langle x, u \rangle, v \rangle = \langle x, u \cup v \rangle & \text{TS3} \\
\langle x, \varnothing \rangle = x & \text{TS4} \\
\langle x, u \rangle \cdot y = x \cdot [u, y] & \text{TRS1} \\
\langle [u, x], v \rangle = [u, \langle x, v \rangle] & \text{TRS2}
\end{array}
$$

TABLE 6. Remaining axioms of BPAS.

The deadlock constant can be added without problems to the above axioms.

| | |
|---|---|
| $x + \delta = x$ | A6 |
| $\delta \cdot x = \delta$ | A7 |
| $\langle \delta, u \rangle = \delta$ | TS6 |

TABLE 7. Signals and deadlock.

An interesting identity that follows with the introduction of $\delta$ is the following:
$$[u, x] = [u, \delta] + x.$$
This equation is indeed very useful for writing efficient process specifications mainly because it allows to a large extent to work with process algebra expressions that are not cluttered with signal insertions.

If we add conditionals over an arbitrary finite Boolean algebra, we need the following extra axioms.

| | |
|---|---|
| $[u, x] \triangleleft \phi \triangleright [v, y] = [u \triangleleft \phi \triangleright v, x \triangleleft \phi \triangleright y]$ | |
| $\phi :\rightarrow \langle x, u \rangle = \langle \phi :\rightarrow x, u \rangle$ | |

TABLE 8. Signal insertion and guarded command.

We formulate the second law of table 8 in terms of the binary operator, since the obvious law for the ternary operator, viz. $\langle x, u \rangle \triangleleft \phi \triangleright \langle y, v \rangle = \langle x \triangleleft \phi \triangleright y, u \triangleleft \phi \triangleright v \rangle$, is *not* valid, since the validity of $\phi$ may change during the execution of the process. A variant for the ternary operator that does work, is the following:
$$\langle x, u \rangle \triangleleft \phi \triangleright \langle y, v \rangle = \langle x \triangleleft \phi \triangleright \delta, u \rangle + \langle \delta \triangleleft \phi \triangleright y, v \rangle$$
As an example, we give a calculation:
$$[u,a] \triangleleft \phi \triangleright \langle b, v \rangle = [u,a] \triangleleft \phi \triangleright [\varnothing, \langle b, v \rangle] = [u \triangleleft \phi \triangleright \varnothing, a \triangleleft \phi \triangleright \langle b, v \rangle] =$$
$$= [\phi :\rightarrow u, (\phi :\rightarrow a) + (\neg \phi :\rightarrow \langle b, v \rangle)] = [\phi :\rightarrow u, (\phi :\rightarrow a) + \langle \neg \phi :\rightarrow b, v \rangle)].$$

2.4 SIGNAL FILTERING AND GLOBAL SIGNAL INSERTION
The signal filtering operation $\cap$ allows to reduce the number of visible signals by filtering them all with a fixed signal. The advantage of the use of signal filtering is to obtain a much simpler signal structure, which can then be specified quite precisely. So if it is hard to find a correct expression for $X$ it may be workable to determine $u \cap X$ for some appropriate signal $u$.

| | |
|---|---|
| $u \cap a = a$ | F1 |
| $u \cap (x + y) = (u \cap x) + (u \cap y)$ | F2 |
| $u \cap (x \cdot y) = (u \cap x) \cdot (u \cap y)$ | F3 |
| $u \cap [v, x] = [u \cap v, u \cap x]$ | F4 |
| $u \cap \langle x, v \rangle = \langle u \cap x, u \cap v \rangle$ | F5 |

TABLE 9. Signal filtering.

Similarly, it is useful to extend $\cup$ over arbitrary processes. The appropriate name for this operation is global signal insertion.

$$
\begin{array}{ll}
u \cup a = [u, \langle a, u \rangle] & \text{GSI1} \\
u \cup (x + y) = (u \cup x) + (u \cup y) & \text{GSI2} \\
u \cup (x \cdot y) = (u \cup x) \cdot (u \cup y) & \text{GSI3} \\
u \cup [v, x] = [u \cup v, u \cup x] & \text{GSI4} \\
u \cup \langle x, v \rangle = \langle u \cup x, u \cup v \rangle & \text{GSI5}
\end{array}
$$

TABLE 10. Global signal insertion.

The extra axioms needed for conditionals are equally straightforward.

$$
\begin{array}{l}
u \cap (x \triangleleft \phi \triangleright y) = (u \cap x) \triangleleft \phi \triangleright (u \cap y) \\
u \cup (x \triangleleft \phi \triangleright y) = (u \cup x) \triangleleft \phi \triangleright (u \cup y)
\end{array}
$$

TABLE 11. Signal filtering, insertion and conditionals.

We give an overview of the signature elements introduced in sections 2.3 and 2.4 in figure 4.

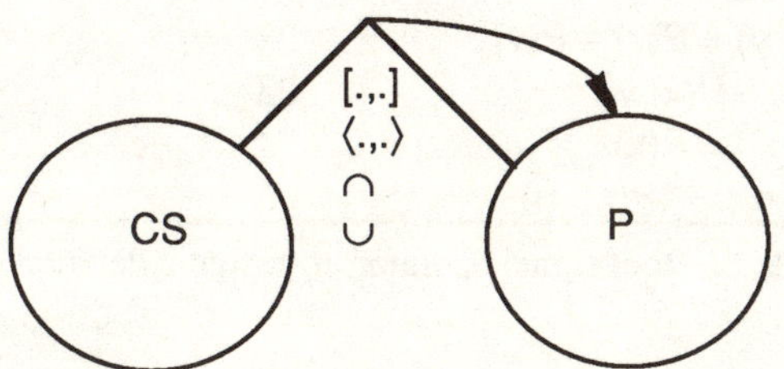


FIGURE 4.

An important application of signals is in the specification of reactive systems. The virtue of the signal mechanism is that it allows to express an asymmetry between action and reaction in the case of reactive systems. The action is an atomic act performed on the initiative of the external environment of a reactive system S. The effect of this action is to change the state of the system and to leave it in a stable resulting state from which it may show one or more signals. The environment can decide itself when to observe these signals, if at all. In the next section we will present a series of examples of process definitions involving signals. Most of these systems may be viewed as (parts of) reactive systems.

Processes with signals constitute a novel concept, at least as seen from the point of view of process algebra. Of course there is nothing new in the intuitions, the novelty lies entirely in the aspect that these matters are wrapped in an algebraic framework. Moreover it turns out that processes with signals allow a natural version of bisimulation semantics. We will exploit the possibilities that are opened by the introduction of signals in processes within the context of process algebra. Processes with signals constitute first and for all a meaningful semantic category. Only secondary is the algebraic treatment of these processes in a specific setting of operators. Nevertheless we hope to have found an algebraic form for the subject that is sufficiently flexible and powerful to be both of technical and conceptual relevance.

## 2.5 ROOT SIGNAL OPERATOR

It is useful to extend the system with operators $S$ and $P$. The operator $S$ determines the root signal of a process. If $S(x) = \emptyset$ we say that $x$ has a *trivial root signal*, otherwise $x$ has a nontrivial root signal. Processes that were studied until now in the context of process algebra always have a trivial root signal. The operator $P$ removes the root signal from its argument, thus obtaining a process with a trivial root signal.

$$
\begin{array}{ll}
S(a) = \emptyset & S1 \\
S(x + y) = S(x) \cup S(y) & S2 \\
S(x{\cdot}y) = S(x) & S3 \\
S([u, x]) = u \cup S(x) & S4 \\
\\
P(a) = a & P1 \\
P(x + y) = P(x) + P(y) & P2 \\
P(x{\cdot}y) = P(x){\cdot}y & P3 \\
P([u, x]) = P(x) & P4 \\
P(\langle x, u \rangle) = \langle P(x), u \rangle & P5
\end{array}
$$

TABLE 12. Root signal operator, root signal deletion operator.

Notice that the equation $S(\langle x, u \rangle) = S(x)$ is derivable from the axioms in table 12:
$$S(\langle x, u \rangle) = S(\langle x, u \rangle{\cdot}y) = S(x{\cdot}[u, y]) = S(x).$$
Also $x = [S(x), P(x)]$ will now be derivable for finite closed process expressions. As a rewrite rule it is useless, however, because it will immediately introduce an infinite loop.

The extra axioms needed for conditionals are straightforward.

$$
\begin{array}{l}
S(x \triangleleft \phi \triangleright y) = S(x) \triangleleft \phi \triangleright S(y) \\
P(x \triangleleft \phi \triangleright y) = P(x) \triangleleft \phi \triangleright P(y)
\end{array}
$$

TABLE 13. Root signals and conditionals.

We can extend the normal form theorem of section 2.2 to BPAS as follows: if $t$ is a term over BPAS, write $t$ in the form $[S(t), P(t)]$. Then $S(t)$ can be written as shown in the lemma in section 2.2, and we can extend the theorem to cover also processes with signals, provided no signals occur in the root, and we also allow terms of the form $\sum_{i \in K} (\phi_i :\rightarrow \langle b_i, u_i \rangle)$ .

## 2.6 EXAMPLES

(i) A traffic light that changes color from green via yellow to red and back to green. As names for the traffic lights one may simply use the natural numbers. For the traffic light with number $n$, the signals are then green($n$) , yellow($n$) and red($n$), the only action is change($n$).

$$TL(n) = [\{green(n)\}, change(n)] \cdot [\{yellow(n)\}, change(n)] \cdot [\{red(n)\}, change(n)] \cdot TL(n).$$

There are various alternatives to this definition which are interesting to display explicitly. For instance one may introduce a data type COLORS = {green, yellow, red} with a function next that permutes the colors as follows: green $\rightarrow$ yellow $\rightarrow$ red $\rightarrow$ green. Then the process TL can be parametrised by a color and one obtains one of the following four recursion equations. In this case we remove the numbers of the traffic lights for brevity.

$$TL(x) = [\{x\}, change] \cdot TL(next(x))$$
$$TL(x) = [\{x\}, change \cdot TL(next(x))]$$
$$TL(x) = [\{x\}, \delta] + change \cdot TL(next(x))$$
$$TL(x) = ([\{x\}, \delta] + change) \cdot TL(next(x))$$

(ii) An electric lamp with power switch. In the equations the names for signals and actions speak for themselves. We will assume that the lamp is indexed by a name $n$ from

NAMES =     {hall1, hall2, hall3, kitchen1, kitchen2,
            living1, living2, living3, living4, staircase, garage, cellar}.

The lamp is parametrised by two additional parameters: L($n$, $x$, $y$), where

   $x \in$ SWITCH = {on, off}, $y \in$ STATUS = {defect, functioning}.

The function sw permutes the elements of SWITCH and the actions switch($n$) correspond to switching on or off lamp $n$. For each $n \in$ NAMES there are two signals: light($n$) and dark($n$). Together these signals constitute the atomic signals. It must be noticed that  the combined signal {light($n$), dark($n$)} will not occur in any process describing a physically meaningful setting.

A recursion equation for the behavior of the lamp with name $n$ is then as follows. Notice that we use a conditional statement on signals, and a guarded command on processes, as discussed in 2.2.

$$
\begin{aligned}
L(n, x, y) = \quad &[\{light(n) \triangleleft (x{=}on \ \& \ y{=}functioning) \triangleright dark(n)\}, \delta] + \\
&switch(n) \cdot L(n, sw(x), y) + \\
&(y{=}functioning) :\rightarrow defect(n) \cdot L(n, x, defect) + \\
&get\_new\_lamp(n) \cdot L(n, x, functioning)
\end{aligned}
$$

We return to this example in sections 3.2 and 3.5.

(iii) A further example is a single lamp with double switch, and the status of the lamp can be controlled from both switches. In these equations the intended meaning of the actions is as follows:

ui =          put switch i in upward position (i=1,2)

di =          put switch i in downward position (i=1,2)

$$L_{0,0} = [\{dark\}, u1] \cdot L_{1,0} + [\{dark\}, u2] \cdot L_{0,1}$$
$$L_{1,0} = [\{light\}, d1] \cdot L_{0,0} + [\{light\}, u2] \cdot L_{1,1}$$
$$L_{0,1} = [\{light\}, u1] \cdot L_{1,1} + [\{light\}, d2] \cdot L_{0,0}$$
$$L_{1,1} = [\{dark\}, d1] \cdot L_{0,1} + [\{dark\}, d2] \cdot L_{1,0}.$$

(iv) A system of two lifts seen from an intermediate floor, say floor 2, has two signals: the signal sur (signal upwards request) indicates that somebody wants (or wanted) to travel in the upward direction, the signal sdr (signal downwards request) indicates that a downward travel has been requested. The signals disappear as soon as a lift has departed in the requested direction. Both lifts move all the way from bottom floor to top floor and back all the time, be it that a request is needed to trigger their action. The actions that can be performed or observed by users at floor 2 are as follows:

up_request (asking for an upward travel),

down_request (asking for a downward travel),

arrive_j_b, (lift j arrives from below),

arrive_j_a (lift j arrives from above),

leave_j_up  (lift j leaves in upward direction),

leave_j_down (lift j leaves in downward direction).

Only the leave actions will influence the signals. In the following specification, the movement of the lifts is taken into account, be it that the internal logic of the lifts is not modeled. For instance, it can happen that  lift 1 leaves in upwards direction twice without having left for the opposite direction in between, whereas all the time the signal sdr was present and lift 2 has not departed in downward direction either. This course of events is not allowed by most lift systems. Let $u \subseteq$ {sdr, sur}.

$$LL_u = [u, \delta] +$$
$$\langle up\_request, u \cup \{sur\}\rangle \cdot LL_{u \cup sur} +$$
$$\langle down\_request, u \cup \{sdr\}\rangle \cdot LL_{u \cup \{sdr\}} +$$
$$\langle (leave\_1\_up + leave\_2\_up), u - \{sur\}\rangle \cdot LL_{u - \{sur\}} +$$
$$\langle (leave\_1\_down + leave\_2\_down), u - \{sdr\}\rangle \cdot LL_{u - \{sdr\}} +$$
$$\langle (arrive\_1\_b + arrive\_1\_a + arrive\_2\_b + arrive\_2\_a), u\rangle \cdot L_u.$$

(v) This example again discusses a lift connecting three floors but now as seen from the inside. The number of lifts involved in the system is of no importance for this specification. In this case the signals are the various sets of requests for travel to floors 0, 1 and 2. These signals for requests are denoted with sr(0), sr(1) and sr(2). Notice that the requests come either from inside the lift or from outside, but in both cases all requests are signalled inside the lift, e.g. by lighted buttons. We see there are 8 possible signals, as there are 8 subsets of the collection of atomic signals.

The actions are:

- req(j) to request transport to floor j for j = 0,1,2;
- move(0,1), move(1,0), move(1,2), move(2,1), move_past(1) denoting the movements of the lift;
- halt(j) denotes the halting of the lift at floor j;
- i_open(j), i_close(j) are the actions of opening and closing the (inner) lift door
      (these actions are controlled by the lift system);
- o_open(j) and o_close(j) are the actions of opening and closing the outer lift door
      (presumably controlled by the users).

There is no time-out mechanism that decides whether or not the outer lift door is going to be opened at all after a halt has been made. The lift has 32 states while moving and 120 states while resting at a floor. The 32 states between floors are organised as follows:

LM(0/1,up, V): the lift is moving between 0 and 1 traveling in an upward direction at the point where it will either continue its travel further without halting at 1 or decide to halt at 1 (be it because of an internal or an external request), moreover V is the signal (i.e. the set of atomic signals) that is visible.

Similarly there are states LM(0/1, down, V), LM(1/2, up, V) and LM(1/2, down,V). Starting from a halted lift, not all moving states can be reached, as the following specification is such, that the lift will only move in a certain direction if there is a request to go in that direction. Thus, only 20 travelling states will actually occur.

The 120 resting states are organised as follows:

for each j ∈ {0, 1, 2},

d ∈ {closed1, inner_door_open1, outer_door_open, inner_door_open2, closed2},

V ⊆ {sr(0), sr(1), sr(2)},

the object (data structure) LR(j, d, V) is a state. Again, not all 120 states will actually occur, as the signal for a given floor is not present when the lift is on that floor and the door is open. Thus, we see that only 84 states can actually occur.

The meaning of these states is as follows: for instance L(2, inner_door_open2, {sr(0), sr(2)}) indicates that the lift rests at floor 2, with its inner door open but its outer again closed and the request signals for floors 0 and 2 on.

Then one needs an equation for each of the 104 (or 152) states of the lift. We will present the whole specification by means of two equations, making heavy use of the guarded command from section 2.2. This model is much simpler than the real situation in most cases. For instance, after halting, the lift does not know anymore in which direction it was travelling.

Let $i \in \{1, 2\}$, $j \in \{0, 1, 2\}$, $b \in \{up, down\}$, $V \subseteq \{sr(0), sr(1), sr(2)\}$, $d \in \{closed1, inner\_door\_open1, outer\_door\_open, inner\_door\_open2, closed2\}$. Then

$L(i/i+1, b, V) = [V, \delta] +$
      $req(0){\cdot}L(i/i+1, b, V{\cup}\{sr(0)\}) +$
      $req(1){\cdot}L(i/i+1, b, V{\cup}\{sr(1)\}) +$
      $req(2){\cdot}L(i/i+1, b, V{\cup}\{sr(2)\}) +$
      $(sr(i){\in}V \;\&\; b{=}down) :{\rightarrow} halt(i){\cdot}LR(i, closed1, V) +$
      $(sr(i+1){\in}V \;\&\; b{=}up) :{\rightarrow} halt(i+1){\cdot}LR(i+1, closed1, V) +$
      $(i{=}0 \;\&\; b{=}up \;\&\; sr(1){\notin}V) :{\rightarrow} move\_past(1){\cdot}L(1/2, up, V) +$
      $(i{=}1 \;\&\; b{=}down \;\&\; sr(1){\notin}V) :{\rightarrow} move\_past(1){\cdot}L(0/1, down, V).$

$LR(j, d, V) = [V, \delta] +$
      $req(0){\cdot}(LR(j, d, V{\cup}\{sr(0)\}) \lhd j{\neq}0 \rhd$
               $(d{=}closed1 :{\rightarrow} LR(j, d, V{\cup}\{sr(0)\} +$
               $d{\in}\{inner\_door\_open1, outer\_door\_open\}) :{\rightarrow} LR(j, d, V) +$
               $d{=}inner\_door\_open2 :{\rightarrow} LR(j, inner\_door\_open1, V) +$
               $d{=}closed2 :{\rightarrow} (LR(0, closed1, \{sr(0)\}{\lhd}V{=}\varnothing{\rhd}LR(j, d, V{\cup}\{sr(0)\})))) +$
      $req(1){\cdot}(LR(j, d, V{\cup}\{sr(1)\}) \lhd j{\neq}1 \rhd$
               $(d{=}closed1 :{\rightarrow} LR(j, d, V{\cup}\{sr(1)\} +$
               $d{\in}\{inner\_door\_open1, outer\_door\_open\}) :{\rightarrow} LR(j, d, V) +$
               $d{=}inner\_door\_open2 :{\rightarrow} LR(j, inner\_door\_open1, V) +$
               $d{=}closed1 :{\rightarrow} (LR(1, closed1, \{sr(1)\}{\lhd}V{=}\varnothing{\rhd}LR(j, d, V{\cup}\{sr(1)\})))) +$
      $req(2){\cdot}(LR(j, d, V{\cup}\{sr(2)\}) \lhd j{\neq}2 \rhd$
               $(d{=}closed1 :{\rightarrow} LR(j, d, V{\cup}\{sr(2)\} +$
               $d{\in}\{inner\_door\_open1, outer\_door\_open\}) :{\rightarrow} LR(j, d, V) +$
               $d{=}inner\_door\_open2 :{\rightarrow} LR(j, inner\_door\_open1, V) +$
               $d{=}closed2 :{\rightarrow} (LR(2, closed1, \{sr(2)\}{\lhd}V{=}\varnothing{\rhd}LR(j, d, V{\cup}\{sr(2)\})))) +$
      $(d{=}closed1) :{\rightarrow} i\_open(j){\cdot}LR(j, inner\_door\_open1, V{-}\{sr(j)\}) +$
      $(d{=}inner\_door\_open1) :{\rightarrow} (o\_open(j){\cdot}LR(j, outer\_door\_open, V) +$
                     $+ i\_close(j){\cdot}LR(j, closed2, V)) +$
      $(d{=}outer\_door\_open) :{\rightarrow} o\_close(j){\cdot}LR(j, inner\_door\_open2, V) +$
      $(d{=}inner\_door\_open2) :{\rightarrow} i\_close(j){\cdot}LR(j, closed2, V) +$
      $(d{=}closed2 \;\&\; j{=}0 \;\&\; (sr(1){\in}V \;or\; sr(2){\in}V)) :{\rightarrow} move(0,1){\cdot}L(0/1, up, V) +$

$(d=closed2 \,\&\, j=1 \,\&\, sr(0){\in}V) :{\to} move(1,0){\cdot}L(0/1, down, V) +$
$(d=closed2 \,\&\, j=1 \,\&\, sr(2){\in}V\,) :{\to} move(1,2){\cdot}L(1/2, up, V) +$
$(d=closed2 \,\&\, j=2 \,\&\, (sr(0){\in}V \text{ or } sr(1){\in}V)) :{\to} move(2,1){\cdot}L(1/2, down, V).$

(vi) An alarm clock. In this example one needs a rather more substantial underlying data type than in the previous examples. We will describe the data type in an informal way but given the data type the description of the clock will be quite precise. Of course there are many formalisms that allow to present a precise description of the data type as well. The atomic signals in this example are:

• The digital time indications measuring time in seconds, collected in the set TIMES. There is a successor function next on TIMES which increases time with one second, counting modulo 24 hours.

• The alarm signal: alarm.

The actions are the following:

| | |
|---|---|
| switch_off_alarm | (takes no time), |
| tick | (a tick of the clock occurs every second), |
| set_alarm(t) | (this action takes 2 seconds). |

The process has a state vector consisting of three attributes:

(a) an element time of TIME, the current time,

(b) a boolean alarm_set that indicates whether or not the alarm has been set

(c) an element a_time of TIME that indicates when (if at all) the alarm must start. The alarm will then be on for two consecutive minutes unless it is switched off before.

Now the state of the clock is given by a record (frame, tuple):

CLOCK(time, alarm_set, a_time).

All actions result in a modification of the (values of the attributes) of the record, and the signals depend directly on the record. In particular, the time signal is just the current value of the first attribute of CLOCK and the signal alarm is on exactly if alarm_set = true and time $\in$ [a_time, a_time + 00.02.00 (modulo 24.00.00)].

Recursion equations for CLOCK are then as follows:

$CLOCK = CLOCK(00.00.00, false, 00.00.00)$

$CLOCK(t, b, t') = [t, \delta] +$

$\quad tick{\cdot}CLOCK(next(t), b, t') +$

$\quad \sum_{r\in TIME} set\_alarm(r){\cdot}CLOCK(next(next(t)), true, r) +$

$\quad switch\_off\_alarm{\cdot}CLOCK(t, false, t') +$

(b=true & t' ≤ t ≤ t' + 00.02.00 (modulo 24.00.00)) :→ [{alarm}, δ].

In the last line, we used the guarded command of 2.2.


# 3. SIGNALS AND PARALLEL COMPOSITION

### 3.1 FREE MERGE

We can extend the system BPAS of section 2 to PAS$^-$, including the free merge (parallel composition without communication) and the left merge with the usual axioms and adding axioms that handle the interaction between the signal insertion operators and the left merge. The superscript $^-$ indicates that there is no feature present that allows the observation of signals, addition of that feature is the subject of section 4. (Of course the extension to ACP will require a modification of the merge expansion axiom by adding an additional term for the communication merge, this will be described in section 5.)

| | |
|---|---|
| $x \parallel y = x \mathbin{\rotatebox[origin=c]{180}{$\mathsf{L}$}} y + y \mathbin{\rotatebox[origin=c]{180}{$\mathsf{L}$}} x$ | M1 |
| $a \mathbin{\rotatebox[origin=c]{180}{$\mathsf{L}$}} x = a{\cdot}x$ | M2 |
| $a{\cdot}x \mathbin{\rotatebox[origin=c]{180}{$\mathsf{L}$}} y = a{\cdot}(x \parallel y)$ | M3 |
| $(x + y) \mathbin{\rotatebox[origin=c]{180}{$\mathsf{L}$}} z = x \mathbin{\rotatebox[origin=c]{180}{$\mathsf{L}$}} z + y \mathbin{\rotatebox[origin=c]{180}{$\mathsf{L}$}} z$ | M4 |
| | |
| $[u, x] \mathbin{\rotatebox[origin=c]{180}{$\mathsf{L}$}} y = [u, x \mathbin{\rotatebox[origin=c]{180}{$\mathsf{L}$}} y]$ | MSI1 |
| $\langle a, u \rangle \mathbin{\rotatebox[origin=c]{180}{$\mathsf{L}$}} x = a{\cdot}(u \cup x)$ | MSI2 |

TABLE 14. PAS$^-$.

### 3.2 EXAMPLES

(i) Simple examples for the application of the free merge can be produced on the basis of the previous series of examples. The simultaneous behavior of traffic lights (see example 2.6.i) with names in {1,...,4} is found by simply merging their process descriptions:

TL(1-4) = TL(1) ∥ [{red(2)}, change(2)]·TL(2) ∥ TL(3) ∥ [{red(4)}, change(4)]·TL(4).

In an intersection where lights 1 and 3 are in the east-west direction, and lights 2 and 4 are in the north-south direction, this system starts out correctly, but as there is no communication between the different traffic lights, the coordination will soon disappear. Some communication mechanism is needed to describe this system correctly. We will return to this example in section 3.5.


(ii) This example refers to the previous example 2.6.ii that introduced a collection of lamps in a private house. The simultaneous behavior of the collection of lamps in the living room is appropriately described by     L(living) = L(living1) ∥ L(living2) ∥ L(living(3) ∥ L(living4).

For the kitchen one obtains: L(kitchen) = L(kitchen1) ∥ L(kitchen2).

For the hall we have L(hall) = L(hall1) ∥ L(hall2). Composing these one obtains:

L(living/kitchen/hall) = L(kitchen) ∥ L(living) ∥ L(hall).

(iii) Examples of derived identities in PAS⁻.

$$[u, \delta] \parallel [v, \delta] = [u \cup v, \delta],$$

$$[u, a] \parallel [v, b] = [u \cup v, a]\cdot[v, b] + [u \cup v, b]\cdot[u, a],$$

$$[u, \langle a, u'\rangle] \parallel [v, \langle b, v'\rangle] = [u \cup v, \delta] + a\cdot[v \& u', \langle b, u' \& v'\rangle] + b\cdot[u \cup v', \langle a, u' \& v'\rangle].$$

### 3.3 SIGNALS AND SYNCHRONOUS COMMUNICATION

The axiom system ACPS⁻ describes the addition of signals to processes with synchronous communication as modeled by ACP.

| | |
|---|---|
| $a \mid b = b \mid a$ | C1 |
| $(a \mid b) \mid c = a \mid (b \mid c)$ | C2 |
| $a \mid \delta = \delta$ | C3 |
| | |
| $x \parallel y = x \mathbin{\underline{\parallel}} y + y \mathbin{\underline{\parallel}} x + x \mid y$ | CM1 |
| $a \mathbin{\underline{\parallel}} x = a\cdot x$ | CM2 |
| $(a\cdot x) \mathbin{\underline{\parallel}} y = a\cdot(x \parallel y)$ | CM3 |
| $(x + y) \mathbin{\underline{\parallel}} z = (x \mathbin{\underline{\parallel}} z) + (y \mathbin{\underline{\parallel}} z)$ | CM4 |
| $[u,x] \mathbin{\underline{\parallel}} y = [u, x \mathbin{\underline{\parallel}} y]$ | MSI1 |
| $\langle a, u\rangle \mathbin{\underline{\parallel}} y = a\cdot(u \cup x)$ | MSI2 |
| | |
| $a \mid (b\cdot x) = (a \mid b)\cdot x$ | CM5 |
| $(a\cdot x) \mid b = (a \mid b)\cdot x$ | CM6 |
| $(a\cdot x) \mid (b\cdot y) = (a \mid b)\cdot(x \parallel y)$ | CM7 |
| $(x + y) \mid z = (x \mid z) + (y \mid z)$ | CM8 |
| $x \mid (y + z) = (x \mid y) + (x \mid z)$ | CM9 |
| $[u, x] \mid y = [u, x \mid y]$ | MSI3 |
| $x \mid [u, y] = [u, x \mid y]$ | MSI4 |
| $\langle a, u\rangle \mid b = \langle a \mid b, u\rangle$ | MSI5 |
| $a \mid \langle b, u\rangle = \langle a \mid b, u\rangle$ | MSI6 |
| $\langle a, u\rangle \mid \langle b, v\rangle = \langle a \mid b, u \cup v\rangle$ | MSI7 |
| $\langle a, u\rangle \mid (b\cdot x) = (a \mid b)\cdot(u \cup x)$ | MSI8 |
| $(a\cdot x) \mid \langle b, u\rangle = (a \mid b)\cdot(u \cup x)$ | MSI9 |
| | |
| $\partial_H(a) = a \qquad$ if $a \notin H$ | D1 |
| $\partial_H(a) = \delta \qquad$ if $a \in H$ | D2 |
| $\partial_H(x + y) = \partial_H(x) + \partial_H(y)$ | D3 |
| $\partial_H(x\cdot y) = \partial_H(x)\cdot\partial_H(y)$ | D4 |
| $\partial_H([u,x]) = [u, \partial_H(x)]$ | DSI1 |
| $\partial_H(\langle x, u\rangle) = \langle \partial_H(x), u\rangle$ | DSI2 |

TABLE 15. ACPS⁻.

The superscript $^-$ indicates that there is no communication by means of observation of signals. Addition of that feature will involve the addition of two more summands to the merge expansion axiom. The remaining axioms of ACPS$^-$ can now be added without leading to any inconsistency, but it is necessary to add some equations that describe the interaction of left merge and communication merge with the node labels. Notice that the axiom CM1 replaces the axiom M1, apart from this the family of axiom systems increases in a monotonic way.

## 3.4 CONDITIONALS

We can extend the results in chapter 2 about conditionals to a setting with parallel composition and communication if we add the following axioms.

$$(x \triangleleft \phi \triangleright y) \mathbin{\parallel} z = (x \mathbin{\parallel} z) \triangleleft \phi \triangleright (y \mathbin{\parallel} z)$$
$$(x \triangleleft \phi \triangleright y) \mid z = (x \mid z) \triangleleft \phi \triangleright (y \mid z)$$
$$x \mid (y \triangleleft \phi \triangleright z) = (x \mid y) \triangleleft \phi \triangleright (x \mid z)$$
$$\partial_H(x \triangleleft \phi \triangleright y) = \partial_H(x) \triangleleft \phi \triangleright \partial_H(y)$$

TABLE 16. Conditionals and parallel composition.

## 3.5 EXAMPLES OF SYSTEM SPECIFICATIONS IN ACPS$^-$

(i) Consider the traffic light system of example 3.2.i. Within ACPS$^-$ it is possible to introduce a control unit that performs the switching of the lights. New actions are needed to deal with switching. In particular we will need names for the switch  actions as performed by the control unit. The actions  set_to(color, number) have an evident meaning. Because the traffic light knows to which color it is set the actions change(n)  are still useful. Having the color as an additional parameter is not strictly needed for programming the control unit either but it serves to get a more readable specification. As new communications we will introduce:

change(n) | set_to(c,n) = c(n).

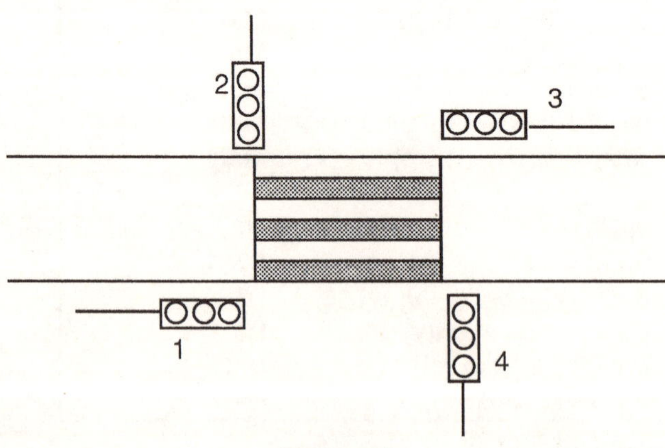

FIGURE 5.

All other communications are $\delta$. The encapsulation set H simply contains all actions that can engage in a nontrivial communication (the change and the set_to actions). Here it is understood that the traffic lights are numbered clockwise. The crossing that will be considered in more detail in this discussion is the one in the direction 2 to 4. We will assume that this crossing is for pedestrians. A picture of the situation is as shown above.

Pedestrians who want to cross from north to south (direction 2 to 4) will wait until the light at 2 is green. A specification for the control unit might be as follows.

```
control =
    (set_to(yellow, 1) ‖ set_to(yellow, 3))·waityellow(1, 3)·
    (set_to(red, 1) ‖ set_to(red, 3))·waitred(1, 3)·
    (set_to(green, 2) ‖ set_to(green, 4))·waitgreen(2, 4)·
    (set_to(yellow, 2) ‖ set_to(yellow, 4))·waityellow(2, 4)·
    (set_to(red, 2) ‖ set_to(red, 4))·waitred(2, 4)·
    (set_to(green, 1) ‖ set_to(green, 3))·waitgreen(1, 3)


loop = control·loop
```

The entire system is obtained by means of an encapsulated merge of the traffic lights and the control unit:

$$\text{SYSTEM} = \partial_H(\text{loop} \parallel \text{TL}(1\text{-}4)).$$

An interesting complication arises if one allows the control unit to receive requests for one of the directions. For instance one may assume that an action read(request(2→4)) is needed to trigger the transition to a state in which the 2 and 4 are green. The program may then look as follows:

```
control2 = (read(request(2→4)) + read(request(4→2)))·control
loop2 = control2·loop2
```

Of course this leads to

$$\text{SYSTEM2} = \partial_H(\text{loop2} \parallel \text{TL}(1\text{-}4)).$$

The third step is to add agents on the scene that want to use the traffic lights. For instance one may consider a pedestrian who wants to use the crossing in the north-south direction. If the pedestrian wants to cross from 2 to 4 (s)he will have to inspect the light at 2, and wait until it is green.

Here is a process describing this person:

```
ped(i) = arrive_at_2(i)·wait(i)·cross(2→4, i)
wait(i) = obs({green(2)}) + put(request(2→4))·wait(i).
```

The action obs({green(2)}) has no special meaning in this specification, be it that in an implementation one would like this action to be implemented in correspondence with its intuitive meaning, namely the observation of the green signal at light 2. In section 5 an observation mechanism will be introduced. Here we assume that put(request(2→4)) and read(request(2→4)) communicate, resulting in an action request(2→4).

H is then extended by the actions read(request(2→4)), yielding H*. The partial order ≤ says that request(2→4) has higher priority than the action put(request(2→4). This order is used as a parameter for the priority operator below. For information on the priority operator see [BBK 86]. For completeness sake we list the axiom system, that makes use of an auxiliary operator ◁ (*unless*). Unfortunately, this binary operator shares a symbol with the ternary conditional operator of 2.2. Appropriate use of brackets is needed to avoid ambiguities.

$$
\begin{array}{ll}
\theta(a) = a & \text{TH1} \\
\theta(x{\cdot}y) = \theta(x){\cdot}\theta(y) & \text{TH2} \\
\theta(x + y) = \theta(x){\triangleleft}y + \theta(y){\triangleleft}x & \text{TH3} \\
a{\triangleleft}b = a \qquad \text{if } \neg(a < b) & \text{P1} \\
a{\triangleleft}b = \delta \qquad \text{if } a < b & \text{P2} \\
x{\triangleleft}(y{\cdot}z) = x{\triangleleft}y & \text{P3} \\
x{\triangleleft}(y + z) = (x{\triangleleft}y){\triangleleft}z & \text{P4} \\
(x{\cdot}y){\triangleleft}z = (x{\triangleleft}z){\cdot}y & \text{P5} \\
(x + y){\triangleleft}z = x{\triangleleft}z + y{\triangleleft}z & \text{P6} \\
\\
\theta([u, x]) = [u, \theta(x)] & \text{THS1} \\
\theta(\langle x, u\rangle) = \langle\theta(x), u\rangle & \text{THS2} \\
x{\triangleleft}[u, y] = x{\triangleleft}y & \text{PS1} \\
x{\triangleleft}\langle y, u\rangle = x{\triangleleft}y & \text{PS2} \\
[u, x]{\triangleleft}y = [u, x{\triangleleft}y] & \text{PS3} \\
\langle x, u\rangle{\triangleleft}y = \langle x{\triangleleft}y, u\rangle & \text{PS4} \\
\\
\theta(x{\triangleleft}\phi{\triangleright}y) = \theta(x){\triangleleft}\phi{\triangleright}\theta(y) & \text{THC} \\
x{\triangleleft}(y{\triangleleft}\phi{\triangleright}z) = (x{\triangleleft}y){\triangleleft}\phi{\triangleright}(x{\triangleleft}z) & \text{PC1} \\
(x{\triangleleft}\phi{\triangleright}y){\triangleleft}z = (x{\triangleleft}z){\triangleleft}\phi{\triangleright}(y{\triangleleft}z) & \text{PC2}
\end{array}
$$

TABLE 17. Priority operator.

This use of the priority operator corresponds precisely to the put mechanism for synchronous but unreliable message passing (see [BW90]). The aim is to allow arbitrarily many put(request(2→4)) actions and to let a communication take place just if the put is a vital one in the sense that it conveys new information to the control unit. A system with three pedestrians about to cross from 2 to 4 is then as follows:

composite_system = θ≤(∂H*(ped(1) ‖ ped(2) ‖ ped(3) ‖ SYSTEM2)).

As was said before, the weakness of this combination is of course that the pedestrian takes no caution to prevent crossing when the light is red. Notice that the action observe_green _at_2 (i) will not be blocked when a green signal at 2 is absent. In the setting of ACPS⁻ that cannot be improved. This example is taken one step further in section 5, however. There, a feature is introduced that allows a proper interaction between the presence of the green signal at 2 and the proper execution of an action observe_green_at_ 2(i).

(ii) Consider once more the CLOCK of example 2.6.v. A plausible user of the clock can execute the following process user = uset_alarm(07.28.00). We assume that uset_alarm(t) can communicate with  the CLOCK action set_alarm(t)  to alarm(t). Let H = {uset_alarm(t), set_alarm(t) | t ∈ TIMES} then

    ∂_H(user ‖ CLOCK)

describes the proper cooperation between user and CLOCK.

(iii) Consider the example of a merge of light behaviours from example 3.2.ii. In this example a person is added to the scene. The person is supposed to enter the house at night and under the assumption that all lights are off. The actions that (s)he can perform are the actions pswitch(n) for n a name in {hall1, hall2, kitchen1, kitchen2, living1, living2, living3, living4}. The communication function works as follows: pswitch(n) | switch(n) = t.

  Here t denotes some action that plays the role of an internal step of the system. The encapsulation set H contains all switch and pswitch actions.The person can for instance execute the following process:

    person = enter_front_door · pswitch(hall1) · enter_living ·
        (pswitch(living1) ‖ pswitch(living2)) · pswitch(living3) ·
          enter_kitchen · pswitch(kitchen1) · leave_kitchen ·
            pswitch(hall1) · enter_living

The combined actions of system and person are then represented by the process expression

    person_in_house = ∂_H(person ‖ L(living/kitchen/hall)).

Let w be the signal light(hall1) & light(hall2) & light(living1) & light(living2) & light(living3) & light(living4) & light(kitchen). Then the following identity can be shown:

  w ∩ ∂_{get_new_lamp(n), defect(n) | n ∈ NAMES}(person_in_house) =
    enter_front_door ·
    t ·
    ({light(hall1)} ∪

```
(enter_living ·
  (((⟨t, {light(living1)}⟩ + ⟨t, {light(living2)}⟩) ·
   ⟨t, {light(living1), light(living2)}⟩ ·
   t ·
   ({light(living1), light(living2), light(living3)} ∪
       enter_kitchen ·
       t ·
       ({light(kitchen)} ∪
           leave_kitchen
       )
   )
 ) ·
 t ·
 ({light(kitchen), light(living1), light(living2), light(living3)} ∪
     enter_living ·
     δ
 ).
```

## 4. PROCESSES WITH FREE MERGE AND OBSERVATION

### 4.1 THE SIGNAL OBSERVATION FUNCTION

In this section an observation mechanism is added to the features available in process algebra. In order to simplify the discussion, the communication mechanism is first left out in order to be added later on, in section 5. It is assumed that some actions $a$ are able to read the signals of processes that are put in parallel with the process executing $a$.

We start out with the observation function as a function $\rho: A_\delta \times AS \to A_\delta$. An action $a \in A$ such that $obs(a, p) = b$ for some $b \in A$, $p \in AS$ is called an *observation action*. The signal observation function satisfies the following axioms:

$$
\begin{array}{|ll|}
\hline
\rho(\delta, p) = \delta & \text{OBS1} \\
\rho(\rho(a, p), q)) = \rho(\rho(a, q), p) & \text{OBS2} \\
\rho(\rho(a, p), p) = \rho(a, p) & \text{OBS3} \\
\hline
\end{array}
$$

TABLE 18. Observation function for atomic signals.

Then, we extend the observation function to a function $\rho: A_\delta \times CS \to A_\delta$ by means of the following additional axioms:

| | |
|---|---|
| $\rho(a, \emptyset) = a$ | OBS4 |
| $\rho(a, \{p\} \cup v) = \rho(\rho(a, p), v)$ | OBS5 |

TABLE 19. Observation function for composite signals.

From the observation function we will manufacture an additional operator on processes denoted with /. This *signal observation operator* describes the inspection of a signal by a process. The inspection of a signal will check whether or not some given atomic signal is contained in it. First, we give axioms for the signal observation operator on atomic actions, i.e. we consider $/: A_\delta \times CS \rightarrow P$. Notice that $a/u$ is always $\delta$, an atomic action or a sum of atomic actions.

| | |
|---|---|
| $\delta/u = \delta$ | O1 |
| $a/\emptyset = a$ | O2 |
| $a/v = \sum_{u \subseteq v} \rho(a, u)$ | O3 |

TABLE 20. Signal observation on atomic actions.

Notice that axioms O2 and O3 imply that $a/v$ always contains $a$ as a summand, and in case we have a trivial observation function ($\rho(a, p) = \delta$ for all $a \in A$), we get $a/v = a$ for all $v \in CS$.

Next we extend signal observation to a operator $/: P \times CS \rightarrow P$. The intuitive meaning of $x/u$ is a process that behaves like $x$ be it that the first action of $x$ is performed as an observation on the signal $u$.

| | |
|---|---|
| $(x + y)/u = x/u + y/u$ | O4 |
| $(x \cdot y)/u = (x/u) \cdot y$ | O5 |
| $[u, x]/v = [u, x/v]$ | O6 |
| $\langle x, u \rangle/v = \langle x/v, u \rangle$ | O7 |
| | |
| $(x \triangleleft \phi \triangleright y)/u = (x/u) \triangleleft \phi \triangleright (y/u)$ | O8 |
| $x/(u \triangleleft \phi \triangleright v) = (x/u) \triangleleft \phi \triangleright (x/v)$ | O9 |

TABLE 21. Signal observation on processes.

Note again that $x/v = x$ for all processes and signals in case of a trivial observation function.

## 4.2. PAS: SIGNAL OBSERVATION AND FREE MERGE

The merge expansion equation M1 of 3.1 has to be modified, in order to obtain an equation for a merge operator that takes signal observation into account as well. The following version of this axiom was suggested in BROUWER [BR 90]. This completes the description of the axiom system PAS.

Notice that this equation reduces to the equation M1 in case of a trivial observation function.

$$x \parallel y = (x/S(y)) \perp\!\!\!\perp y + (y/S(x)) \perp\!\!\!\perp x \qquad \text{OM}$$

TABLE 22. Free merge with observation.

In order to support the intuition for the observation mechanism we will now describe several examples of observation functions.

### 4.3 EXAMPLES OF SIGNAL OBSERVATION FUNCTIONS

In all examples below we specify the observation function $\rho: A_\delta \times AS \to A_\delta$. Thus, $p,q$ denote elements of $AS$.

i. The observation actions all have the form $obs(p)$ with $p$ an element of $AS$. These actions represent the intention to observe a signal $p$. There is a special atomic action $yes$ which is the result (confirmation) of a successful observation. The action $yes$ is not an observation action. The signal observation function then works as follows.

$$\rho(obs(p), p) = yes$$
$$\rho(obs(p), q) = \delta \qquad \text{if } p \neq q$$
$$\rho(a, p) = \delta \qquad \text{for all actions } a \text{ not of the form } obs(p).$$

In connection with this observation function one will use a form of encapsulation which shields off all observations except $yes$. In this way seen from outside the encapsulation context, only succesful observations will be visible.

In this format it suffices to indicate which non-trivial observations exist and to omit the complementary definition of the observations that lead to $\delta$.

Notice that an unsuccessful observation cannot be affirmed by an atomic action: if we put $\rho(obs(p), q) = no$ in case $p \neq q$, we get $obs(p) / \{p,q\} = yes + no$, which is certainly counter-intuitive.

ii. The second example takes the same sets of signals and observations as model (i) except for the addition of a label $p$ to the act of confirmation of successful observation, which thus becomes $yes(p)$. The definition is as follows.

$$\rho(obs(p), p) = yes(p)$$
$$\rho(obs(p), q) = \delta \qquad \text{if } p \neq q$$

All other observations lead to $\delta$ as in the previous example. This kind of observation allows one to keep track of the observed signals after encapsulation. The advantage being that no abstraction through renaming is involved and that a potential source of non-determinism is thereby removed. Notice that non-determinism in specifications is generally due to the level of abstraction rather than to the intrinsic non-determinism. Almost every system becomes non-deterministic when viewed at a sufficiently high level of abstraction.

iii. This example takes an element $u$ of $CS$ as the parameter of the observation actions. The observation $test(u)$ succeeds if some atomic signal $p$ in $u$ is present. In this case the observation function can be defined as follows.

$$\rho(test(u), p) = yes \qquad if\ p \in u$$
$$\rho(test(u), q) = \delta \qquad if\ p \notin u.$$

iv. We consider the generalization of the format in (i) above to actions $obs(u)$ with $u$ an element of $CS$. These actions represent the intention to observe all atomic signals in the composite signal $u$. The action $obs(\varnothing)$ will occur when all signals in $u$ have been observed.

$$\rho(obs(u), p) = obs(u-\{p\})$$

v. The last example generalizes (ii) as in the previous format. Action $obs(u,v)$ expresses that signals in $u$ still have to be observed, and signals in $v$ have been observed. We start with an action $obs(u,\varnothing)$ and complete success is then given by $obs(\varnothing,u)$. We call this observation function the *standard observation function*. The definition is as follows.

$$\rho(obs(u,v), p) = obs(u-\{p\}, v\cup(u\cap\{p\})).$$

## 5. ACPS: SYNCHRONOUS COMMUNICATION AND OBSERVATION

### 5.1 AXIOMS

Now we will combine the above theories $ACPS^-$ and PAS to obtain ACPS (ACP with signals and signal observation) by changing the axiom of the merge. The purpose of the following equations is to ensure that all operators except $+$, $\cdot$, $[.,.]$ and $\langle.,.\rangle$ can be eliminated from finite process expressions by means of left to right term rewriting. Notice that the original case of ACP can be viewed similarly with the understanding that all operators except $+$ and $\cdot$ can be eliminated in that case.

Besides taking the axioms for $ACPS^-$ and PAS together while taking the sum of their merge expansion axioms, three axioms are needed that ensure that observation and communication will not interfere. The purpose of these axioms is to guarantee that merge will be associative. It is possible that more general schemes exist, but in any case having an observation that communicates at the same time seems to be a rather fancy feature. In table 23, $a,b \in A$, $p \in AS$.

| | |
|---|---|
| $(\exists p \in AS \ \rho(a, p) \neq \delta) \Rightarrow a \mid b = \delta$ | OBS6 |
| $\rho(a, p) \mid b = \delta$ | OBS7 |
| $\rho(a \mid b, p) = \delta$ | OBS8 |

TABLE 23. Observation and communication.

The new expansion axiom for merge with communication is then as follows:

| | |
|---|---|
| $x \parallel y = (x/S(y)) \mathbin{\rlap{\rule[-.5ex]{1.2ex}{.1ex}}\kern.2ex L\kern-.6ex L} y + (y/S(x)) \mathbin{\rlap{\rule[-.5ex]{1.2ex}{.1ex}}\kern.2ex L\kern-.6ex L} x + x \mid y$ | OCM |

TABLE 24. Merge with observation and communication.

In case of a trivial observation function, this expression reduces to axiom CM1 of 3.3 and the interaction between processes works exactly as in ACP. ACP is therefore the special case of ACPS if the signal observation function vanishes everywhere. Similarly PA is the special case of ACP if the communication function vanishes everywhere. Obviously there is a situation where the communication function vanishes everywhere but the signal observation function may assume non-trivial values. In such cases one may omit the summand with the  communication merge and we get PAS.


## 5.2  EXAMPLES

(i) Let us reconsider the alarm clock of example 3.5.ii. We are now able to allow the user of the clock to observe the alarm and to switch off the clock subsequently. Let the user be described by the following process:

$$\text{user1} = \text{uset\_alarm}(07.00.00) \cdot \text{obs}(\{\text{alarm}\}, \varnothing) \cdot \text{uswitch\_off\_alarm}$$

Here we assume that uswitch_off_alarm communicates with switch_off_alarm to off_alarm. The standard signal observation from section 4.3.v is used.

Finally the encapsulation set H contains all actions uset_alarm(t)  as well as the action obs({alarm}, $\varnothing$). Then the cooperation of user and clock is given by:

$$\partial_H(\text{user1} \parallel \text{CLOCK}).$$

The specification allows but does not require the user to observe the alarm. If it is essential that the user hears the alarm a priority operator is needed that gives a higher priority to obs($\varnothing$, {alarm}) than to tick. A second step introduces a slightly more complex behavior for the user:

$$\text{user2} = \text{uset\_alarm}(07.00.00) \cdot \text{sleep} \cdot \text{user2loop}$$

$$\text{user2loop} = \text{obs}(\{\text{alarm}\}, \varnothing) \cdot \text{uswitch\_off\_alarm} + \text{wakeup} \cdot \text{sleep} \cdot \text{user2loop}$$

A third step is to make the behavior of the user again more complex, for instance if the user wakes up before the alarm has gone off (s)he may decide to sleep again without switching off the alarm provided $t < 06.45.00$. Here we need an observation for the time signal. We again use the standard signal observation function.

A new version user3 of the user then turns into:

$$user3 = uset\_alarm(07.00.00)\cdot sleep\cdot user3loop$$
$$user3loop = obs(\{alarm\}, \varnothing)\cdot uswitch\_off\_alarm +$$
$$wakeup\cdot(\sum_{r\in\{00.00.00,..,06.45.00\}} obs(\{r\},\varnothing)\cdot sleep\cdot user3loop \quad +$$
$$\sum_{r\in\{06.45.01,..,07.02.00\}} obs(\{r\},\varnothing)\cdot uswitch\_off\_alarm)$$

Of course the observations $obs(\{t\}, \varnothing)$ must be added to the encapsulation set H. Then the cooperation between user and alarm clock is again given by

$$\partial_H(user3 \parallel CLOCK).$$

Of course substantially more involved types of user behavior can be modelled using the primitives of ACPS.


(ii) Returning to the example of the traffic lights (3.5.i), using the signal observation function of 4.3.iv, one adds the actions $obs(\{green(k)\})$ to H. By simply working in the setting of ACPS the composite_system of example (3.5.i) will now work properly.


(iii) In this example we will elaborate once more on the example of the person in a house with various light switches (see 3.5.iii). The point here is that the person may want to inspect whether or not a light that was switched on indeed functions correctly. If not (s)he will replace the bulb by a new one. The actions that must be introduced here are inspections of the signals of the various lamps. We use the standard observation function. Recall that the atomic signals are in this case of the form light(n) and dark(n).

The action pswitch(n) can now be replaced, whenever it is assumed to switch the lamp on, by a more involved non-atomic process as follows:

$$pswitch(n)\cdot(obs(\{light(n)\},\varnothing) + obs(\{dark(n)\},\varnothing)\cdot put\_new\_lamp(n))$$

Of course the communication function has to be extended. In particular one needs a communication for the new lamp actions, for instance:

$$put\_new\_lamp(n) \mid get\_new\_lamp(n) = new\_lamp(n)$$

A modification of the person process from (3.5.iii) to incorporate this additional feature is as follows:

```
person* = enter_front_door · pswitch(hall1) ·
    (obs({light(hall1)},∅) + obs({dark(hall1)},∅)·put_new_lamp(hall1)) ·
        enter_living ·
    (pswitch(living1) ‖ pswitch(living2)) · pswitch(living3) ·
    (     (obs({light(living1)},∅) + obs({dark(living1)},∅)·put_new_lamp(living1)) ‖
          (obs({light(living2)},∅) + obs({dark(living2)},∅)·put_new_lamp(living2)) ‖
          (obs({light(living3)},∅) + obs({dark(living3)},∅)·put_new_lamp(living3)
    ) ·
    enter_kitchen · pswitch(kitchen1) ·
    (obs({light(kitchen)},∅) + obs({dark(kitchen)},∅)·put_new_lamp(kitchen)) ·
    leave_kitchen · pswitch(hall1) · enter_living
```

Composing this process **person*** with L(living/kitchen/hall) in the system ACPS will indeed produce a process that shows appropriate interaction of the person with its environment. Notice however that there are many different plausible behaviours of the person. There seems to be no universal generic behavior for **person** even in this very simple case. Obviously other factors that bear no relation to the switching of light influence the behavior of the person. Nevertheless it is a fair hypothesis to assume that the person behaves for some short time as a process. Thus locally (in time) **person** (or **person***) is a process in the sense of process algebra, but globally a much more complex stucturing mechanism is needed. One can imagine, however, that it is possible to formally incorporate in process algebra a small knowledge base which structures the decision taking process of the person.


## 6. FURTHER EXTENSIONS OF ACPS

Having the mechanisms of ACPS available it is natural to formulate many additional features on top of it which will increase expressive power without leading to semantical difficulties. In this section some of such mechanisms will be reviewed.


### 6.1 STATE OPERATORS THAT GENERATE SIGNALS

Let us assume that a state operator in the sense of [BB 88] is given by a domain S and functions act: $A_\delta \times S \to A_\delta$ and eff: $A \times S \to S$. The expression $\lambda_s(x)$ with $s \in S$ denotes process $x$ working on the state space S with the current state being $s \in S$.

We can assume that there is an additional function sig: $S \to CS$ which determines for each state the signal that is produced by that state. The absence of signals is modeled by taking sig(s) $= \varnothing$ of course. Now the six equations for the state operator are as shown below.

$$\begin{array}{ll}
\lambda_s(\delta) = [sig(s), \delta] & \text{LS1} \\
\lambda_s(a) = [sig(s), act(a, s)] & \text{LS2} \\
\lambda_s(a \cdot x) = [sig(s), act(a, s) \cdot \lambda_{eff(a, s)}(x)] & \text{LS3} \\
\lambda_s(x + y) = \lambda_s(x) + \lambda_s(y) & \text{LS4} \\
\lambda_s([u, x]) = [u, \lambda_s(x)] & \text{LS5} \\
\lambda_s(\langle x, u \rangle) = \langle \lambda_s(x), u \rangle & \text{LS6}
\end{array}$$

TABLE 25. State operator generating signals.

Using a state operator that generates signals one can define signaling processes in such a way that the recursion equations need not contain any signal at all, thus considerably optimizing the notation. We will illustrate this in two examples.

EXAMPLE 1.

Let D be a finite alphabet of data, and let ST(D) be the collection of finite sequences over D. The empty sequence is denoted by $\emptyset$ and adding an element d to the list x results in push(d, x). The atomic signals are as follows:

   top(d)      for $d \in D$,

   empty.

ST(D) will be the state space for a process that represents a stack over D. The signal function sig is de fined by $sig(\emptyset) = \{empty\}$, $sig(push(d, x)) = \{top(d)\}$. The atomic actions are:

   push_int(d), push(d) for $d \in D$ (the suffix int denotes an *intended* action),

   pop_int, pop.

The functions act and eff are given by:

   act(push_int(d), x) = push(d) (the act function transforms an intended action into an actual action),

   act(pop_int, x) = pop,

   eff(push_int(d), x) = push(d, x) (the eff function gives the resulting contents of the stack),

   $eff(pop\_int, \emptyset) = \emptyset$,

   eff(pop_int, push(d, x)) = x.

(For act only those cases are given where act will not lead to $\delta$.). The behavior of a stack over D is the given by the following process definition.

   $stack(D) = \lambda_\emptyset(ST\_INT)$

   $ST\_INT = (\sum_{d \in D} push\_int(d) + pop\_int) \cdot ST\_INT$.

EXAMPLE 2.

In this example two buffers A and B with data from the finite set D are maintained in the state. Both buffers have length $k > 1$. The process to be defined allows to read data in both buffers in a concurrent mode. For both buffers A and B there are two atomic signals: open_A indicates that there is still room in A, closed_A indicates that A has been filled (likewise for B). When both buffers have been loaded the action comp compares the contents of the buffers. The comparison will send value true if the buffers were equal and false otherwise. Thereafter the buffers are made empty again and the process restarts. We will describe the system in a top-down fashion, first explaining the overall architecture and then completing the details. As a notational convention we have adopted moreover that intended actions (which have not yet been processed by the state operator but for which this is envisaged in the design) are marked with the suffix int.

$$\text{SYSTEM} = \lambda_{\langle\varnothing,\varnothing\rangle}\circ\partial_H (X)$$
$$X = (\text{int\_A} \parallel \text{int\_B})\cdot X.$$

The state consists of a pair of buffers A and B. Initially both are empty. The signals produced by a state $\langle\alpha, \beta\rangle$ are as follows.

$$
\begin{aligned}
\text{sig\_A}(\langle\alpha, \beta\rangle) = \quad &\{\text{open\_A}\} &&\text{if length}(\alpha) < k, \\
&\{\text{closed\_A}\} &&\text{if length}(\alpha) = k. \\
\text{sig\_B}(\langle\alpha, \beta\rangle) = \quad &\{\text{open\_B}\} &&\text{if length}(\beta) < k, \\
&\{\text{closed\_B}\} &&\text{if length}(\beta) = k.
\end{aligned}
$$
$$\text{sig}(x) = \text{sig\_A}(x) \cup \text{sig\_b}(x).$$

The processes int_A and int_B are defined by:
$$\text{int\_A} = \sum_{d\in D} \text{read\_int\_A}(d)\cdot\text{int\_A} + \text{comp\_int\_A}$$
$$\text{int\_B} = \sum_{d\in D} \text{read\_int\_B}(d)\cdot\text{int\_B} + \text{comp\_int\_B}$$

Communication is defined as follows:
$$\text{comp\_int\_A} \mid \text{comp\_int\_B} = \text{comp\_int},$$
$$H = \{\text{comp\_int\_A}, \text{comp\_int\_B}\}$$

The next step is to explain the effect function for those actions that pass the encapsulation operator:

$$
\begin{aligned}
\text{eff}(\text{read\_int\_A}(d), \langle\alpha, \beta\rangle) = \quad &\alpha * d &&\text{if length}(\alpha) < k \\
&\alpha &&\text{otherwise} \\
\text{eff}(\text{read\_int\_B}(d), \langle\alpha, \beta\rangle) = \quad &\beta * d &&\text{if length}(\beta) < k \\
&\beta &&\text{otherwise} \\
\text{eff}(\text{comp\_int}, \langle\alpha, \beta\rangle) = \quad &\varnothing
\end{aligned}
$$

Finally the action function must be specified:

$$
\begin{aligned}
\text{act}(\text{read\_int\_A}(d), \langle\alpha, \beta\rangle) = \quad &\text{read\_A}(d) &&\text{if length}(\alpha) < k \\
&\delta &&\text{otherwise} \\
\text{act}(\text{read\_int\_B}(d), \langle\alpha, \beta\rangle) = \quad &\text{read\_B}(d) &&\text{if length}(\beta) < k
\end{aligned}
$$

$$\text{act(comp\_int, } \langle \alpha, \beta \rangle) = \begin{cases} \delta & \text{otherwise} \\ \text{write(true)} & \text{if lenght}(\alpha) = k \text{ and } \alpha = \beta \\ \text{write(false)} & \text{if length}(\alpha) = \text{length}(\beta) = k \text{ and } \alpha \neq \beta, \\ \delta & \text{otherwise.} \end{cases}$$

The use of the state operator in this example is hard to avoid because of the parallel reading of data that must be used simultaneously later on. This issue is worked out in [VE 90].


## 6.2 STATE OPERATOR WITH CONDITIONS

In case we have conditions over an arbitrary Boolean algebra $\mathbb{B}_n$ with $n$ generators, we need a function eval: $\mathbb{B}_n \times S \to \mathbb{B}_n$ ($S$ the set of states) satisfying the following axioms. With the help of this function we can give an axiom for the state operator on conditional expressions.

| | |
|---|---|
| $\text{eval(s, true)} = \text{true}$ | E1 |
| $\text{eval(s, false)} = \text{false}$ | E2 |
| $\text{eval}(s, \phi \wedge \psi) = \text{eval}(s, \phi) \wedge \text{eval}(s, \psi)$ | E3 |
| $\text{eval}(s, \phi \vee \psi) = \text{eval}(s, \phi) \vee \text{eval}(s, \psi)$ | E4 |
| $\text{eval}(s, \neg\phi) = \neg\text{eval}(s, \phi)$ | E5 |
| | |
| $\lambda_s(x \triangleleft \phi \triangleright y) = \lambda_s(x) \triangleleft \text{eval}(\phi, s) \triangleright \lambda_s(y)$ | LC |

TABLE 26. State operator and conditions.


## 6.3 INFINITE NUMBER OF ATOMIC SIGNALS

This extension is a rather plausible one. Indeed there is no particular reason why the set of atomic signals must be finite as long as all composite signals that are used in a process are finite combinations of atomic signals. For instance let the natural numbers be written in unary form by means of $0$ and a function $S$, act as atomic signals. Then process count(k) for $k \in \mathbb{N}$ can be defined as follows:

$$\text{count}(k) = [\{S^k(0)\}, t] \cdot (\text{count}(k + 1) + \text{stop}).$$

Assume the presence of an observation action above(10) that allows the following observations (in the format of 4.3.i): $\text{obs(above(10), } S^{k+10}(0)) = \text{yes}$.

Let the only non-trivial communication be stop | yes = end, then with $H = \{\text{above(10), stop}\}$, the process $\partial_H(\text{count(0)} \parallel \text{above(10)})$ denotes a process that will count until some value above 9 and then terminate. Under the assumption that there are only finitely many atomic actions, the use of these signals taken from an arbitrary data type is only limited as far as process interaction within process algebra is concerned. Nevertheless, one can imagine that an external observer can indeed distinguish all signals so that the complex signals are useful to model unintended external behavior.

## 6.4 AN ALTERNATIVE NOTATION FOR ROOT SIGNAL INSERTION

An alternative to the use of the binary root signal insertion operator [.,.] is to introduce the unary deadlock signal mapping $\delta(u)$ which is defined by $\delta(u) = [u, \delta]$.

In fact root signal insertion can be expressed using this operation:

$$[u, x] = x + \delta(u).$$

Because the deadlock signal mapping and root signal insertion are expressible with respect to one another it is possible to provide an axiomatisation of the algebra of communicating processes with signals using $\delta(.)$ rather than [.,.]. In terms of the complexity of the signature that leads to a substantial simplification, because a unary operator is a simpler concept than a binary one. The main reason, however, not to use the deadlock signal operator instead of root signal insertion is that it would lead to a set of axioms which is considerably harder to read than the given equations.

## 7. BISIMULATION SEMANTICS FOR PROCESSES WITH SIGNALS

### 7.1 PROCESS GRAPHS WITH NODE LABELS

Many possible semantic worlds exist for ACP. It is impossible and unnecessary to review all models of importance here. The bisimulation model being the initial algebra semantics of ACP stands out as the most important model in our view, in spite of its ignorance of the concept of true concurrency and it not being fully abstract. We will indicate how a bisimulation semantics can be obtained for processes with signals. This new notion of bisimulation specialises to the original form in the case of processes with trivial signals only.

We assume that a finite set $A$ of atoms is given as well as a finite set $AS$ of signals. We assume $A$ and $AS$ to be disjoint. Outside $A$ we have an additional constant $\delta$. A process in the setting of this simple form of bisimulation semantics is a finitely branching, directed, acyclic and rooted graph, with nodes labeled by a signal in $CS$ (a subset of $AS$) and edges labeled by actions in $A$. Terminal nodes in the graph may have an additional label $\delta$ besides the signal label from $CS$. This indicates that the branch terminates in deadlock. Absence of the label $\delta$ at a terminal node implies proper and correct termination.

The interpretation of the operators is as follows.

- $\delta$ corresponds to the single node graph with two labels: $\varnothing$ and $\delta$.
- The atomic action $a$ corresponds to a graph with two nodes and one connecting edge. Both nodes are labeled with $\varnothing$ and the edge is labeled with $a$.

- Given graphs $g$ and $h$, their sum is obtained by identifying the root nodes and taking the union of the root signals. Only if both roots have label $\delta$, the new root obtains a label $\delta$ as well.

- The product $g \cdot h$ of two graphs $g$ and $h$ is obtained from $g$ by attaching a copy of $h$ at each terminal node of $g$ that has no label $\delta$. The signal labels of the terminal nodes of $g$ in $g \cdot h$ are recomputed by taking the union in each case with the root signal of $Q$.

- The graph $[u, g]$ is obtained from $g$ by combining its root signal with $u$, and the graph $\langle g, u \rangle$ is found from $g$ by combining each signal label of a properly terminating node of $g$ with $u$.

- $S(g)$ is the root signal of $g$ and $P(g)$ results from $g$ by replacing the root signal of $g$ by $\varnothing$.

- The merge of process graphs $g$ and $h$ without communication corresponds to taking the cartesian product of the two graphs. The signal of a pair $(p, q)$ of nodes is just the combination of the signals of the individual nodes. A transition $(p, q) \xrightarrow{a} (p', q')$ exists if either $p \xrightarrow{a} p'$ and $q = q'$ or $q \xrightarrow{a} q'$ and $p = p'$. In the case of communication the diagonal transitions $(p, q) \xrightarrow{c} (p', q')$ are added whenever there is a non-trivial communication $c$ between actions $a$ and $b$ such that $p \xrightarrow{a} p'$ and $q \xrightarrow{b} q'$. The left merge is constructed by first defining the merge and then removing all initial transitions that correspond to first move of the second argument of the left merge. Similarly a communication merge is defined starting from the merge by removing al initial transitions that are not communications. Of course both in the case of the left merge and in the case of the merge there may result inaccessible parts of the graph after deletion of the relevant initial transitions. These superfluous parts of the process graphs of course have to be removed  afterwards.

Finally the presence of signal observations must be taken into account. The merge of two processes in the case of ACPS starts with constructing their merge in the sense of ACPS⁻. The domain of the new graph is then obtained but more edges have to be added in order to take the succesful observations into account. If $q = q'$, $p \xrightarrow{a} p'$ and $a / S(q) \xrightarrow{b} \surd$, or $p = p'$, $q \xrightarrow{a} q'$ and $a / S(p) \xrightarrow{b} \surd$, then a transition $(p, q) \xrightarrow{b} (p', q')$ is added.

Equally interesting and informative as these graph constructions is an operational semantics based on actions and labeled transitions between expressions over the free syntax of ACPS. Such an operational semantics is given in section 8.


7.2 BISIMULATION FOR PROCESS GRAPHS WITH NODE LABELS

A *bisimulation* between graphs $g$ and $h$ is a bisimulation in the sense usual for process algebra with the additional restriction that if a bisimulation $R$ relates a pair $(s, t)$ of states of $g$ and $h$ the labels of the nodes $s$ and $t$ must coincide. In this way a model of BPAS is obtained. It can be shown that BPAS provides a complete axiomatisation for bisimulation congruence on processes with signals. The proof uses the second canonical form as described below and is postponed until after the discussion of these canonical forms. Without proof we state the following properties of bisimulation semantics on processes with signals.

(i)  In the initial algebra of ACPS the merge operator is associative.

(ii) The initial algebra of ACPS is an expansion of the initial algebra of ACP, whence ACPS is a conservative enrichment of ACP (at least as far as identities between finite closed process expressions are concerned).

(iii) The initial algebra of ACPS is isomorphic to the bisimulation model with appropriate definitions of all operators other than + and $\cdot$.

(iv) ACPS constitutes (on finite closed process expressions) a terminating and confluent term rewriting system modulo its permutative identities for +.

(v) The notion of a regular process is defined just as in the case without signals. A process graph is regular if it bisimulates with a finite graph. All regular processes with signals can be specified by means of finite linear systems of equations. All operations on processes described in this paper transform regular processes into regular processes. The algebra of regular processes exists just as well for ACPS as it does for ACP.

(vi) ACPS allows a projective limit model in very much the same way as ACP does.


The equations of ACPS first and for all express an intuition. The virtue of these equations is to be consistent in bisimulation semantics, secondly and not of less importance to allow the transformation of each finite process expression in each of several canonical forms. These canonical forms are discussed below. Finally an aim is to show the axioms complete with respect to bisimulation semantics, at least in the case of finite processes. A third objective is to anticipate for guarded recursive definitions and to enforce that finite projections of processes defined with guarded recursion can be transformed to finite process expressions without recursion. These finite process expressions can then be transformed into one of the canonical forms.


### 7.3 CANONICAL FORMS FOR ACPS

The canonical forms for finite processes are less easily established in the presence of signals and observations. First of all one observes that each process graph can be unfolded into a bisimilar tree. The canonical forms are all derived from the tree representation. Consider the following classes of finite closed process expressions.


(i) CANP1.

- For $u$ in CS, $[u, \delta]$ is in CANP1.

- If $n, m, k \geq 0$ with $n + m + k > 0$, and $a_1,...,a_n$, $b_1,...,b_m$, $c_1,...,c_k$ are atomic actions different from $\delta$; if moreover $u, u_1,..., u_m$ are signals in CS and $X_1,...,X_n$ are in CANP1 then

$$[u, a_1 \cdot X_1 + ... + a_n \cdot X_n + \langle b_1, u_1 \rangle + ... + \langle b_m, u_m \rangle + ... + c_1 + ... + c_k]$$

is an expression in CANP1.

In CANP1 the number of occurrences of $\langle .,. \rangle$ is minimised. Moreover each node label is displayed at exactly one position in the expression. Closing brackets of $[.,.]$ are moved to the last possible position. This type of canonical form arises from the most obvious translation of processes as labeled trees into process expressions. These canonical forms are needed if one is to understand the equations for / above. CANP1 contains exactly the normal forms of expressions according to BPAS.

(ii) CANP2.
- For $u$ in CS, $[u, \delta]$ is in CANP2.
- If $n, m, k \geq 0$ with $n + m + k > 0$, and $a_1,...,a_n$, $b_1,...,b_m$, $c_1,...,c_k$ are atomic actions different from $\delta$; if moreover $u, u_1,..., u_m$ are signals in CS and $X_1,...,X_n$ are in CANP2 then

$$[u, a_1]\cdot X_1 + ... + [u, a_n]\cdot X_n + [u, \langle b_1, u_1 \rangle] + ... + [u, \langle b_m, u_m \rangle] + ... + [u, c_1] + ... + [u, c_k]$$

is an expression in CANP2.

The expressions in CANP2 have the disadvantage that the root label of each subexpression is duplicated for every branch of the subexpression. The advantage of this kind of canonical form lies in the simple definition of finite projections and the corresponding option to consider infinite expressions as limits of finite expressions and to manufacture a projective limit model in the style of [BK 84].

(iii) CANP3.
- For $u$ in CS, $[u, \delta]$ is in CANP3.
- If $n, m, k \geq 0$ with $n + m + k > 0$, and $a_1,...,a_n$, $b_1,...,b_m$, $c_1,...,c_k$ are atomic actions different from $\delta$; if moreover $u, v_1,..., v_n, u_1,..., u_m$ are signals in CS and $X_1,...,X_n$ are in CANP3 such that for $X_i$ the root signal is $v_i$ then

$$[u, \langle a_1,v_1 \rangle]\cdot X_1 +...+ [u, \langle a_n,v_n \rangle]\cdot X_n + [u, \langle b_1,u_1 \rangle] +.. + [u, \langle b_m,u_m \rangle] +.. + [u, c_1] +.. + [u, c_k]$$

is an expression in CANP3.

The expressions in CANP3 have the disadvantage that the root label and terminal label of each subexpression is duplicated for every branch of the subexpression. The advantage is that this form is useful if one is to translate the signal observation mechanism into the more primitive communication mechanism.

(iv) CANP4.
Let CANP4* be the following class of expressions:
- $\delta$ is in CANP4*.
- If $n, m, k \geq 0$ with $n + m + k > 0$, and $a_1,...,a_n$, $b_1,...,b_m$, $c_1,...,c_k$ are atomic actions different from $\delta$; if moreover $v_1,..., v_n, u_1,..., u_m$ are signals in CS and $X_1,...,X_n$ are in CANP4* then

$$\langle a_1, v_1 \rangle\cdot X_1 + ... + \langle a_n, v_n \rangle\cdot X_n + \langle b_1, u_1 \rangle + ... + \langle b_m, u_m \rangle + ... + c_1 + ... + c_k$$

is an expression in CANP4*.

Now CANP4 consists of all expressions of the form $[u, X]$ with $X$ an expression in CANP4*. Clearly the point of CANP4 is to minimise the use of $[.,.]$.

(v) CANP5.

If $n, m, k \geq 0$ and $a_1,...,a_n$, $b_1,...,b_m$, $c_1,...,c_k$ are atomic actions different from $\delta$; if moreover $u, u_1,..., u_m$ are signals in CS and $X_1,...,X_n$ are in CANP5 then

$$[u, \delta] + a_1 \cdot X_1 + ... + a_n \cdot X_n + \langle b_1, u_1 \rangle + ... + \langle b_m, u_m \rangle + ... + c_1 + ... + c_k$$

is an expression in CANP5.

These normal forms depend entirely on the presence of $\delta$. Clearly these forms are very simple and in particular this form helps in the design of recursion equations for parametrised processes with signals. Of course this canonical form cannot be used in absence of the deadlock constant.

### 7.4 COMPLETENESS OF THE AXIOMS FOR BISIMULATION IN THE CASE OF FINITE PROCESSES

Using the third canonical form, a completeness proof for bisimulation congruence for processes with signals can be given. We will consider the simplest case without deadlock. We also have to assume that the set of atomic signals, AS, is finite. One introduces a new set of atomic actions, viz. for each atomic action $a$ and signals $u, v$ we have $[u, \langle a, v \rangle]$ as a new atomic action. Let B be the alphabet of these new actions. Let CANP3 be the collection of process expressions over B that is in the third canonical form.

Let P and Q be two bisimilar process expressions with signals. Using the equations of BPAS both expressions can be transformed into a CANP3 form, this results in P' and Q'. Now observe that two processes in third canonical form bisimulate in the new sense if and only if they bisimulate as processes over B in the usual sense. Therefore P' and Q' are bisimilar as processes over B. Because BPA is a complete axiomatisation for bisimulation semantics on finite processes P' and Q' can be proven equal by means of axioms of BPA and considering them as processes over B. It turns out however that the axioms of BPA are just as valid for P' and Q' seen as CANP3 expressions. Indeed the axioms of BPA when applied as rewriting rules to CANP3 expressions do not lead outside that class.

### 7.5 REMARK

Models in the presence of conditonal expressions can be given similarly, be it that transitions will be labeled by pairs of actions and (enabling) conditions, and a notion of bisimulation must be adapted to that setting. As an example, we would have $\phi{:}\to (\neg\psi {:}\to a{\cdot}(b + c)) \xrightarrow{a, \phi \wedge \neg \psi} (b + c)$.

We will not do so at this point. Instead, we will provide a structural operational semantics involving signals in 8.2.

## 8. AN OPERATIONAL SEMANTICS FOR ACPS

### 8.1 RULES FOR ACP AND ACPS

An operational semantics of ACP is nothing new and can be retrieved from several sources, such as [VGL 87]. We prefer the approach in which only the operators of ACP that genuinely represent program constructions are provided with an operational semantics. The motivation for this style is the point of view that the auxiliary operators are there to facilitate the algebra but that the primary operators have an importance that far exceeds that of the algebra. Therefore the operational semantics of the syntax for process expressions should be accessible also to readers who have no knowledge of the auxiliary operators and their sometimes not quite straightforward role.

In the case of ACPS one assumes that the communication function is given as a partial function on the atomic actions (i.e. not for $\delta$) and that the signal observation function is given as a total operator on atomic actions and atomic signals. Each of these operations must satisfy the required axioms.

In these circumstances the proper atomic actions and the operators $+$, $\cdot$, $[.,.]$, $\langle.,.\rangle$, $\parallel$, $\partial_H$ represent appropriate system description mechanisms, all others are auxiliary operators used to make the algebraic specification possible or concise. We need the notation $\sqrt{}(u)$ to indicate that a process terminates in a proper terminal state with node label $u$. As an abbreviation for $\sqrt{}(\emptyset)$ we use the simpler notation $\sqrt{}$. These expressions are itself not a process expression and will not occur in complicated ways as a subexpression of large process expression. In the rules below $X$ and $Y$ denote arbitrary process expressions.

$$a \xrightarrow{a} \sqrt{}$$

$$\frac{X \xrightarrow{a} X'}{X + Y \xrightarrow{a} X'} \qquad \frac{X \xrightarrow{a} \sqrt{}(u)}{X + Y \xrightarrow{a} \sqrt{}(u)}$$

$$\frac{Y \xrightarrow{a} Y'}{X + Y \xrightarrow{a} Y'} \qquad \frac{Y \xrightarrow{a} \sqrt{}(u)}{X + Y \xrightarrow{a} \sqrt{}(u)}$$

$$\frac{X \xrightarrow{a} X'}{X \cdot Y \xrightarrow{a} X' \cdot Y} \qquad \frac{X \xrightarrow{a} \sqrt{}(u)}{X \cdot Y \xrightarrow{a} [u, Y]}$$

$$\frac{X \xrightarrow{a} X'}{\langle X, u \rangle \xrightarrow{a} \langle X', u \rangle} \qquad \frac{X \xrightarrow{a} \sqrt{}(v)}{\langle X, u \rangle \xrightarrow{a} \sqrt{}(v \cup u)}$$

$$\frac{X \xrightarrow{a} X'}{[u, X] \xrightarrow{a} X'} \qquad \frac{X \xrightarrow{a} \sqrt{}(v)}{[u, X] \xrightarrow{a} \sqrt{}(v)}$$

TABLE 27. Rules for BPAS.

$$\frac{X \xrightarrow{a} X'}{X \| Y \xrightarrow{a} X' \| Y} \qquad \frac{X \xrightarrow{a} \sqrt{}(u)}{X \| Y \xrightarrow{a} u \cup Y}$$

$$\frac{Y \xrightarrow{a} Y'}{X \| Y \xrightarrow{a} X \| Y'} \qquad \frac{Y \xrightarrow{a} \sqrt{}(u)}{X \| Y \xrightarrow{a} u \cup X}$$

TABLE 28. Additional rules for PAS$^-$.

$$\frac{X \xrightarrow{a} X',\, Y \xrightarrow{b} Y',\, a \mid b = c}{X \| Y \xrightarrow{c} X' \| Y'} \qquad \frac{X \xrightarrow{a} \sqrt{}(u),\, Y \xrightarrow{b} Y',\, a \mid b = c}{X \| Y \xrightarrow{c} u \cup Y'}$$

$$\frac{X \xrightarrow{a} X',\, Y \xrightarrow{b} \sqrt{}(u),\, a \mid b = c}{X \| Y \xrightarrow{c} u \cup X'} \qquad \frac{X \xrightarrow{a} \sqrt{}(u),\, Y \xrightarrow{b} \sqrt{}(v),\, a \mid b = c}{X \| Y \xrightarrow{c} \sqrt{}(u \cup v)}$$

$$\frac{X \xrightarrow{a} X',\, a \notin H}{\partial_H(X) \xrightarrow{a} \partial_H(X')} \qquad \frac{X \xrightarrow{a} \sqrt{}(u),\, a \notin H}{\partial_H(X) \xrightarrow{a} \sqrt{}(u)}$$

TABLE 29. Additional rules for ACPS$^-$.

$$\frac{X \xrightarrow{a} X',\, \rho(a,u) = b \neq \delta,\, u \subseteq S(Y)}{X \| Y \xrightarrow{b} X' \| Y} \qquad \frac{X \xrightarrow{a} \sqrt{}(v),\, \rho(a,u) = b \neq \delta,\, u \subseteq S(Y)}{X \| Y \xrightarrow{b} v \cup Y}$$

$$\frac{Y \xrightarrow{a} Y',\, \rho(a,u) = b \neq \delta,\, u \subseteq S(X)}{X \| Y \xrightarrow{b} X \| Y'} \qquad \frac{Y \xrightarrow{a} \sqrt{}(v),\, \rho(a,u) = b \neq \delta,\, u \subseteq S(X)}{X \| Y \xrightarrow{b} v \cup X}$$

TABLE 30. Additional rules for ACPS.

In the two next rules $E$ is a (guarded) system of recursion equations that contains the equation $X = s$. $\langle X \mid E \rangle$ denotes a process that is a solution for this equation. $\langle s \mid E \rangle$ denotes a process as follows: $s$ is a process expression with some free variables in $E$, for these process variables the unique solution of these variables in $E$ is substituted.

$$\frac{\langle s \mid E \rangle \xrightarrow{a} Y}{\langle X \mid E \rangle \xrightarrow{a} Y} \qquad \frac{\langle s \mid E \rangle \xrightarrow{a} \sqrt{}(u)}{\langle X \mid E \rangle \xrightarrow{a} \sqrt{}(u)}$$

TABLE 31. Rules for recursion.

## 8.2 CONDITIONALS AND STATE OPERATOR

In the presence of conditionals over an arbitrary finite Boolean algebra $\mathbb{B}_n$, we will label the transitions with pairs $a,\phi$, where $a \in A$ is an atomic action (so $a \neq \delta$) and $\phi$ a non-false expression over $\mathbb{B}_n$ ($\phi \neq \text{false}$). We obtain the following rules.

$$a \xrightarrow{a,\text{true}} \sqrt{}$$

$$\frac{X \xrightarrow{a,\phi} X'}{X + Y \xrightarrow{a,\phi} X'} \qquad \frac{X \xrightarrow{a,\phi} \sqrt{}(u)}{X + Y \xrightarrow{a,\phi} \sqrt{}(u)}$$

$$\frac{Y \xrightarrow{a,\phi} Y'}{X + Y \xrightarrow{a,\phi} Y'} \qquad \frac{Y \xrightarrow{a,\phi} \sqrt{}(u)}{X + Y \xrightarrow{a,\phi} \sqrt{}(u)}$$

$$\frac{X \xrightarrow{a,\phi} X'}{X \cdot Y \xrightarrow{a,\phi} X' \cdot Y} \qquad \frac{X \xrightarrow{a,\phi} \sqrt{}(u)}{X \cdot Y \xrightarrow{a,\phi} [u,Y]}$$

$$\frac{X \xrightarrow{a,\phi} X'}{\langle X,u \rangle \xrightarrow{a,\phi} \langle X',u \rangle} \qquad \frac{X \xrightarrow{a,\phi} \sqrt{}(v)}{\langle X,u \rangle \xrightarrow{a,\phi} \sqrt{}(v \cup u)}$$

$$\frac{X \xrightarrow{a,\phi} X'}{[u,X] \xrightarrow{a,\phi} [u,X']} \qquad \frac{X \xrightarrow{a,\phi} \sqrt{}(v)}{[u,X] \xrightarrow{a,\phi} \sqrt{}(v)}$$

$$\frac{X \xrightarrow{a,\phi} X', \ \phi \wedge \psi \neq \text{false}}{X \triangleleft \psi \triangleright Y \xrightarrow{a,\phi \wedge \psi} X'} \qquad \frac{X \xrightarrow{a,\phi} \sqrt{}(u), \ \phi \wedge \psi \neq \text{false}}{X \triangleleft \psi \triangleright Y \xrightarrow{a,\phi \wedge \psi} \sqrt{}(u)}$$

$$\frac{Y \xrightarrow{a,\phi} Y', \ \phi \wedge \neg \psi \neq \text{false}}{X \triangleleft \psi \triangleright Y \xrightarrow{a,\phi \wedge \neg \psi} Y'} \qquad \frac{Y \xrightarrow{a,\phi} \sqrt{}(u), \ \phi \wedge \neg \psi \neq \text{false}}{X \triangleleft \psi \triangleright Y \xrightarrow{a,\phi \wedge \neg \psi} \sqrt{}(u)}$$

TABLE 32. Rules for BPAS with conditionals.

Adding the rules for parallel composition is straightforward. It is interesting to look at the rules for the state operator. This works as shown in table 33.

With this extended operational semantics involving conditions on the arrows comes a new definition for bisimulation. Instead of just requiring matching actions, we also require matching conditions. The following defintion starts from the set of valuations of the generators of the boolean algebra, i.e. all mappings $v: \{\theta_1,...,\theta_n\} \to \text{BOOL}$. Each such mapping naturally extends to a mapping $v: \mathbb{B}_n \to \text{BOOL}$. Then we say that a relation $\approx$ on a transition system is a *bisimulation* when the following holds:

i.   if $X \approx Y$ then $S(X) = S(Y)$

314

$$\frac{X \xrightarrow{a,\phi} X', \; act(a,s)=b\neq\delta, \; eval(s,\phi)=\psi\neq false}{\lambda_s(X) \xrightarrow{b,\psi} \lambda_{eff(a,s)}(X')}$$

$$\frac{X \xrightarrow{a,\phi} \surd(u), \; act(a,s)=b\neq\delta, \; eval(s,\phi)=\psi\neq false}{\lambda_s(X) \xrightarrow{b,\psi} \surd(u)}$$

TABLE 33. Rules for state operator.

ii.  if $X \approx Y$ and $X \xrightarrow{a,\phi} X'$, then for all valuations $v$ with $v(\phi) = \text{true}$, there is a condition $\psi$ with $v(\psi) = \text{true}$ and an expression $Y'$ such that $Y \xrightarrow{a,\psi} Y'$ and $X' \approx Y'$

iii.  if $X \approx Y$ and $Y \xrightarrow{a,\phi} Y'$, then for all valuations $v$ with $v(\phi) = \text{true}$, there is a condition $\psi$ with $v(\psi) = \text{true}$ and an expression $X'$ such that $X \xrightarrow{a,\psi} X'$ and $X' \approx Y'$

iv.  if $X \approx Y$ and $X \xrightarrow{a,\phi} \surd(u)$ then for all valuations $v$ with $v(\phi) = \text{true}$, there is a condition $\psi$ with $v(\psi) = \text{true}$ such that $Y \xrightarrow{a,\psi} \surd(u)$

v.  if $X \approx Y$ and $Y \xrightarrow{a,\phi} \surd(u)$ then for all valuations $v$ with $v(\phi) = \text{true}$, there is a condition $\psi$ with $v(\psi) = \text{true}$ such that $X \xrightarrow{a,\psi} \surd(u)$.

We call two expressions $X, Y$ bisimilar if there is a bisimulation relating $X$ and $Y$. We claim that transition systems modulo bisimulation form a model for our theory. It is an open question whether our axiomatization is complete for this model.

8.3 STANDARD SIGNAL OBSERVATION FUNCTION

In ACPS it is quite often convenient to have a fixed signal observation function in mind. This was already mentioned in 4.3 and here, we give an operational semantics for some of the functions described there. First the case 4.3.i. Then, the rules in table 30 specialize to the following form.

$$\frac{X \xrightarrow{obs(p)} X', \; p\in S(Y)}{X\Vert Y \xrightarrow{yes} X'\Vert Y} \qquad \frac{X \xrightarrow{obs(p)} \surd(u), \; p\in S(Y)}{X\Vert Y \xrightarrow{yes} u\cup Y}$$

$$\frac{Y \xrightarrow{obs(p)} Y', \; p\in S(X)}{X\Vert Y \xrightarrow{yes} X\Vert Y'} \qquad \frac{Y \xrightarrow{obs(p)} \surd(u), \; p\in S(X)}{X\Vert Y \xrightarrow{yes} u\cup X}$$

TABLE 34. Signal observation function of 4.3.i.

As another example, we can consider the format 4.3.iv. This looks as follows.

$$\frac{X \xrightarrow{obs(\underline{u})} X',\ w \subseteq u \cap S(Y)}{X \| Y \xrightarrow{obs(u-w)} X' \| Y} \qquad \frac{X \xrightarrow{obs(\underline{u})} \sqrt{(v)},\ w \subseteq u \cap S(Y)}{X \| Y \xrightarrow{obs(u-w)} v \cup Y}$$

$$\frac{Y \xrightarrow{obs(\underline{u})} Y',\ w \subseteq u \cap S(X)}{X \| Y \xrightarrow{obs(u-w)} X \| Y'} \qquad \frac{Y \xrightarrow{obs(\underline{u})} \sqrt{(v)},\ w \subseteq u \cap S(X)}{X \| Y \xrightarrow{obs(u-w)} v \cup X}$$

TABLE 35. Signal observation function of 4.3.iv.

# 9. ABSTRACTION

In this section, we will consider extensions of process algebra with signals to the case where internal, non-observable actions (so-called $\tau$-actions) are present. We will consider two semantics of such an extension:

i.   branching bisimulation equivalence of [GW 89], [BW 90];

ii.  weak bisimulation equivalence (or observational equivalence) of Milner [MI 80, MI 89], [BK 85].

In both cases, the constant $\tau$ is the result of abstraction, as given by the abstraction operator $\tau_I$, that renames all actions from I into $\tau$. In the next table, $I \subseteq A$, $a \in A \cup \{\tau\}$ (so always $\tau_I(\delta) = \delta$).

| | | |
|---|---|---|
| $\tau_I(a) = a$ | if $a \notin I$ | TI1 |
| $\tau_I(a) = \tau$ | if $a \in I$ | TI2 |
| $\tau_I(x + y) = \tau_I(x) + \tau_I(y)$ | | TI3 |
| $\tau_I(x \cdot y) = \tau_I(x) \cdot \tau_I(y)$ | | TI4 |
| $\tau_I([u, x]) = [u, \tau_I(x)]$ | | TSI1 |
| $\tau_I(\langle x, u \rangle) = \langle \tau_I(x), u \rangle$ | | TSI2 |
| $\tau_I(x \triangleleft \phi \triangleright y) = \tau_I(x) \triangleleft \phi \triangleright \tau_I(y)$ | | TIC |

TABLE 36. Abstraction operator.

## 9.1 BRANCHING BISIMULATION

Graphs now have edges labeled by elements of $A \cup \{\tau\}$, and nodes labeled by an element of CS. Endnodes may have an additional label $\delta$. If s and t are two nodes in a process graph g, then we put $s \Rightarrow t$ if there is a (possibly empty) path from s to t containing only $\tau$-edges.

R: $g \Leftrightarrow_b h$, relation R between nodes of g and nodes of h is a *branching bisimulation* between g and h iff

i.   the roots of g and h are related by R

ii.  if $R(s,t)$ and the atomic signal p is present in s, then there is a node t' in h and a path $t \Rightarrow t'$ in h such that $R(s,t')$ and signal p is present in t'

iii. if R(s,t) and the atomic signal p is present in t, then there is a node s' in g and a path s $\Rightarrow$ s' in g such that R(s',t) and signal p is present in s'

iv. if R(s,t) and s $\xrightarrow{a}$ s' is an edge in g, then either a=$\tau$ and R(s',t), or there are nodes t*, t' in h and a path t $\Rightarrow$ t* $\xrightarrow{a}$ t' in h such that R(s,t*) and R(s',t')

v. if R(s,t) and t $\xrightarrow{a}$ t' is an edge in h, then either a=$\tau$ and R(s,t'), or there are nodes s*, s' in g and a path s $\Rightarrow$ s* $\xrightarrow{a}$ s' in g such that R(s*,t) and R(s',t')

vi. if R(s,t) then s is an endnode in g iff t is an endnode in h.


We say graphs g and h are *branching bisimilar*, g $\underline{\leftrightarrow}_b$ h iff there is an R: g $\underline{\leftrightarrow}_b$ h.

A branching bisimulation R is called *rooted* iff in addition

vii. if root(g) $\xrightarrow{a}$ s is an edge in g (a$\in$ A$\cup\{\tau\}$), then there is a node t in h and an edge root(h) $\xrightarrow{a}$ t in h such that R(s,t)

viii. if root(h) $\xrightarrow{a}$ t is an edge in h (a$\in$ A$\cup\{\tau\}$), then there is a node s in g and an edge root(g) $\xrightarrow{a}$ s in g such that R(s,t).

We say graphs g and h are rooted branching bisimilar , g $\underline{\leftrightarrow}_{rb}$ h iff there is a rooted R: g $\underline{\leftrightarrow}_b$ h.


Now we claim that the set of process graphs modulo rooted branching bisimulation forms a model for BPAS. In this model, moreover the following $\tau$-law is valid (a $\in$ A$\cup\{\tau\}$). This law is taken from [BW 90]. See also [GW 89].

$$a \cdot (\tau \cdot (x + y) + x) = a \cdot (x + y) \qquad \text{BE}$$

TABLE 37. Branching bisimulation with signals.


We leave as an open problem whether BPAS + BE constitutes a *complete* axiomatization of this model. In order to extend these results to theories with parallel composition, extra laws are not needed, since all laws that hold for atomic actions also hold for the constant $\tau$. Thus, we can use the axiomatization of PAS$^-$ in table 12, ACPS$^-$ in table 14, PAS in tables 16-19 and ACPS in tables 20-21, with in all cases a,b,c $\in$ A$\cup\{\delta,\tau\}$. All we need in addition is an extra requirement on the communication and observation function. In table 38, a $\in$ A, p $\in$ AS.

$$\begin{array}{ll} \tau \mid a = \delta & \text{C4} \\ \rho(\tau, p) = \delta & \text{OBS4} \end{array}$$

TABLE 38. $\tau$ with communication and observation function.


9.2 WEAK BISIMULATION

We give the definition of weak bisimulation in the following.

R: g $\underline{\leftrightarrow}_\tau$ h, relation R between nodes of g and nodes of h is a $\tau$-bisimulation between g and h iff

i.   the roots of g and h are related by R

ii.   if R(s,t) and the atomic signal p is present in s, then there is a node t' in h and a path t $\Rightarrow$ t' in h such that R(s,t') and p is present in t'

iii.  if R(s,t) and the atomic signal p is present in t, then there is a node s' in g and a path s $\Rightarrow$ s' in g such that R(s',t) and p is present in s'

iv.   if R(s,t) and s $\xrightarrow{a}$ s' is an edge in g (a$\neq\tau$), then there is a node t' in h and a path t $\Rightarrow$ $\xrightarrow{a}$ $\Rightarrow$ t' in h such that R(s',t')

v.    if R(s,t) and t $\xrightarrow{a}$ t' is an edge in h (a$\neq\tau$), then there is a node s' in g and a path s $\Rightarrow$ $\xrightarrow{a}$ $\Rightarrow$ s' in g such that R(s',t')

vi.   if R(s,t) and s $\xrightarrow{\tau}$ s' is an edge in g, then there is a node t' in h and a path t $\Rightarrow$ t' in h such that R(s',t')

vii.  if R(s,t) and t $\xrightarrow{\tau}$ t' is an edge in h, then there is a node s' in g and a path s $\Rightarrow$ s' in g such that R(s',t')

viii. if R(s,t) then s is an endnode in g iff t is an endnode in h.


We say graphs g and h are $\tau$-bisimilar, g $\underline{\leftrightarrow}_\tau$ h iff there is an R: g $\underline{\leftrightarrow}_\tau$ h.

A $\tau$-bisimulation R is called *rooted* iff in addition

ix.   if root(g) $\xrightarrow{\tau}$ s is an edge in g, then there is a node t in h and a path root(h) $\xrightarrow{\tau}$ $\Rightarrow$ t in h such that R(s,t)

x.    if root(h) $\xrightarrow{\tau}$ t is an edge in h, then there is a node s in g and a path root(g) $\xrightarrow{\tau}$ $\Rightarrow$ s in g such that R(s,t).

We say graphs g and h are weakly bisimilar or rooted $\tau$-bisimilar, g $\underline{\leftrightarrow}_{r\tau}$ h iff there is a rooted R: g $\underline{\leftrightarrow}_\tau$ h.


Now we claim that the set of process graphs modulo rooted $\tau$-bisimulation forms a model for BPAS. In this model, moreover the following three $\tau$-laws are valid (a $\in$ A). They are variants of the $\tau$-laws of Milner [MI 80].

$$
\begin{array}{ll}
u \subseteq S(x) \Rightarrow a \cdot [u, \tau] \cdot x = a \cdot x & \text{TO1} \\
u \subseteq S(x) \Rightarrow \tau \cdot x + x = [u, \tau \cdot x] & \text{TO2} \\
a \cdot (\tau \cdot x + y) = a \cdot (\tau \cdot x + y) + a \cdot x & \text{TO3}
\end{array}
$$

TABLE 39. Observational equivalence with signals (a$\neq\tau$).

We leave as an open problem whether BPAS + TO1-3 constitutes a *complete* axiomatization of this model.

Unfortunately, two of the earlier laws are not valid anymore, viz. the laws S3 and P3 of table 12 in 2.5. They have to be replaced by the following laws (a $\in$ A).

| | |
|---|---|
| $S(ax) = \emptyset$ | S3* |
| $S(\tau x) = S(x)$ | TS |
| $P(ax) = ax$ | P3* |
| $P(\tau x) = \tau \cdot P(x)$ | TP |

TABLE 40. $\tau$ and root signals.

In order to extend these results to theories with parallel composition, we need extra laws, as not all laws for atomic actions will hold for $\tau$. In order to obtain PAS$^-$ with internal action, we have the following obvious laws (see also [BK 85]).

| | |
|---|---|
| $\tau \mathbin{\underline{\parallel}} x = \tau \cdot x$ | TM1 |
| $\tau x \mathbin{\underline{\parallel}} y = \tau \cdot (x \parallel y)$ | TM2 |
| $\langle \tau, u \rangle \mathbin{\underline{\parallel}} x = \tau \cdot (u \cup x)$ | TMO |

TABLE 41. $\tau$ and free merge.

Adding communication and encapsulation necessitates the following laws (see also [BK 85]).

| | |
|---|---|
| $\tau \mid x = \delta$ | TC1 |
| $x \mid \tau = \delta$ | TC2 |
| $\tau x \mid y = x \mid y$ | TC3 |
| $x \mid \tau y = x \mid y$ | TC4 |
| $\langle \tau, u \rangle \mid x = \delta$ | TCO1 |
| $x \mid \langle \tau, u \rangle = \delta$ | TCO2 |
| $\partial_H(\tau) = \tau$ | D0 |

TABLE 42. $\tau$ and communication.

Finally, adding observations leads to the following additional axioms.

| | |
|---|---|
| $\tau/u = \delta$ | OT1 |
| $\tau x/u = x/u$ | OT2 |

TABLE 43. $\tau$ and observation.

## 10. EXAMPLES

In this section we discuss a number of examples of the use of signals and observations.

### 10.1 QUEUE

A specification of a (FIFO) queue can be given as follows. We have a given finite data set D, and the following specification has variables indexed by sequences over D. $\varepsilon$ is the empty sequence, and concatenation of a sequence and an element is denoted by *. Signals are empty, nonempty.

$$Q_\varepsilon = [\{empty\}, \sum_{d \in D} enqueue(d) \cdot Q_d]$$

$$Q_{\sigma * d} = [\{nonempty\}, dequeue(d)] \cdot Q_\sigma + \sum_{e \in D} enqueue(e) \cdot Q_{e * \sigma * d} \ .$$

## 10.2 BAG

The bag is indexed by multi-sets. Signals are again empty, nonempty.

$$B_\emptyset = [\{empty\}, \sum_{d \in D} inbag(d) \cdot B_{\{d\}}]$$

$$V \neq \emptyset \ \Rightarrow \ B_V = [\{nonempty\}, \sum_{d \in D} outbag(d) \cdot B_{V - \{d\}}] + \sum_{d \in D} inbag(d) \cdot B_{V \cup \{d\}}.$$

## 10.3 STACK

We use conventions as above. We give a number of alternatives. First a stack without signals.

$$S_\varepsilon^1 = \sum_{d \in D} push(d) \cdot S_d^1$$

$$S_{\sigma * d}^1 = pop \cdot S_\sigma^1 + top(d) \cdot S_\sigma^1 + \sum_{e \in D} push(e) \cdot S_{\sigma * d * e}^1.$$

Next, we add a signal showing the top of the stack.

$$S_\varepsilon^2 = \sum_{d \in D} push(d) \cdot S_d^2$$

$$S_{\sigma * d}^2 = pop \cdot S_\sigma^2 + top \cdot [\{show(d)\}, S_{\sigma * d}^2] + \sum_{e \in D} push(e) \cdot S_{\sigma * d * e}^2.$$

In the third specification, we add signals empty, nonempty, and also allow actions top, pop in case the stack is empty. If this happens, an error signal is emitted, and no further action is possible.

$$S_\varepsilon^3 = [\{empty\}, \sum_{d \in D} push(d) \cdot S_d^3] + \langle top, \{error\}\rangle + \langle pop, \{error\}\rangle$$

$$S_{\sigma * d}^3 = pop \cdot S_\sigma^3 + top \cdot [\{show(d)\}, S_{\sigma * d}^3] + [\{nonempty\}, \sum_{e \in D} push(e) \cdot S_{\sigma * d * e}^3].$$

The fourth specification has a state of underflow, when an empty stack is popped. A subsequent push leads out of the error situation.

$$S^4_\varepsilon = [\{empty\}, \sum_{d \in D} push(d) \cdot S^4_d] + \langle top, \{error\}\rangle + pop \cdot U^4$$

$$U^4 = [\{underflow\}, \sum_{d \in D} push(d) \cdot S^4_\varepsilon] + \langle top, \{error\}\rangle + pop \cdot U^4$$

$$S^4_{\sigma*d} = pop \cdot S^4_\sigma + top \cdot [\{show(d)\}, S^4_{\sigma*d}] + [\{nonempty\}, \sum_{e \in D} push(e) \cdot S^4_{\sigma*d*e}].$$

The fifth stack keeps functioning, when a pop or top is executed on an empty stack.

$$S^5_\varepsilon = [\{empty\}, \sum_{d \in D} push(d) \cdot S^5_d] + top \cdot [\{show(\bot)\}, S^5_\varepsilon] + pop \cdot [\{error\}, S^5_\varepsilon]$$

$$S^5_{\sigma*d} = pop \cdot S^5_\sigma + top \cdot [\{show(d)\}, S^5_{\sigma*d}] + [\{nonempty\}, \sum_{e \in D} push(e) \cdot S^5_{\sigma*d*e}].$$

In the sixth specification, a pop or top executed on an empty stack leads to an irrecoverable error state, but actions can still be executed.

$$S^6_\varepsilon = [\{empty\}, \sum_{d \in D} push(d) \cdot S^6_d] + top \cdot ERROR + pop \cdot ERROR$$

$$ERROR = [\{error\}, (top + pop + \sum_{d \in D} push(d))] \cdot ERROR$$

$$S^6_{\sigma*d} = pop \cdot S^6_\sigma + top \cdot [\{show(d)\}, S^6_{\sigma*d}] + [\{nonempty\}, \sum_{e \in D} push(e) \cdot S^6_{\sigma*d*e}].$$

### 10.4 COMMUNICATING BUFFERS

In this example we will give an example where both observation and communication play a role. We consider a one element buffers, that always signals its contents on the output port. It receives a communication when the contents are read out. On the input port, it tries to read an item. When it has succeeded in doing this, it sends a communication acknowledging this. The buffer $B^{ij}$ has input port $i$ and output port $j$. The signal $^jd$ means that message $d$ is offered at port $j$.

$$B^{ij} = [\{^j\bot\}, \delta] + obs(^j\bot) \cdot B^{ij} + \sum_{d \in D} obs(^jd) \cdot [\{^jd\}, s_i(ack)] \cdot B^{ij}_d$$

$$B^{ij}_d = [\{^jd\}, \delta] + r_j(ack) \cdot B^{ij}.$$

Now we connect two buffers together. We have the communication $r_2(ack) \mid s_2(ack) = c_2(ack)$, and we use the signal observation function of section 4.3.ii. We define

$$X = \partial_H(B^{12} \parallel B^{23}),$$

where the encapsulation set is $H = \{obs(^2\bot), r_2(ack), s_2(ack)\} \cup \{obs(^2d) : d \in D\}$. Some calculations result in the following recursive specification:

$$X = [\{^2\bot, {}^3\bot\}, \delta] + obs(^1\bot)\cdot X + \sum_{d \in D} obs(^1d)\cdot X_1^d$$

$$X_1^d = [\{^2d, {}^3\bot\}, \delta] + s_1(ack)\cdot X_2^d + yes(^2d)\cdot[\{^2d, {}^3d\}, s_1(ack)]\cdot X_3^d$$

$$X_2^d = [\{^2d, {}^3\bot\}, yes(^2d)]\cdot X_3^d$$

$$X_3^d = [\{^2d, {}^3d\}, c_2(ack)]\cdot X_4^d$$

$$X_4^d = [\{^2\bot, {}^3d\}, \delta] + r_3(ack)\cdot X + obs(^1\bot)\cdot X_4^d + \sum_{e \in D} obs(^1e)\cdot X_5^{de}$$

$$X_5^{de} = [\{^2e, {}^3d\}, \delta] + r_3(ack)\cdot X_1^e + s_1(ack)]\cdot[\{^2e, {}^3d\}, r_3(ack)]\cdot X_2^e.$$

Let us now abstract from the set of actions $I = \{obs(^1\bot), s_1(ack), c_2(ack)\} \cup \{yes(^2d) : d \in D\}$. This leaves only the input actions $obs(^1d)$ and the output action $r_3(ack)$. We get the following specification for $Y = \tau_I(X)$:

$$Y = \tau_I(X) = [\{^2\bot, {}^3\bot\}, \delta] + \sum_{d \in D} obs(^1d)\cdot Y_1^d$$

$$Y_1^d = \tau_I(X_1^d) = \tau_I(X_2^d) = [\{^2d, {}^3\bot\}, \tau]\cdot Y_3^d$$

$$Y_3^d = \tau_I(X_3^d) = [\{^2d, {}^3d\}, \tau]\cdot Y_4^d$$

$$Y_4^d = \tau_I(X_4^d) = [\{^2\bot, {}^3d\}, \delta] + r_3(ack)\cdot Y + \sum_{e \in D} obs(^1e)\cdot Y_5^{de}$$

$$Y_5^{de} = \tau_I(X_5^{de}) = [\{^2e, {}^3d\}, \delta] + r_3(ack)\cdot Y_1^e.$$

Having hidden the actions at port 2, we can proceed by also hiding the signals at port 2. We do this by means of the signal filtering operator of 2.4. We put $u = \{^3d : d \in D\} \cup \{^3\bot\}$, and derive the following specification for $Z = u \cap Y$:

$$Z = u \cap Y = u \cap \tau_I(X) = [\{^3\bot\}, \delta] + \sum_{d \in D} obs(^1d)\cdot Z_1^d$$

$$Z_1^d = u \cap Y_1^d = [\{^3\bot\}, \tau]\cdot Z_3^d$$

$$Z_3^d = u \cap Y_3^d = u \cap Y_4^d = [\{^3d\}, \delta] + r_3(ack)\cdot Z + \sum_{e \in D} obs(^1e)\cdot Z_5^{de}$$

$$Z_5^{de} = u \cap Y_5^{de} = [\{^3d\}, \delta] + r_3(ack)\cdot Z_1^e.$$

## ACKNOWLEDGEMENTS

We thank Wouter Brouwer for many comments and improvements. Frits Vaandrager pointed out an error in a previous version of equation MSI2 (section 3.3) as well as a difficulty in a first version of the observation function. Henk Goeman, Sjouke Mauw, Hans Mulder and Gert Veltink also contributed by suggesting improvements. Our treatment of conditionals benefitted from correspondence with Tony Hoare.


## REFERENCES

[AB 84] G. AUSTRY & G. BOUDOL, *Algèbre de processus et synchronisation,* TCS 30, 1984, pp. 91-131.

[BB 88] J.C.M. BAETEN & J.A. BERGSTRA, *Global renaming operators in concrete process algebra,* Information & Computation 78 (3), 1988, pp. 205-245.

[BBK 86] J.C.M. BAETEN, J.A. BERGSTRA & J.W. KLOP, *Syntax and defining equations for an interrupt mechanism in process algebra,* Fundamenta Informaticae IX (2), 1986, pp. 127-168.

[BBMV 89] J.C.M. BAETEN, J.A. BERGSTRA, S. MAUW & G.J. VELTINK, *A process specification formalism based on static COLD,* report CS-R8930, CWI Amsterdam 1989. To appear in Proc. METEOR Workshop Methods based on Formal Specifications, Mierlo 1989 (L. Feijs & J.A. Bergstra, eds.), Springer LNCS.

[BW90] J.C.M. BAETEN & W.P. WEIJLAND, *Process algebra,* Cambridge Tracts in TCS 18, Cambridge University Press 1990.

[BE 88a] J.A. BERGSTRA, *Process algebra for synchronous communication and observation,* Report P8815, Programming Research Group, University of Amsterdam 1988. *Revised version,* report P8815b, 1989.

[BE 88b] J.A. BERGSTRA, *ACP with signals,* in: Algebraic and Logic Programming (J. Grabowski, P. Lescanne & W. Wechler, eds.), Springer LNCS 343, 1988, pp. 11-20.

[BK 84] J.A. BERGSTRA & J.W. KLOP, *Process algebra for synchronous communication,* Information and Control 60 (1/3), 1984, pp. 109-137.

[BK 85] J.A. BERGSTRA & J.W. KLOP, *Algebra of communicating processes with abstraction,* Theoretical Computer Science 37 (1), 1985, pp. 77-121.

[BR 90] W.S. BROUWER, *Stable signals and observation in a process specification formalism,* M.Sc. thesis, University of Amsterdam 1990.

[VGL 87] R.J. VAN GLABBEEK, *Bounded nondeterminism and the approximation induction principle in process algebra,* in: Proc. STACS 87 (F.J. Brandenburg, G. Vidal-Naquet & M. Wirsing, eds.), LNCS 247, Springer Verlag 1987, pp. 336-347.

[GW 89] R.J. VAN GLABBEEK & W.P. WEIJLAND, *Branching time and abstraction in bisimulation semantics. Extended abstract,* in: Information Processing 89, IFIP World Congress, San Francisco (G.X. Ritter, ed.), North-Holland, Amsterdam 1989, pp. 613-618.

[GP 90] J.F. GROOTE & A. PONSE, *Process algebra with guards,* report CS-R9069, CWI Amsterdam 1990.

[HE 88] M. HENNESSY, *Algebraic theory of processes,* MIT Press 1988.

[HO 85] C.A.R. HOARE, *Communicating sequential processes,* Prentice Hall 1985.

[HHJ+ 87] C.A.R. HOARE, I.J. HAYES, HE JIFENG, C.C. MORGAN, A.W. ROSCOE, J.W. SANDERS, I.H. SORENSEN, J.M. SPIVEY & B.A. SUFRIN, *Laws of programming,* Comm. of the ACM 30 (8), 1987, pp. 672-686.

[MV 90] S. MAUW & G.J. VELTINK, *A process specification formalism,* Fund. Inf. XII, 1990, pp. 85-139.

[MI 80] R. MILNER, *A calculus of communicating systems,* Springer LNCS 92, 1980.

[MI 89] R. MILNER, *Communication and concurrency,* Prentice Hall 1989.

[VA 90] F.W. VAANDRAGER, *Process algebra semantics for POOL,* in: Applications of Process Algebra (J.C.M. Baeten, ed.), Cambridge Tracts in TCS 17, Cambridge University Press 1990, pp. 173-236.

[VE 90] C. VERHOEF, *On the register operator,* report P9003, Programming Research Group, University of Amsterdam 1990.

[VR 86] J.L.M. VRANCKEN, *The algebra of communicating processes with empty process,* Report FVI 86-01, Programming Research Group, University of Amsterdam 1986.

# FUNCTIONAL SPECIFICATION
## OF
# TIME SENSITIVE COMMUNICATING SYSTEMS[1]


Manfred Broy

Institut für Informatik

Technische Universität München

Postfach 20 24 20, D-8000 München 2, FRG


## Abstract


A formal model and a logical framework for the functional specification of time sensitive communicating systems and their components is outlined. The specification method is modular w.r.t. sequential composition, parallel composition, and communication feedback. Nondeterminism is included by underspecification. The application of the specification method to timed communicating functions is demonstrated. Abstractions from time are studied. In particular a rational is given for the chosen concepts of the functional specification technique. The relationship between system models based on nondeterminism and system models based on explicit time notions is investigated. Forms of reasoning are considered. The alternating bit protocol is used as a running example.


---


[1]This work was supported by the ESPRIT BASIC RESEARCH ACTION "DEMON" and by the Sonderforschungsbereich 342 "Werkzeuge und Methoden für die Nutzung paralleler Architekturen"

# 1. Introduction

The behaviour of distributed systems that communicate by exchanging messages can be represented by functional models. In such a modelling systems that operate in some environment are described by functions that map input communications onto output communications. Communications of a system can be internal, i.e. between the components of a system and therefore not necessarily important nor even observable from the environment, or external, i.e. lead to an exchange of messages between the system and its environment. In the later case we speak also of input and output of the system. The causal or logical relationship between input messages and output messages is called the (extensional) behaviour of a system. A simple way for representing the behaviour of systems is a trace. A trace is a sequence of communication actions providing the input/output history of a particular instantiation of a system. Since the specification of systems in terms of traces is often technically inelegant and traces do not emphasize compositional forms and the conceptual difference between input and output we rather use functional models of the behaviour of systems instead of traces.

In a functional specification of a component being part of a distributed system, called an *agent* in the following, interaction and communication have to be considered essential. The behaviour of system components can be understood by mappings that relate the input received from the environment to output sent to the environment (cf. [Kahn, MacQueen 77], [Broy 85]). Formally this can be modelled by prefix monotonic functions mapping histories ("streams") of input communications onto histories of output communications. Using streams and stream processing functions for modelling interactive systems has been proposed many times before in the literature (see also data flow approaches).

Prefix monotonicity reflects a basic property of communicating systems: if we have observed a finite sequence of output messages for a corresponding finite sequence of input messages, then if we observe additional input (thus the old input sequence is a prefix of the extended one) we may just observe additional output (thus the old output sequence is a prefix of the extended one). Prefix monotonicity provides a notion of causality between input and output. It reflects the stepwise consumption of input and production of output and guarantees the existence of least fixpoints, which is mandatory for giving meaning to communication feedback loops.

A specification of a system component is understood as a predicate that characterizes the component's behaviour. Technically such a behaviour is described by a class of prefix monotonic functions. For every input x and every choice of a function f from this class a system behaviour with output f(x) is obtained.

It seems methodologically fruitful to specify systems property-oriented: the functional behaviour of a system is described by a number of properties represented by logical formulas. The properties can be classified into safety properties and liveness properties. Safety properties describe characteristics that can be falsified by finite observations. Accordingly liveness properties correspond to characteristics that can be falsified only in infinite observations.

Agents the behaviours of which depend on the timing of the input messages can be modelled also along the lines described above by introducing particular time messages. An interesting question that we want to discuss in the following concerns the relationship between models of communicating systems including explicit notions of time and abstractions from time by the concept of nondeterminism.

In our functional setting a system component is called nondeterministic, if there are several output streams possible for a given input stream. Nondeterminism generally brings a number of complications for the semantic treatment of communicating systems. Nevertheless, since nondeterminism is an important methodological concept, it should be incorporated into a specification formalism. The reasons are in particular as follows:

- the abstraction from certain operationally relevant details (such as time, storage etc.) keeps the specification more manageable,

- leaving open certain properties in specifications that are not relevant or should be determined only later in the design process allows to avoid overspecification and to obtain additional flexibility.

Accordingly also nondeterministic specifications of systems with time dependent behaviour are of interest.

Nondeterminism of systems is easily incorporated into functional specifications by underspecification: there may be several functions that meet the predicate specifying an agent and any of them can be chosen for generating system behaviours. Strictly speaking it does not make any difference whether such a choice is done throughout the design of a system component (removing underspecification) or left open in the implementation as nondeterminism in the execution.

However, for particular agents the i/o behaviour cannot be modelled simply by a set of prefix monotonic functions at particular levels of abstraction. Examples are the nonstrict merge of message streams or system components that produce output until there is input on one of its input lines. For those agents certain in principle possible behaviours should only be chosen for input histories with certain properties. In particular, when abstracting away explicit time, there may exist conflicts between time oriented choice conditions and the requirement of monotonicity. These conflicts can be solved by either introducing an explicit notion of time into the proposed model or by considering so-called *input choice specifications*. Input choice specifications allow to express liveness properties of agents that depend on certain time properties of input histories. For such agents the nondeterministic choice of a behaviour function may depend on the timing of the specific input. Input choice specifications try to handle even such time dependent agents by nondeterminism without including necessarily explicit time notions into the semantic model.

The presented approach is powerful enough to deal with the specification of liveness properties that are considered especially difficult such as fairness or responsiveness in the absence of input. The particular formal model for specification has been chosen carefully for avoiding anomalies (such as the one described in [Brock, Ackermann 81]).


## 2. Basic Structures

Given a set M of messages a stream over M is a finite or infinite sequence of elements from M. By $M^*$ we denote the finite sequences over the set M. We denote the finite sequence consisting of the elements $x_1, ..., x_n$ by $\langle x_1 ... x_n \rangle$ or by $\langle x_{i+1}: i < n \rangle$. $M^*$ includes the empty sequence which is denoted by $\langle \rangle$.

By $M^\infty$ we denote the infinite sequences over the set M. $M^\infty$ can be understood to be represented by the total mappings from the natural numbers $\mathbb{N}$ into M. We denote the infinite sequence consisting of the elements $x_1, x_2, \dots$ by $\langle x_{i+1}: i < \infty \rangle$.

We denote the set of streams over the set M by $M^\omega$. Formally we have

$$M^\omega =_{\text{def}} M^* \cup M^\infty.$$

For simplicity we write also $a^n$ for $\langle x_{i+1}: i < n \rangle$ with $x_{i+1} = a$ for all i, $0 \leq i < n$. Similarly we write $S^n$ for the set of all streams $\langle x_{i+1}: i < n \rangle$ with $x_{i+1} \in S$ for all i, $0 \leq i < n$.

Streams can be understood to represent the history of communications (on channels) between components of interactive systems. We introduce a number of functions on streams that are useful in system descriptions.

For every stream s we may define its *length*. The length of a stream is infinite or a natural number. It will be denoted by #s. Formally we have

$$\#: M^\omega \to \mathbb{N} \cup \{\infty\}$$

with

$$\#\langle x_{i+1}: i < n \rangle = n.$$

For a stream s we write s.i to denote the i-th element of the stream s.

A classical operation on sequences is the *concatenation* which we denote by ˆ. The concatenation is a function that takes two sequences (say s and t) and produces a sequence as result starting with s and continuing with t . If s is infinite then the result of concatenating s with t yields s again. Formally we have for the concatenation the following functionality:

$$\hat{\,}. : M^\omega \times M^\omega \to M^\omega$$

Its precise meaning is defined for streams $s = \langle s_{i+1}: i < n \rangle$ and $t = \langle t_{i+1}: i < m \rangle$ by

$$s \hat{\,} t = \langle \text{ if } i < n \text{ then } s_{i+1} \text{ else } t_{i-n+1} \text{ fi}: i < n+m \rangle .$$

On the set $M^\omega$ of streams we define a *prefix ordering* $\sqsubseteq$. We write $s \sqsubseteq t$ for streams s and t if s is a *prefix* of t. Formally we have for streams $s = \langle s_{i+1}: i < n \rangle$ and $t = \langle t_{i+1}: i < m \rangle$

$$s \sqsubseteq t \quad \text{iff} \quad n \leq m \wedge \forall i, i < n: s_{i+1} = t_{i+1} .$$

The prefix ordering defines a partial ordering on the set $M^\omega$ of streams. If $s \sqsubseteq t$, then we also say that s is an *approximation* of t. The set of streams ordered by $\sqsubseteq$ is even complete in the sense that every directed set $S \subseteq M^\omega$ of streams has a *least upper bound* denoted by lub S. A nonempty set S which is subset of a partially ordered set is called *directed*, if

$$\forall x, y \in S: \exists z \in S: x \sqsubseteq z \wedge y \sqsubseteq z .$$

With the technique of least upper bounds of directed sets of finite streams we are able to describe infinite streams. Infinite streams are also of interest as (and can also be described by) fixpoints of prefix monotonic functions. Note that the streams associated with feedback loops in interactive systems correspond to such fixpoints.

A *stream processing function* is a function

$$f: M^\omega \to N^\omega$$

that is *prefix monotonic* and *continuous*. The function f is called *monotonic*, if for all streams s and t we have

$$s \sqsubseteq t \Rightarrow f.s \sqsubseteq f.t .$$

For better readability we often write for the function application f.x instead of f(x). The function f is called *continuous*, if for all directed sets $S \subseteq M^\omega$ of streams we have

$$\text{lub } \{f.s: s \in S\} = f.\text{lub } S .$$

If a function is continuous, then its results for infinite input can be already predicted from its results on all finite approximations of the input.

By $\bot$ we denote the pseudo element which represents the result of diverging computations. We write $M^\bot$ for $M \cup \{\bot\}$. Here we assume that $\bot$ is not an element of M. On $M^\bot$ we define also a simple partial ordering by:

$$x \sqsubseteq y \quad \text{iff} \quad x = y \vee x = \bot$$

We use the following functions on streams

$$\text{ft}: M^\omega \to M^\bot,$$

$$\text{rt}: M^\omega \to M^\omega,$$

$$.\&.: M^\bot \times M^\omega \to M^\omega.$$

They are defined as follows: the function ft selects the first element of a stream, if the stream is not empty:

$$\text{ft.}\langle s_{i+1}: i < n \rangle = \begin{cases} \bot & \text{if } n = 0 \\ s_1 & \text{if } n > 0 . \end{cases}$$

The function rt deletes the first element of a stream, if the stream is not empty:

$$\text{rt.}\langle s_{i+1}: i < n \rangle = \begin{cases} \langle \rangle & \text{if } n = 0 \\ \langle s_{i+2}: i < n\text{-}1 \rangle & \text{if } n > 0 . \end{cases}$$

We sometimes use iterated applications of the function rt. Then we write:

$$\text{rt}^0.s = s,$$

$$\text{rt}^{k+1}.s = \text{rt}^k(\text{rt.}s),$$

$$\text{rt}^\infty.s = \langle \rangle.$$

The function & appends an element to a stream (as first element), if the element is defined (i.e. $\neq \bot$):

$$x \,\&\, \langle s_{i+1}: i < n \rangle = \begin{cases} \langle \rangle & \text{if } x = \bot \\ \langle s_i: i < n+1 \rangle & \text{if } x \neq \bot \text{ with } s_0 = x . \end{cases}$$

The definition of this function has been chosen carefully to keep it monotonic and continuous w.r.t. the prefix ordering. However, there are also nonformal reasons for taking this definition: as soon as one tries to send the value x of an expression as the first member of a stream the computation of which never terminates (i.e. $x = \perp$) the transmission of the first element can never take place and the resulting stream is empty.

The properties of the functions can be expressed by the following equations:

$$\perp \,\&\, s = \langle\rangle, \qquad\qquad rt.\langle\rangle = \langle\rangle,$$

$$ft(x \,\&\, s) = x, \qquad\qquad x \neq \perp \Rightarrow rt(x \,\&\, s) = s,$$

$$\langle\rangle\hat{\ }s = s = s\hat{\ }\langle\rangle, \qquad\qquad x \neq \perp \Rightarrow (x \,\&\, s)\hat{\ }r = x \,\&\, (s\hat{\ }r),$$

$$\#\langle\rangle = 0, \qquad\qquad x \neq \perp \Rightarrow \#(x \,\&\, s) = 1 + \#s.$$

Sometimes it is useful to work with a filter function on streams. Given a set $S \subseteq M$ and a stream $x \in M^\omega$ we write $S©x$ for the substream of $x$ consisting only of the elements of $x$ contained in $S$. Formally we define

$$S©\langle\rangle = \langle\rangle,$$

$$S©(d \,\&\, x) = d \,\&\, (S©x) \qquad\qquad \text{if } d \in S,$$

$$S©(d \,\&\, x) = S©x \qquad\qquad \text{if } \neg(d \in S).$$

We denote the function space of (n,m)-ary prefix continuous stream processing functions by:

$$[(M^\omega)^n \to (M^\omega)^m]$$

The prefix ordering $\sqsubseteq$ induces an ordering on tuples of streams and on this function space, too. For convenience we introduce the concept of a *resumption* $f_x$ of the stream processing function f for the input stream x. It is specified by

$$f_x(s) = rt^i(f(x\hat{\ }s)) \qquad\qquad \text{where} \quad i = \#f.x \ .$$

The prefix monotonicity of a stream processing function is essential for its interpretation as representation of the meaning of a communicating device. Monotonicity w.r.t. the prefix ordering basically is expressed by the following property

$$f(s\hat{\ }r) = f.s\hat{\ }f_s(r) \ .$$

This equation shows that for a prefix monotonic function f the output of $f(s\hat{\ }r)$ is concatenated from the output f.s produced by f on input s and the output produced by the resumption $f_s$ on the further input r. Thus every prefix of the input determines a prefix of the output.

Note that the operations ft, rt, and & are prefix monotonic and continuous, but the concatenation $\hat{\ }$ as defined above is not prefix monotonic.

Prefix monotonicity reflects a characteristic property of interactive systems: communicated data cannot be changed after being shown as output. If f.s is the stream that results as output from the input of the stream s then if we continue with the input r, i.e. give after all the input $s\hat{\ }r$, then we get some additional output $f_s(r)$, i.e. obtain after all the output $(f.s)\hat{\ }f_s(r)$. In addition monotonicity gives the formal platform for handling communication in feedback loops: feedback is translated to fixpoint equations, which are known to have solutions, if the involved functions are monotonic.

For composing stream processing functions we may use the three classical forms of composition namely sequential and parallel composition and feedback. Let g be a (n,m)-ary function and h be a (n',m')-ary stream processing function.

We denote *parallel composition* of g and h by g‖h. g‖h is a (n+n',m+m')-ary stream processing function. Parallel composition can be visualized by a diagram:

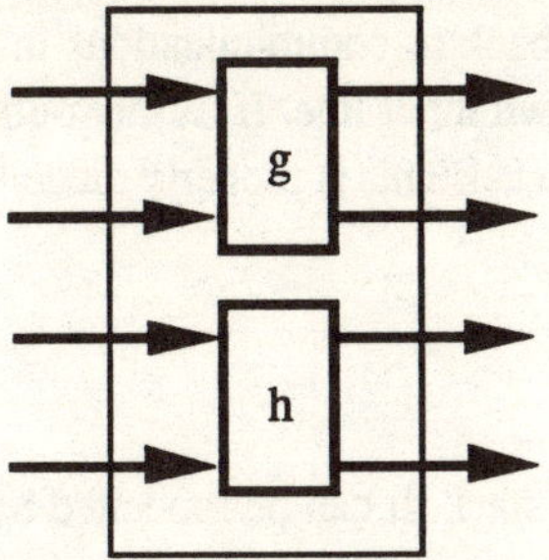

We define for $x \in (M^\omega)^n$, $z \in (M^\omega)^{n'}$

(g‖h).(x,z) = (g.x, h.z) .

Let m = n'; we denote *sequential composition* by g∘h. g∘h is a (n,m')-ary stream processing function. Sequential composition can be visualized by a diagram:

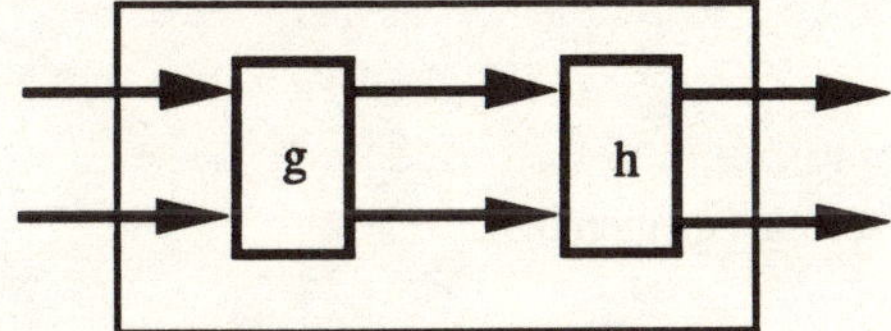

We define for $x \in (M^\omega)^n$:

(g∘h).x = h.g.x .

We define the feedback operator $\mu^k$ for a (n+k,m)-ary stream processing function f where m ≥ k. $\mu^k f$ is a (n,m)-ary stream processing function derived from f by k feedback loops. This can be visualized by a diagram:

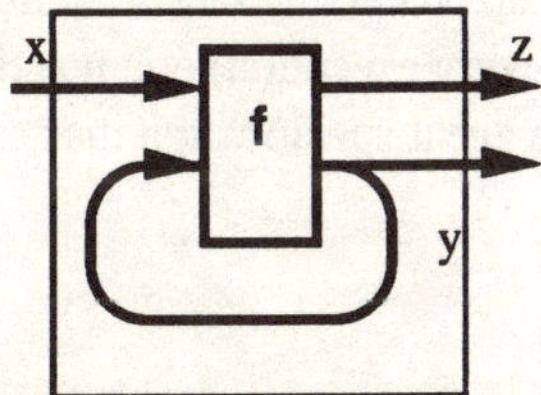


We define for $x \in (M^\omega)^n$

$(\mu^k f).x = \textbf{fix } \lambda\, z, y: f(x, y)$

where $z \in (M^\omega)^{m-k}$, $y \in (M^\omega)^k$. By **fix** we denote the least fixpoint operator. We write $\mu f$ for $\mu^1 f$.

In a more readable version we may specify the μ-operator as follows: we have

$(z, y) = (\mu^k f).x,$

if (z, y) is the least fixpoint of the equation

$(z, y) = f(x, y).$

Formally $(\mu^k f).x$ denotes a fixpoint relative to x, i.e. for every given tuple x of streams we obtain (z, y) by a fixpoint construction.

Note that the fixpoint concept reflects the characteristics of feedback of communications in an appropriate manner. The output of the agent f is sent back to its own input line. If further output requires more input than available this way no more output is generated. This is properly modelled by the assumption that the fixpoint is the least one.


## 3. System Specification

A simple example for a specification of an agent is an interactive stack. It can be modelled by a function

$$f: (D \cup \{?\})^\omega \to D^\omega.$$

Here data messages in the input stream are understood as to be pushed onto the stack and messages "?" are understood as requests for data. We specify the behaviour of the interactive stack by the formula (for $x \in D^*, d \in D, y \in (D \cup \{?\})^\omega$):

$$f.x = \langle\rangle,$$

$$f(x^\smallfrown\langle d\rangle^\smallfrown\langle ?\rangle^\smallfrown y) = d \ \& \ f(x^\smallfrown y).$$

Similarly a queue can be specified by replacing the second equation by:

$$f(\langle d\rangle^\smallfrown x^\smallfrown\langle ?\rangle^\smallfrown y) = d \ \& \ f(x^\smallfrown y).$$

Note that in both cases f is not uniquely determined by the given equation, since nothing is said about the result for an input of the form ? & y.

An agent with n input lines and m output lines accepts n streams as input and produces m streams as output. It is called *(n,m)-ary agent*. An agent specification describes the behaviour of an agent.

For a deterministic agent its behaviour is represented by a stream processing function. For a nondeterministic agent its behaviour is specified by a *class* of stream processing functions. Several instances of an agent may occur in a system. In every instance (every execution) one function is chosen from that class for determining one particular output. An agent specification thus can be represented by a predicate

$$Q: [(M^\omega)^n \to (M^\omega)^m] \to \mathbb{B}$$

where $\mathbb{B} = \{1, 0\}$ represents the set of truth values.

A simple example for a nondeterministic agent is a fork that receives a stream as input and produces two streams as output by splitting the input stream:

$$\text{FORK}.f = \forall \ x, y, z, d: f.x = (y, z) \Rightarrow \#(\{d\}\copyright x) = \#(\{d\}\copyright y) + \#(\{d\}\copyright z).$$

Another basic example for a nondeterministic agent is the merge agent that nondeterministically merges two streams. It can be specified by the predicate MERGE:

$$\text{MERGE}.f \equiv \forall \ x, y: \exists \ oracle \in \{1, 0\}^\omega: f(x, y) = sched(x, y, oracle) \wedge \#oracle = \infty$$

$$\textbf{where} \ \forall \ a, b, c: \quad sched(a, b, 0 \ \& \ c) = ft.a \ \& \ sched(rt.a, b, c) \wedge$$

$$sched(a, b, 1 \ \& \ c) = ft.b \ \& \ sched(a, rt.b, c) .$$

A function f with MERGE.f need not to be fair and it may be strict, i.e. we may have (even if $x = \langle\rangle$ and $y \neq \langle\rangle$) :

$f(x, y) = x$ .

A version of a specification of a (on infinite input) fair merge agent reads as follows:

$\text{FAIR\_MERGE}.f \equiv \forall\, x, y: \exists\, \text{oracle} \in \{1, 0\}^\omega: f(x, y) = \text{sched}(x, y, \text{oracle}) \wedge$

$$\#(\{1\}\copyright\text{oracle}) = \infty \wedge \#(\{0\}\copyright\text{oracle}) = \infty .$$

Again we may have

$f(x, y) = x$

for finite x, but not for infinite ones. A specification of a nonstrict and fair version of merge is given later.

Stream processing functions

$$f. (M^\omega)^n \to (M^\omega)^m$$

can be seen as a special case of an agent specification, i.e. as a deterministic agent specification. For notational convenience we write

$f$

also for a deterministic agent specification Q with $Q.h = (f = h)$.

The combining forms for stream processing functions carry over to system specifications by pointwise application. Let Q and Q' be specifications for (n, m)-ary and (n',m')-ary agents resp. We define:

$(Q\|Q').f \equiv_{df} \exists\, g, h: Q.g \wedge Q'.h \wedge f = g\|h,$

$(Q{\circ}Q').f \equiv_{df} \exists\, g, h: Q.g \wedge Q'.h \wedge f = g{\circ}h,$

$(\mu^k Q).f \equiv_{df} \exists\, g: Q.g \wedge f = \mu^k g.$

A decomposition of specifications into safety and liveness properties as suggested in [Lamport 83] and [Schneider 87] may help in the understanding of an agent's behaviour. It can be explained as follows:

- *safety properties* correspond to characteristics of a system that can be falsified by certain finite observations. A finite observation for an input x consists of some finite output which is an approximation for some possibly infinite totally correct output produced for input x. This observation does include the causality between input and observed output. For every prefix $x' \sqsubseteq x$ we may observe which is the maximal prefix $y' \sqsubseteq y$ that can be caused by input x' (when assuming that the output is a prefix of y).

- *liveness properties* correspond to characteristics of a system that can be falsified only by infinite observations.

Since a specification of an agent should provide enough information by its safety properties to determine the safety properties of composed systems in which the agent is used as component we do not specify a simple input/output relation, but a set of prefix continuous (and thus monotonic) functions. This way the causality between input and output in the sense of safety properties is properly indicated. This will be demonstrated by an example below. Each of the functions specifies information about system behaviours in the sense of possible courses of computation.

The observation of the causality between input and output as being part of the safety properties is a subtle point and therefore requires further explanation. Let us assume that we may observe for certain output that it was produced before particular input had been provided. Mathematically then an observation is a partially ordered set of events labelled by input and output actions. For simplicity we represent finite observations by finite traces of messages M labelled by "in" for input and "out" for output. Then a (finite) trace is a (finite) sequence t

$$t \in (I \cup O)^\omega$$

where I and O denote the set of input and output actions resp.

$$I = \{in.m: m \in M\},$$

$$O = \{out.m: m \in M\} .$$

By

$$strip: (I \cup O)^\omega \to M^\omega$$

we denote the function that derives the pure message stream from a given trace. It is specified by the following equations:

$$strip.\langle\rangle = \langle\rangle ,$$

$$strip(in.m \ \& \ t) = strip(out.m \ \& \ t) = m \ \& \ strip.t .$$

For defining formally observations by traces we introduce a predicate obs. A trace t is called a *partial observation* for a continuous stream processing function

$$f: M^\omega \to M^\omega ,$$

if we have

$$obs(t, f)$$

where the predicate

$$obs: (I \cup O)^\omega \times (M^\omega \to M^\omega) \ \to \ \mathbb{B}$$

specifies that the sequence of actions in t corresponds to possible observations of a sequence of input actions for the function f. Formally the predicate obs is specified by the most liberal (weakest) predicate that fulfils the following equations:

$$obs(\langle\rangle, f) \equiv true,$$

$$obs(in.m \ \& \ t, f) \equiv obs(t, g) \qquad\qquad where \ \forall \ x: g.x = f(m \ \& \ x),$$

$$obs(out.m \ \& \ t, f) \equiv (ft.f.\langle\rangle = m \land obs(t, rt \circ f))$$

Note that

$$obs(t, f) \land f \sqsubseteq g \Rightarrow obs(t, g) .$$

A trace t is called *total observation* for f, if obs(t, f) and f.strip(I©t) = strip(O©t).

**Lemma**: Alternative characterisation of traces

(a) t is partial observation for f, iff

$$\forall\ t': t' \sqsubseteq t \Rightarrow strip(O©t') \sqsubseteq f.strip(I©t') .$$

(b) t is total observation for f, iff in addition

$$f.strip(I©t) = strip(O©t) .$$

**Proof:** (a) By induction on the length of t. If t = ‹›, then the proposition (a) trivially holds. Assume t is finite and nonempty and the formula holds for finite streams of length #rt.t. For t we obtain:

$$obs(t, f) = obs(rt.t, g)$$

where

$$ft.t \in I \Rightarrow \forall\ x: g.x = f(ft.strip.t\ \&\ x),$$

$$ft.t \in O \Rightarrow \forall\ x: ft.strip.t\ \&\ g.x = f.x .$$

For rt.t and g we apply the induction hypothesis and obtain (a) for t and f.

This completes the proof for finite traces by induction on the number of elements in t. To infinite traces the proof is extended by the continuity of f and g. Note that we have for a chain of finite traces t.i :

$$(\forall\ i: obs(t.i, f)) \Rightarrow obs(lub\ \{t.i: i \in \mathbb{N}\}, f) .$$

(b) is just the definition. ◊

Every predicate Q on functions defines a property of functions and also specifies a set of finite and infinite observations represented by finite and infinite resp. traces. We say that (a trace) t *is an observation about the property* Q, if

$$\exists\ f: Q.f \wedge obs(t, f) .$$

We write then also obs(t, Q).

We say that property Q is falsified for a specification R, if there is a trace t such that

$$obs(t, R) \wedge \neg\ obs(t, Q)$$

i.e. if we can make an observation about R that does not have property Q.

For an agent specification Q and some input $x \in (M^\omega)^n$ an output $y \in (M^\omega)^m$ is called *correct*, if there exists a stream processing function f such that

$$Q.f \wedge f.x = y .$$

Partial correctness corresponds to safety properties and finite observations. For an input x partial correctness corresponds to observations restricted to finite prefixes of the output streams, but it includes observations of maximal output for approximations (prefixes) of x. For modelling this we introduce the notion of output finite function. A function f is called *output finite*, if its output always is a tuple of finite streams, i.e. if

$$f: (M^\omega)^n \to (M^*)^m$$

An output finite prefix monotonic function f is called *partially correct* or *safe* for the specification Q, if we have:

$$\exists\, g\colon Q.g \wedge f \sqsubseteq g$$

Generally a function is called *safe* w.r.t. Q if all its output finite approximations are safe. According to this definition, if for a function f which is partially correct for Q we have $f.x = y$ where y is infinite, this does not imply that there exists a function g with Q.g and $g.x = y$. It does only indicate that for all finite prefixes y' of y there exists a function g with Q.g and $y' \sqsubseteq g.x$.

**Lemma**: The function f is safe w.r.t. Q iff all finite observations w.r.t. f are observations for Q, i.e.

$$\forall\, t \in (I \cup O)^*\colon \text{obs}(t, f) \Rightarrow \text{obs}(t, Q)\,.$$

**Proof:** We have for finite traces t that

$$\text{obs}(t, f)$$

is equivalent to the proposition: there exists a prefix monotonic output finite function h with $h \sqsubseteq f$ such that

$$\text{obs}(t, h)$$

Let $x = \text{strip}(I \copyright t)$; then

$$h \sqsubseteq g$$

and therefore

$$\text{obs}(t, h) \Rightarrow \text{obs}(t, g)$$

since f is safe w.r.t. Q iff there exists such a g with Q.g.                    ◊

By this definition of safety we can be sure that the following property of compositionality holds: if f is partially correct w.r.t. Q, then $\mu^k f$ is partially correct w.r.t $\mu^k Q$. The same holds for parallel and sequential composition. Also w.r.t. liveness our approach is compositional.

There are two reasons why we work with sets of functions rather than set-valued functions or relations when modelling nondeterminism in communicating systems:

- we can use the classical functional calculus for proving properties of specifications,

- for determining the meaning of feedback loops least fixpoints have to be considered, which cannot be characterized uniquely in general in the framework of set-valued functions or relations.

We illustrate the second remark by an example.

**Example**: *Agent specification with identical input/output relations*

We consider two (1,1)-ary agents. Let the continuous stream processing functions

$$g, h, i, j\colon [\{1\}^\omega \to \{1\}^\omega]$$

be specified (for all $x, y \in \{1\}^\omega$) by the following equations

$$g.\langle\rangle = h.\langle\rangle = \langle 1\rangle, \qquad\qquad i.\langle\rangle = j.\langle\rangle = \langle\rangle,$$

$$g.\langle 1\rangle = j.\langle 1\rangle = \langle 1\ 1\rangle, \qquad\qquad h.\langle 1\rangle = i.\langle 1\rangle = \langle 1\rangle,$$

$$y = \langle 1\ 1\rangle^\smallfrown x \Rightarrow g.y = h.y = i.y = j.y = \langle 1\ 1\rangle.$$

We define the specifying predicates Q1 and Q2 by

$\quad$ Q1.f $\equiv$ (f = g $\vee$ f = i),

$\quad$ Q2.f $\equiv$ (f = h $\vee$ f = j).

Obviously for every input the sets of totally correct output of Q1 and Q2 coincide. The behaviours $\mu$Q1 and $\mu$Q2 for the agent specifications Q1 and Q2 under feedback are given by the least fixpoints z of functions f with Q1.f and Q2.f resp. The least fixpoint of the functions i and j is ‹›. The least fixpoints of g is ‹1 1›. The least fixpoints of h is ‹1›. So $\mu$Q1 is distinct from $\mu$Q2. Note that for Q1 the trace

$\quad$ ‹out.1  in.1  out.1  in.1›

is a possible observation, but not for Q2. $\hfill \Diamond$

For explaining the adequacy of the given approach let us for a moment refer to operational considerations: for an agent specification Q given some input x every finite prefix y of f.x for some function f with Q.f represents some approximation for the totally correct output f.x . The stream y can be seen as the output produced for the input x after a certain finite amount of computation time. It corresponds to a "finite observation". If we "wait long enough" without adding any further input then the output should be increased eventually as long as "more" output is guaranteed by all functions f with Q.f and y $\sqsubseteq$ f.x.

## 4. The Alternating Bit Protocol

For illustrating the proposed specification techniques we consider the alternating bit protocol. The alternating bit protocol can be graphically represented by the following diagram:

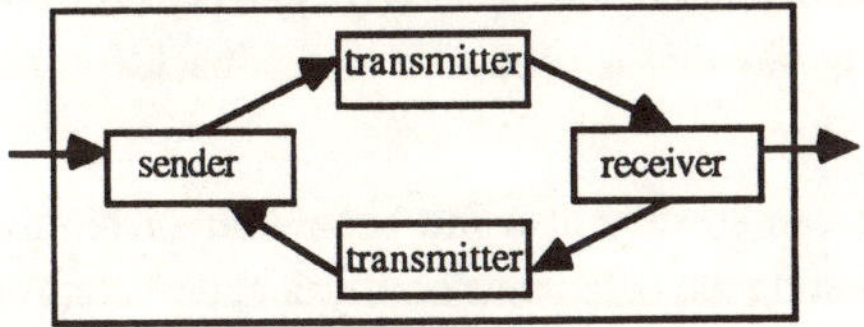


From this diagram the number of input lines and output lines for the involved agents are obvious. The diagram represents the following composed specification

$\quad$ AB = $\mu$(CS $\circ$ CT $\circ$ CR $\circ$ (ID $\|$ CT)) $\circ$ CP

where CP specifies the projection function

$\quad$ CP.f $\equiv \forall$ x, y: f(x, y) = x,

and ID specifies the identity function

$\quad$ ID.f $\equiv \forall$ x: f.x = x.

The function that is to be realized by the alternating bit protocol is simply the identity.

We use for the specification of the sender and the receiver the following auxiliary functions:

$$\text{alt: } \{1, 0\}^\omega \rightarrow \{1, 0\}^\omega.$$

The function alt records the alternations from 1 to 0 in its input stream :

$$\text{alt.}\langle b\rangle = \langle b\rangle,$$

$$\text{alt}(b \;\&\; b \;\&\; x) = \text{alt}(b \;\&\; x),$$

$$\text{alt}(\neg b \;\&\; b \;\&\; x) = \neg b \;\&\; \text{alt}(b \;\&\; x).$$

The continuity of alt implies $\text{alt}(1^\infty) = \langle 1\rangle$. The function (let D denote the set of data to be transmitted)

$$\text{res}: (D \times \{1, 0\})^\omega \to D^\omega$$

produces as output those data that are labelled by an alternating bit:

$$\text{res.}\langle\rangle = \langle\rangle,$$

$$\text{res.}\langle(d, a)\rangle = \langle d\rangle,$$

$$\text{res}((d, a) \;\&\; (d, a) \;\&\; x) = \text{res}((d, a) \;\&\; x),$$

$$\text{res}((d, a) \;\&\; (d', a') \;\&\; x) = d \;\&\; \text{res}((d', a') \;\&\; x) \qquad \text{if } a \neq a' \lor d \neq d'.$$

The continuity of res implies $\text{res}((d,1)^\infty) = \langle d\rangle$. The function

$$\text{ack}: (D \times \{1, 0\})^\omega \to \{1, 0\}^\omega$$

drops the data and produces a stream of bits:

$$\text{ack.}\langle\rangle = \langle\rangle,$$

$$\text{ack}((d, a) \;\&\; x) = a \;\&\; \text{ack.}x.$$

We may specify the agent receiver by the following predicate CR:

$$\text{CR.f} \equiv \forall\, x: f.x = (\text{res.}x, \text{ack.}x).$$

Note that the receiver in the specification above yields pairs of result streams $(r, y)$ where r is the output of the system and y is the stream of alternating bits that is produced as feedback for the sender.

For a specification technique it is essential that we can specify also the behaviour of highly (unbounded) nondeterministic components for representing the behaviour of physical devices. As an illustration for such components we specify the transmitter. We give a specification by the predicate CT:

$$\text{CT.f} \equiv \forall\, x: \exists\, s \in \mathbb{B}^\omega: f.x = h(x, s) \land \#(1\copyright s) = \infty$$

$$\textbf{where } \forall\, x, s: h(x, 1\&s) = \text{ft.}x \;\&\; h(\text{rt.}x, s) \land h(x, 0\&s) = h(\text{rt.}x, s).$$

A further example for a specification is the sender. The sender should send its current message again and again as long as there is no (correct) feedback from the receiver. However, there is a clear conflict between the prefix monotonicity and the intended behaviour: if there is never feedback the message should be repeated infinitely often, if there is correct feedback the message should be sent no longer (i.e. at most finitely often after all). We may try to give a specification CS for the sender as follows:

$$\text{CS.f} \equiv \forall\, x, y: \text{res.f}(x,y) \sqsubseteq x \land \#\text{res.f}(x,y) = \min(\#x, 1+\#\text{alt.}y) \land \text{ft.ack.f}(x,y) \sqsubseteq 1 \land$$

$$(\#x > \#\text{alt.}y \Rightarrow \#f(x,y) = \infty).$$

This specification asserts in particular by the last clause that if not enough acknowledgements are sent back then the first nonacknowledged message is repeated infinitely often.

From CS.f we may conclude

$$f(d\&d'\&x, \diamond) = (d, 1)^\infty,$$

$$\#(\{(d',0)\} \copyright f(d\&d'\&x, 1\&y)) \geq 1.$$

These formulas show that there does not exist a prefix monotonic function that fulfils the predicate CS. This problem has to do with timing. Intuitively we may say that the sender repeats its message *as long as* it does not get the expected acknowledgement. However, without explicit time information we cannot express the notion "as long as" in the considered model. So we have either to introduce explicit time considerations (see below) or to regard the specification either as contradictory w.r.t. monotonicity or we have to give up the requirement of monotonicity which makes it difficult to ensure the existence of least fixpoints. Least fixpoints, however, are badly needed to give meaning to feedback loops such as the one occurring in the alternating bit protocol.

For solving this trade off we switch to real time models and later to so-called input choice specifications of the form introduced in the following sections.

## 5. Agents with Time Dependent Behaviour

For many applications of communicating systems the timing of their input history is essential: the behaviour of a communicating agent may critically depend on information about the relative or absolute timing of received messages. Such a situation can be formally modelled by introducing a notion of time explicitly into a semantic model. We work in the following with a fairly simple but nevertheless sufficiently expressive model of time by using a special element $\sqrt{}$ called "tick" or "timeout" for representing the situation that no message has arrived at some input line (or no message has been produced at some output line) within a certain time interval. The fact that no actual message has arrived within such an interval of time can be seen as the specific information represented by $\sqrt{}$.

We use $\sqrt{}$ as a message and write $M^{\sqrt{}}$ for $M \cup \{\sqrt{}\}$. A *timed stream* is an element from the set

$$(M^{\sqrt{}})\omega .$$

It represents the communication over a channel with additional time information, i.e. about the time that it takes between the transmission of two messages on some input or output line. Then a *time annotated (or timed for short) (n, m)-ary stream processing function* f is a continuous function of the following functionality:

$$f: [((M^{\sqrt{}})\omega)^n \to ((M^{\sqrt{}})\omega)^m]$$

In principle timed streams and time annotated stream processing functions can be specified by the same techniques as introduced and used for streams without explicit time information above.

**Example**: *Specification of a time sensitive function*

A time sensitive stream processing function f that produces some input message as output only if the input is repeated within a certain amount of time can be specified by the following equations (let $x \in (M^{\sqrt{}})\omega$, $a, b \in M$, $a \neq b$):

$$f.\diamond = \diamond,$$

$$f.\langle a \rangle = \langle \sqrt{} \rangle,$$

$$f(\sqrt{} \ \& \ x) = \sqrt{} \ \& \ f.x,$$

$$f(a \ \& \ a \ \& \ x) = \sqrt{} \ \& \ a \ \& \ f.x,$$

$$f(a \ \& \ \sqrt{} \ \& \ a \ \& \ x) = \sqrt{} \ \& \ \sqrt{} \ \& \ a \ \& \ f.x,$$

$$f(a \ \& \ b \ \& \ x) = \sqrt{} \ \& \ f(b \ \& \ x),$$

$$f(a \ \& \ \sqrt{} \ \& \ b \ \& \ x) = \sqrt{} \ \& \ \sqrt{} \ \& \ f.(b \ \& \ x),$$

$$f(a \ \& \ \sqrt{} \ \& \ \sqrt{} \ \& \ x) = \sqrt{} \ \& \ \sqrt{} \ \& \ \sqrt{} \ \& \ f.x.$$

Note that we have $\#f.x = x$ for all input streams x. The function f can be understood as a simplified version of a login function with a password: if not the correct password is given within a certain amount of time the login is aborted. $\lozenge$

From every timed stream $s \in (M^{\sqrt{}})^{\omega}$ we may derive its time information free stream of actual messages by $M\copyright s$. Note that an infinite timed stream $s \in (M^{\sqrt{}})^{\omega}$ provides a complete timing information about the communication history on an input or output line.

Again we may think about traces of timed agents. Note that in principle time ticks in the input stream may refer to an interval of different time duration than a time tick in an output stream. Such assumptions, however, may lead to conceptual anomalies when considering feedback. The concept of time ticks can be applied for different modelings of time such as quantitative or relative time models. The different models can be reflected by different additional properties that are assumed for the stream processing functions (see next section).

One of the most interesting aspects in connection with models for interaction including timing aspects are questions of abstracting away time. So in the following a short explanation is given how streams with explicit timing informations are related to streams without explicit timing information. A timed partial stream $s \in (M^{\sqrt{}})^{*}$ of length n can be understood to provide incomplete information about the communication history on a channel just representing the behaviour on the channels in the first n time intervals. In contrast to this a nontimed partial stream $s \in M^{*}$ of length n represents information more ambiguously. It may correspond to the complete information about the communication history on an input channel where after all only a finite number of messages is sent (i.e. it is the abstraction from an infinite timed stream s' by s = M\copyright s') or it may just represent the behaviour on the channels in the first n+k time intervals (i.e. it is the abstraction from a finite timed stream s' of length n+k by s = M\copyright s'). The identification of these two essentially different situations into one when abstracting away time information is the reason for the problems with monotonicity (and continuity) in the sender example above. The sender should behave differently on a finite timed stream (which of course contains only a finite number of messages) of acknowledgement feedbacks without the expected bit (where it should send only a "sufficient" number of messages) in contrast to its behaviour on an infinite timed stream without the expected bit (where it should send an infinite number of messages).

If an infinite timed stream s contains only a finite number of proper messages, i.e. if

$$\#(M\copyright s) < \infty,$$

then after a finite prefix the stream s consists of an infinite stream of $\sqrt{}$. However, there is a remarkable difference between s and $M\copyright s$ (and also between s and every finite stream $s' \in (M^{\sqrt{}})^{\omega}$ with $M\copyright s' = M\copyright s$). The stream s is a total element w.r.t. prefix ordering, while $M\copyright s$ (and similarly s') is a partial element, i.e. an element representing incomplete information about the input.

Note that there does not exist a prefix continuous function end_of_transmission with

$$\text{end\_of\_transmission.s} = \text{true} \qquad \text{iff } s = \sqrt{\infty}.$$

Clearly we can never test algorithmically if end_of_transmission is valid, because we cannot predict the future.

However, we may define functions that for instance repeat a message as often as there is no new actual message in the next time intervals. Such a repeater (rep.m) starting with message m can be defined on timed streams as follows (for $x \in M$):

$$(\text{rep.m}).(\sqrt{} \ \& \ s) = m \ \& \ (\text{rep.m}).s,$$

$$(\text{rep.m}).(x \ \& \ s) = x \ \& \ (\text{rep.x}).s \ .$$

Of course there is no way to define a monotonic function g.m in analogy to the repeater on the level of nontimed streams such that

$$(\text{rep.m}).s = (\text{g.m}).(M©s).$$

This simple observation throws some light on the difference between modelling systems with or without explicit time notions. In an infinite stream s including time ticks with $\#(M©s) < \infty$ (i.e. s is the representation of the partial stream M©s with full time information) we cannot test (by monotonic predicates) whether the stream M©s is empty, but we can test again and again whether a stream is not continued by actual messages and according to this react to this situation by producing more and more output. We get this way a weak (stepwise) test for the availability of further input on timed streams which does not exist (by a monotonic predicate) for the nontimed partial stream M©s.

The behaviour of a timed agent with n input lines and m output lines can be specified by a predicate over the space of timed stream processing functions. A nontimed agent may be obtained as an abstraction from a timed agent. If the timing information for the input streams does not influence the actual messages in the output streams, but only their timing, then an agent is called *time insensitive*. A more formal definition of time insensitivity is given later.

The use of time notions in specifications may make specifications longer and more detailed, since we have to deal with all kinds of time considerations, but it may also make the specification of certain agents simpler, since using the explicit notion of time may allow to express certain liveness aspects more directly.

**Example**: *Timed specifications for the alternating bit example*

Assuming time information in the input streams the sender specification of the alternating bit example above for instance is rather simple. We give such a specification by the predicate TCS for the sender:

$$\text{TCS.f} \equiv (\text{Q.1}).\text{f} \quad \textbf{where} \ \ \forall \ \text{f, b: } (\text{Q.b}).\text{f} \equiv \forall \ x, y, a:$$

$$f(a \ \& \ x, \langle\rangle) = \langle(a,b)\rangle \ \wedge$$

$$f(a \ \& \ x, \sqrt{} \ \& \ y) = (a,b) \ \& \ f(a \ \& \ x, y) \ \wedge$$

$$f(a \ \& \ x, \neg b \ \& \ y) = (a,b) \ \& \ f(a \ \& \ x, y) \ \wedge$$

$$f(a \ \& \ x, b \ \& \ y) = (a,b) \ \& \ g(x, y) \qquad \textbf{where} \ \ (\text{Q.}\neg b).g$$

In contrast to the time free specification above there do not arise any problems with monotonicity here. Of course, we may also give more nondeterministic versions of the sender (repeating a message an arbitrary - but finite - number of times on nonpositive feedback).

For being able to use such a timed specification of the sender for the alternating bit protocol we need to have of course also specifications dealing with explicit time notions for the transmitter and the receiver. We may specify the agent receiver by the timed agent specification TCR:

$$\text{TCR.f} \equiv \forall\ x: f.x = h(x, 1)$$

$$\textbf{where}\ \forall\ x, b, a:\quad h(\diamond, b) = (\diamond, \diamond),$$

$$h(\sqrt{}\ \&\ x, b) = (\sqrt{}\ \&\ r, \sqrt{}\ \&\ y)\ \textbf{where}\ (r, y) = h(x, b),$$

$$h((a,b)\ \&\ x, b) = (a\&r, b\&y)\ \textbf{where}\ (r, y) = h(x, \neg b),$$

$$h((a,\neg b)\ \&\ x, b) = (r, \neg b\&y)\ \textbf{where}\ (r, y) = h(x, b).$$

We give a specification with explicit time notions for the transmitter by the predicate TCT:

$$\text{TCT.f} \equiv \forall\ x: \exists\ s \in \mathbb{B}^{\omega}: f.x = h(x, s) \wedge \#(1\copyright s) = \infty$$

$$\textbf{where}\ \forall\ x, s: h(x, 1\&s) = ft.x\ \&\ h(rt.x, s) \wedge h(x, 0\&s) = \sqrt{}\ \&\ h(rt.x, s)\ .$$

We may also give a more nondeterministic version of the transmitter where a lost message may lead to an arbitrary but finite (and nonempty) number of time ticks. ◊

The specifications of the timed agents given here are rather simple and straightforward. In the next section we study the abstraction from time for agents with time dependent behaviour.


## 6. Properties of Agents with Time Dependent Behaviour: Abstracting from Time

In principle the message $\sqrt{}$ can be used and treated in specifications like any other message. However, looking at timed systems we may assume additional properties that are characteristic for timing. We may assume that for every agent it takes some time until arriving input messages lead to certain output messages. This can be modelled by time ticks in the output streams.

If we understand a timed agent as a function or a set of functions mapping input histories including time information onto output histories, then it is suggestive to add the following assumptions:

- infinite input histories are mapped on infinite output histories, i.e. on a complete timing of the output streams,

- finite input histories of length n lead to output histories of length $\geq n+1$.

The second assumption implies the first by the continuity of the functions.

According to these assumptions for an agent for every output stream its first k elements are determined by the first k elements of the input streams. A timed stream processing function f is called *pulse driven* (or also *time synchronous*), if we have (for all i, $1 \leq i \leq m$, $x \in ((M^{\sqrt{}})^{\omega})^n$):

$$\#(f.x).i \geq \min\ \{1+\#x.j: 1 \leq j \leq n\}\ .$$

This formula expresses that for each output channel at least the first k output elements are determined by the first k input elements.

We may even ask for the time, the *delay*, that it takes until messages received at input lines show effects at output lines. The most remarkable observations, however, can be stated as follows

(a) Pulse driven agents produce infinite output streams on infinite input streams.

(b) The streams that are components of least fixpoints of pulse driven functions with are infinite.

(c) Parallel and sequential composition of pulse driven agents as well as feedback lead to pulse driven agents.

Observation (b) allows to conclude that fixpoints of pulse driven functions are unique. Observation (a) shows that for pulse driven functions every input history with a complete timing is mapped onto an output history with a complete timing.

Pulse driven functions in particular form an adequate functional model for switching circuits.

Studying communicating systems by abstracting away time information is an important issue. Therefore it is useful to classify those timed functions and timed agents that do not use timing information in an essential way. In general, however, the behaviour of communicating systems may depend on timing. The order (the relative timing) in which messages are received on different input lines may influence the behaviour of systems. Nevertheless for some agents this timing is not relevant for the produced actual messages or it influences the produced actual messages not in an essential way: this can be modelled by nondeterminism. Some other situations cannot be mapped so simply onto nondeterminism. The resp. agents are called *time critical*.

A timed stream processing function f is called *time insensitive*, if the time information on the input streams does not influence the produced message streams at all, i.e. if

$$M©x = M©x' \implies M©f.x = M©f.x'.$$

Here we write for $x \in (S^\omega)^n$ simply $M©x$ instead of $(M©x_1, ..., M©x_n)$.

A time insensitive timed stream processing function f determines uniquely a nontimed stream processing function g by the formula

$$g(M©x) = M©f.x .$$

Moreover g is prefix monotonic and continuous, if f is so. This is immediately seen by the fact that $(M^\omega)^n \subseteq ((M^\surd)^\omega)^n$, and, since $M©x = x$ for $x, y \in (M^\omega)^n$ with $x \sqsubseteq y$ we have:

$$g.x = M©f.x \sqsubseteq M©f.y = g.y .$$

Here we used the fact that $\lambda\, x: M©x$ is prefix monotonic. Note moreover that it is continuous.

We define a time abstraction function

$$\varphi: ((M^\surd)^\omega)^n \to (M^\omega)^n$$

that deletes the ticks in a stream by

$$\varphi.x = M©x .$$

A timed function f is time insensitive if $\varphi$ is homomorphic w.r.t. a function g called the *time abstraction* for f where:

$$\varphi.f.x = g.\varphi.x$$

or in a more combinatorial notation if

$$f \circ \varphi = \varphi \circ g.$$

For a time insensitive function abstracting away time does not introduce any nondeterminism. We obtain a uniquely determined function on nontimed pure message streams.

Functions specified by the predicate TCR characterize timed functions that are time insensitive and their time abstractions coincide with the functions specified by CR given for nontimed functions above.

Given a (n, m)-ary timed agent specification T the relation

$$W: (M^\omega)^n \times (M^\omega)^m \to \mathbb{B}$$

defined by

$$W(x, y) \equiv \exists\, x' \in ((M\cup\{\surd\})^\omega)^n,\ f: T.f \wedge x = M\copyright x' \wedge y = M\copyright f.x',$$

is called *time free input/output relation*.

For keeping specifications abstract we would like to be able to abstract away time aspects in specifications. This is certainly not reasonable for arbitrary timed agents or stream processing functions. However, for certain agents it is possible.

Of course we require some properties for the abstraction mapping. These requirements are classical:

(a)    the abstraction allows to deduce the time free input/output relation,

(b)    the abstraction is *modular* w.r.t. parallel and sequential composition as well as w.r.t. feedback.

The requirement (b) essentially means that the abstraction function is required to be a homomorphism w.r.t. the compositional forms. We define the time abstraction function $\Phi$ for time insensitive stream processing functions f by

$$\Phi.f =_{df} f' \quad\quad if \quad\quad\quad f \circ \varphi = \varphi \circ f'$$

$\Phi$ is modular, i.e. we have

$$\Phi(h\|g) = \Phi.h \,\|\, \Phi.g,$$

$$\Phi(h\circ g) = \Phi.h \circ \Phi.g,$$

$$\Phi(\mu^k f) = \mu^k\, \Phi.f.$$

as long as f, h, g are time insensitive. The proof is straightforward.

Similar to the definition of time insensitive functions we define time insensitivity of agents. Let the agent be specified by

$$T: [((M^{\surd})^\omega)^n \to ((M^{\surd})^\omega)^m] \to \mathbb{B}$$

T is called time insensitive if there exists an agent specification

$$Q: [(M^\omega)^n \to (M^\omega)^m] \to \mathbb{B}$$

such that

$$T \circ \varphi = \varphi \circ Q.$$

If a stream processing function is not time insensitive, then abstracting away time can be seen as introducing some kind of nondeterminism.

Given a (not necessarily time insensitive) timed function f, a nontimed function g is called a *strong time abstraction for f*, and we write

$$g \textbf{ abs } f,$$

if for every input x to g there exists a total timing y such that the behaviour of g on x coincides with the time abstraction of the behaviour of f on y, i.e.

$$\forall \, x \in (M^{\omega})^n \colon \exists \, y \in ((M^{\sqrt{}})^{\infty})^n \colon x = M©y \wedge (\varphi \circ g)|_{y\downarrow} = (f \circ \varphi)|_{y\downarrow} \, .$$

Here we use the following abbreviations: $f|_S$ denotes the restriction of the function f to S (where S is a subset of the domain of f) and:

$$y\downarrow = \{x \colon x \sqsubseteq y\} \, .$$

A timed function f is called *weakly time insensitive*, if for every input $y \in (M^{\sqrt{}})^{\omega}$ there exists a prefix continuous strong time abstraction g for f such that:

$$g(M©y) = M©f.y \, .$$

When abstracting from time we may represent the behaviour (apart from timing aspects) of a weakly time insensitive function by its set of strong time abstractions. For a time insensitive function all its strong time abstractions coincide.

**Example:** *Weakly time insensitive functions*

Let us consider a function that reproduces its input messages provided there is "sufficient time" between two messages:

$$f \colon (M^{\sqrt{}})^{\omega} \to (M^{\sqrt{}})^{\omega}$$

is specified by (let $x \in M$, i.e. $x \neq \sqrt{}$, $y \in M^{\sqrt{}}$):

$$f.\langle\rangle = \langle\rangle,$$

$$f.\langle y \rangle = \langle\sqrt{}\rangle,$$

$$f(\sqrt{} \ \& \ s) = \sqrt{} \ \& \ f.s,$$

$$f(x \ \& \ y \ \& \ s) = \sqrt{} \ \& \ x \ \& \ f.s.$$

f is not time insensitive, since

$$M©f.\langle x \ \sqrt{} \ x \ \sqrt{}\rangle = \langle x \ x \rangle \neq \langle x \rangle = M©f.\langle x \ x \rangle \, .$$

Nevertheless f is weakly time insensitive. All strong time abstractions for f are functions that fulfil the nontimed transmitter specification. ◊

The transmitter specification TCT defines functions that are weakly time insensitive. The sender specification TCS by the timed agent above is not weakly time insensitive. Since the input of an infinite stream $\sqrt{}^{\infty}$ of time ticks leads to an infinite number of copies of the message (a,b) in the output stream, there are due to monotonicity no strong time abstractions for the timed functions specified by TCS.

In a weakly time insensitive function the output may depend on the timing of the input, but only in a very weak ("nondeterministic") form. In particular, the function f must not react to infinite streams of time ticks by an infinite number of actual messages (but with different messages in reaction to proper messages), since this would generally be in conflict with the monotonicity requirement.

A further example for a timed function that is not weakly time insensitive is the repeater rep.m. We have:

$$(\text{rep.m}).(\sqrt{\infty}) = m^{\infty}$$

and (for $a \in M$):

$$(\text{rep.m}).(\sqrt{n} \,\hat{}\, \langle a \rangle \hat{}\, \sqrt{\infty}) = (m^n)\hat{}(a^{\infty})$$

which contradicts the required monotonicity for strong time abstractions.

A nontimed agent specified by the predicate Q on nontimed stream processing functions is called *time abstraction* for T and we denote Q by $\Phi.T$, if for every nontimed input x for Q every nontimed behaviour shown by Q is a reflection of a timed behaviour of T:

$$Q.g \Rightarrow \exists\, f: T.f \wedge g \textbf{ abs } f .$$

This generalizes the notion of time insensitivity to timed agents. Q is called *full time abstraction* for T, if for every timed input y to T all behaviours possibly generated by T are mirrored by strong time abstractions specified by Q:

$$T.f \Rightarrow \forall\, y \in ((M^{\sqrt{}})^{\infty})^n: \exists\, g: g \textbf{ abs } f \wedge Q.g \wedge (\varphi \circ g)|_y{\downarrow} = (f \circ \varphi)|_y{\downarrow} .$$

In a time insensitive nondeterministic agent specified by the predicate T either all functions f with T.f are time insensitive or the time sensitivity is not observable due to the nondeterminism in T. Technically speaking whenever we have a timed input y and some behaviour function f with T.f, then for every timed input y' with $M©y = M©y'$ there is a function f' with T.f' such that $M©f.y = M©f'.y'$. So for the output $M©f.y$ for the function f for which we only know T.f it cannot be predicted more by knowing the timing of y.

The requirement of the existence of strong time abstractions g is very strict. Since we require from a strong time abstraction g that g is continuous and for every input x the output g.x is the time abstraction $M©f.y$ for an output f.y of a function f with T.f for some infinite timed input y. Monotonicity (which is part of continuity) enforces that for certain distinct infinite streams y and y' with $M©y \sqsubseteq M©y'$ we have functions f and f' such that

$$T.f \wedge T.f' \wedge (\varphi \circ g)|_y{\downarrow} = (f \circ \varphi)|_y{\downarrow} \wedge (\varphi \circ g)|_{y'}{\downarrow} = (f' \circ \varphi)|_{y'}{\downarrow} .$$

This requirement is rather strong. In particular it implies

$$M©f.y \sqsubseteq M©f'.y'$$

which is not at all true if for instance for $n = 1$ both $M©f.y$ and $M©f'.y'$ are infinite and distinct. So for certain agents such as fair timed merge such strong time abstractions do not exist.

Moreover time insensitivity of agents T and T' does not allow to conclude time insensitivity of

$$T \circ T',$$

since the timing in the output f.x of functions f with T.f on input x may be very specific such that for the function f' with T'.f' only very particular output f'.f.x is produced, due to the timing of f.x. So we cannot conclude time insensitivity for $T \circ T'$ from the time insensitivity of T and T'.

Nevertheless we may introduce a further notion that expresses that for given input y all timings of the output are possible. An agent specification T is called *time unspecific*, if for every infinite input y, every pulse driven function f and f'' with T.f and $(f \circ \varphi)|_{(f.y)}{\downarrow} = (f'' \circ \varphi)|_{(f.y)}{\downarrow}$ there exists a function f' such that

$$T.f' \wedge f'|(f.y)\!\downarrow = f''|(f.y)\!\downarrow \, .$$

This means that the timing in the output is not influenced by the timing in the input. If an agent is time unspecific then the timing in the input does not influence the data messages in the output nor the timing in the output.


## 7. Input Choice Specifications

Even for not weakly time insensitive agents we may look for a nondeterministic nontimed modelling under certain assumptions. For certain time sensitive agents such as the sender in the example of the alternating bit protocol the choice of the (function generating its) output may depend on the timing of the input. In particular a infinite stream of time ticks may be mapped onto an infinite stream of proper messages. Looking for strong time abstractions for those function we run into conflicts with monotonicity. However, particular time abstractions for such functions can be represented at the level of nondeterminism by an input choice specification.

Formally an *input choice specification* of an (n,m)-ary agent is given by a predicate

$$R: [(M^{\omega})^n \rightarrow (M^{\omega})^m] \times (M^{\omega})^n \rightarrow \mathbb{B}.$$

The proposition R is used to specify the behaviour of an agent. R(f, x) is supposed to express that f and also all g with $g \sqsubseteq f$ are partially correct stream processing functions w.r.t. the specified agent, i.e. f reflects correctly safety properties. Moreover, for the input x the output f.x is correct output, i.e. correct also w.r.t. liveness conditions. In other words R(f, x) is supposed to stand for the following two logical statements:

- f is a function that is safe (fulfils the safety requirements),

- the output f.x is live (fulfils the liveness requirements) for input x and the behaviour (the causality between input and output) indicated by f.

For a programming language that allows to write programs satisfying such specifications cf. [Broy 86] and [Broy 87a]. For a logical calculus for relating the programs written in such a language to input choice specifications see [Broy 87b].

Every agent specification

$$Q: [(M^{\omega})^n \rightarrow (M^{\omega})^m] \rightarrow \mathbb{B}$$

can be seen as an input choice agent specification

$$Q': [(M^{\omega})^n \rightarrow (M^{\omega})^m] \times (M^{\omega})^n \rightarrow \mathbb{B}$$

by

$$Q'(f, x) = \exists \, f': Q.f' \wedge f \sqsubseteq f' \wedge f.x = f'.x \, .$$

For notational convenience we write often Q instead of Q'.

Seen as a time abstraction of an agent T specifying a set of timed pulse driven functions g we may understand R(f, x) as the logical statement "f is a partially correct time abstraction w.r.t. some timed function g with T.g and f.x is totally correct with respect to a timed stream y with complete time information, i.e. y is infinite and $x = M\copyright y$ holds". This way input choice specifications provide a concept for a particular abstraction from time for functions that are time sensitive in a characteristic manner.

For composing input choice specifications of system components we may again use the three classical concepts of sequential and parallel composition and feedback. For input choice specifications R and R' we define the compositional forms as follows:

$$(R\|R').(f, (x, z)) \equiv \exists\ g, h: R(g, x) \wedge R'(h, z) \wedge f = g\|h,$$

$$(R \circ R').(f, x) \equiv \exists\ g, h: R(g, x) \wedge R'(h, g.x) \wedge f = g \circ h,$$

$$(\mu^k R).(f, x) \equiv \exists\ y, z, g: R(g, (x,y)) \wedge (z, y) = f.x \wedge f = \mu^k g.$$

We again write $\mu R$ for $\mu^1 R$.

For an input choice specification R and an input $x \in (M^\omega)^n$ an output $y \in (M^\omega)^m$ is called *correct*, if there exists a stream processing function f such that

$$R(f, x) \wedge f.x = y.$$

An output finite function g is called *partially correct* or *safe* for the specification R, if for all input elements x:

$$\exists\ f: R(f, x) \wedge g \sqsubseteq f.$$

Generally a function f is called *safe*, if all its approximations by output finite functions are safe.

We call two input choice specifications R and R' *relationally equivalent*, if for every input x the sets of totally correct output for R and R' coincide. We then write

$$R \sim R'.$$

Unfortunately the equivalence relation $\sim$ is not a congruence.

**Example:** *The equivalence $\sim$ is not a congruence.*

We consider two (1,1)–ary agents. Let the continuous stream processing functions

$$f, g, h: [\mathbb{B}^\omega \to \mathbb{B}^\omega]$$

be defined (for all $x \in \mathbb{B}^\omega$) by the following equations

$$f.x = \langle 1\ \ 1 \rangle, \qquad\qquad g.\langle\rangle = \langle 1 \rangle, \qquad\qquad\qquad h.\langle\rangle = \langle\rangle,$$

$$g(1\&x) = h(1\&x) = \langle 1\ \ 0 \rangle,$$

$$g(0\&x) = h(0\&x) = \langle 1\ \ 0 \rangle.$$

We define the specifying predicates R1 and R2 by

$$R1(q, x) \equiv (q = f \vee (x \neq \langle\rangle \wedge q = g)),$$

$$R2(q, x) \equiv (q = f \vee (x \neq \langle\rangle \wedge q = h)).$$

If we consider feedback $\mu R1$ and $\mu R2$ for the agents R1 and R2 we define its meaning by the least fixpoints z of functions q with $R(q, z)$. The least fixpoints of f and g are $\langle 1\ \ 1 \rangle$ and $\langle 1\ \ 0 \rangle$ resp. and fulfil R1. The least fixpoint of f also fulfils R2. So R1 exhibits two behaviours under feedback while R2 exhibits only one, since the least fixpoint of h is $\langle\rangle$ which is ruled out since it does not satisfy the input choice condition. ◊

The example in particular shows that under feedback relationally equivalent agents may lead to agents that are not relationally equivalent.

This trivial example shows further aspects of safety properties. If we give a partial input x (in our example ‹›) to an agent and we observe some finite output y, which of course has to be an approximation for some totally correct output, then this may indicate that some particular decisions have been taken inside the agent (in the sense of nondeterministic choices). If we observe in our example the output ‹1› for the input ‹›, then agent R1 still may be free to choose between f and g, while R2 certainly has already chosen f. Partial output may indicate that certain choices have taken place. This is relevant for determining the behaviour of agents within feedback loops. Note that the chosen examples correspond directly to the example provided in [Brock, Ackermann 81].

The power of this specification method can be demonstrated by the specification of an agent which represents fair nonstrict merge:

$$\text{FAIR\_NONSTRICT\_MERGE}(f, (x, y)) \equiv \text{FAIR\_MERGE.f} \wedge \#f(x, y) = \#x + \#y.$$

The sender specification for the alternating bit protocol now can be written as an input choice specification as follows:

$$
\begin{aligned}
CS(f, (x', y')) \equiv \quad (\forall\ x, y: \quad &\text{res.f}(x, y) \sqsubseteq x\ \wedge \\
&\#y < \infty \Rightarrow \#\text{res.f}(x,y) \leq 1+\#\text{alt.y}\ \wedge \\
&\#y = \infty \Rightarrow \#\text{res.f}(x,y) \leq \#\text{alt.y})\ \wedge \\[6pt]
&(\#x' > \#\text{alt.y}' \Rightarrow \#f(x', y') = \infty\ )\ \wedge \\
&\#\text{res.f}(x', y') = \min(\#x', 1+\#\text{alt.y}')\ .
\end{aligned}
$$

The first part of the specification defines the safety properties (which are just properties for f) and the second part defines the liveness properties.


## 8. Safety and Liveness

The concept of input choice specification is strongly based on the notion of safety and liveness. An input choice specification

$$R: [(M^{\omega})^n \to (M^{\omega})^m] \times (M^{\omega})^n \to \mathbb{B}$$

may in particular always be decomposed into safety and liveness properties.

Partial correctness or safety properties of communicating systems indicate which communications may occur as output caused by particular input. The predicate

$$S: [(M^{\omega})^n \to (M^{\omega})^m] \to \mathbb{B}$$

is called the *safety predicate* for the specification R, if a function f is partially correct w.r.t. R iff S.f. According to our definition of partial correctness if f is lub of a chain of functions fulfilling S, then S.f. The safety predicate is downward closed and closed w.r.t. lubs of chains of functions. Note that S specifies more than just a relation between input and output: as shown in section 3 additional information about the causality between input, output and nondeterministic choice is provided by S, too.

For the specification R the predicate

$$L: [(M^{\omega})^n \to (M^{\omega})^m] \times (M^{\omega})^n \to \mathbb{B}$$

is called the *liveness predicate* for R, if

$$L(f, x) \equiv (S.f \Rightarrow R(f, x))\ .$$

The liveness predicate indicates for a partially correct function, if for some input history enough output is provided. Obviously an input choice specification R can be reconstructed from its safety and its liveness predicates if we assume $R(f, x) \Rightarrow S.f$.

## 9. Abstracting from Time (continued)

Similar to the definition of time insensitive functions we define time insensitivity of agents described by input choice specifications. Let the agent be specified by

$$T: [((M^{\sqrt{}})^{\omega})^n \to ((M^{\sqrt{}})^{\omega})^m] \to \mathbb{B}$$

T is called *time insensitive on infinite timed streams* if there exists an input choice agent specification

$$Q: [(M^{\omega})^n \to (M^{\omega})^m] \times (M^{\omega})^n \to \mathbb{B}$$

such that

$$\#x = \infty \Rightarrow (T \circ \varphi).x = (\varphi \circ Q).x \, .$$

In a time insensitive nondeterministic agent specified by the predicate T either all functions f with T.f are time insensitive or the time sensitivity is not observable due to the nondeterminism in T. Technically speaking like in the case of general time abstractions whenever we have a timed input y and some behaviour function f with T.f, then for every timed input y' with $M©y = M©y'$ there is a function f' with T.f' such that $M©f.y = M©f'.y'$. So the output $M©f.y$ for f for which we only know T.f cannot be predicted by the timing of y.

The agent specified by TNFM is time insensitive. This also means that the associated input choice specification fulfils the requirements that will be given for input choice specifications below.

When considering abstractions from time the difference between timed agents and their time abstractions becomes crucial, if timed agents behave essentially differently on timed partial i.e. finite streams s and timed total i.e. infinite streams t with $M©s = M©t$. A special problem arises in particular for completely timed streams s and r (i.e. $\#s = \#r = \infty$) where $M©s \sqsubseteq M©r$. Without assuming $s \sqsubseteq r$ we cannot assume for timed stream processing functions f any particular relationship between f.s and f.r and therefore nothing specific can be said in general about the relationship between $M©f.s$ and $M©f.r$ (in particular $M©f.s \sqsubseteq M©f.r$ does not hold, in general).

From $M©s \sqsubseteq M©r$ we may only conclude that there exists a finite time transformation t such that for some finite stream v we have

$$v \sqsubseteq t.s \wedge v \sqsubseteq r \wedge M©v = M©s.$$

In general time abstraction leads to a loss of monotonicity or putting it the other way around strong time abstractions do not exist. An abstraction from time can be obtained also for certain not (in sense defined so far) time sensitive agents by input choice specifications as follows.

Let

$$f: ((M^{\sqrt{}})^{\omega})n \to ((M^{\sqrt{}})^{\omega})m$$

be an arbitrary time dependent function. A prefix continuous function

$$g: (M^{\omega})^n \to (M^{\omega})^m$$

is called *safe time abstraction* for f, if for every $x \in (M^\omega)^n$ there exists $y \in ((M^{\sqrt{}})^\infty)^n$ such that

$$\forall x \in (M^\omega)^n: \exists y \in ((M^{\sqrt{}})^\infty)^n: x = M\copyright y \wedge (\varphi \circ g)|_y{\downarrow} \sqsubseteq (f \circ \varphi)|_y{\downarrow} .$$

We then write  g **abs_safe** f. For $x \in M^\omega$ the function g is called *life for f and input x*, if in addition for $y \in ((M^{\sqrt{}})^\infty)^n$ we have

$$(x = M\copyright y \Rightarrow (\varphi \circ g)|_y{\downarrow} = (f \circ \varphi)|_y{\downarrow}) .$$

We then write (g, x) **abs_live** (f, y). Let T be a specifying predicate for a timed agent with n input lines and m output lines. We define the input choice specification R called the *time abstraction* from T as follows:

$$R(g, x) \equiv_{df} \exists f: T.f \wedge g \text{ \textbf{abs\_safe} } f \wedge \exists y \in (M^{\sqrt{}})^\infty)^n: (g, x) \text{ \textbf{abs\_live} } (f, y) .$$

For a timed agent specification T an input choice specification R is called a time abstraction if

$$T.f \equiv \forall y \in ((M^{\sqrt{}})^\infty)^n: \exists g: R(g, M\copyright y) \wedge (g, M\copyright y) \text{ \textbf{abs\_live} } (f, y) .$$

Time abstractions in the form of input choice specifications always exist. However, in the time abstraction in general the choice of a particular behaviour that may be influenced by the timing using the specification T cannot be influenced any more.

$\Phi$ therefore provides a time abstraction for time insensitive and time unspecific specifications mapping them on input choice specifications. Of course for arbitrary time dependent functions f and f' (or agents T and T') time abstraction is not modular in the following sense: let R and R' be the input choice specifications such that R(g, x) (and R'(g, x)) if g is a safe time abstraction for f (and for f' resp.) that is life w.r.t. to input x. For the safe time abstraction h for f ∘ f' that is life w.r.t. x we have

$$(R \circ R')(h, x),$$

but from (R ∘ R')(h, x) we cannot conclude that h is a safe time abstraction for f ∘ f' that is life w.r.t. x. The individual use of time information in the composition of f and f' cannot be expressed sufficiently at the level of nondeterminism if the involved agents are time sensitive.

## 10. Functional System Description: Techniques for the Specification of Agents

In this section we give a short survey on techniques for the functional description and specification of interactive systems.

From a theoretical point of view in our approach to the functional specification of agents an agent is specified by a predicate characterizing a set of stream processing functions or by an input choice specification, also characterizing a set of stream processing functions and the liveness properties for given input streams or tuples of input streams. From a methodological point of view we are interested in more specific techniques for specifying agents. In this section we give a brief survey on a number of such possibilities for specifying agents.

### 10.1 Specifications by Direct Predicates

A simple way to specify agents are logical formulas that talk about stream processing functions. First we may write (conditional) equations. In simple cases we write specifications of functions by recursive equations. An example for a specification of an agent by a recursive equation is the elementwise application of functions on message to a stream. The elementwise addition by 1 can be expressed by a function

$$add\_one: N^\omega \to N^\omega$$

with the equation

$$add\_one.x = 1+ft.x \ \& \ add\_one.rt.x \ .$$

The addition of successive numbers for instance can be expressed by the function

$$add\_successive\_elements: N^\omega \to N^\omega$$

with the equation

$$add\_successive\_elements.x = (ft.x+ft.rt.x) \ \& \ add\_successive\_elements.rt.rt.x \ .$$

Such specifications of stream-processing functions are operational as long as expressions on the right-hand sides of the equations are formed of functions with constructive descriptions. Then they can be understood as programs.

We may also specify functions by equations for streams. This a special case of (recursive) equations for functions. Consider the function

$$numbers: \ N^\omega \to N^\omega$$

with the equation

$$numbers.x = ft.x \ \& \ add\_one.numbers.x \ .$$

Here the recursive call of the function numbers on the righthand side shows the same parameters as the left-hand side.

Further examples for direct specifications have been given above such as the specification of interactive stacks or queues.

Apart from simple equations there are many ways for writing logical formulas characterising properties of a stream processing function f. In particular the functions length "#", the filter function "©" and the prefix ordering " ⊑ " often are useful here.


## 10.2 State-based Specifications

Our functional approach to the specification of systems is closely related to state based approaches. A stream processing function can be understood as a special rather abstract form of an automaton.

States are a fundamental concept for the specification of agents. For a given system or system component a state describes the set of possible behaviours. From a functional view there are two ways for using states for specifying agents:

- *Concrete states* can be used as parameters for functions.

- *Abstract states* specify a set of stream processing functions.

Concrete states can be in particular used as an auxiliary concept to specify abstract states. This will be demonstrated below.

## 10.2.1 Concrete States

Any set V can be used as a concrete state. We then consider stream processing functions of the form:

$$h : V \rightarrow [(M^{\omega})^n \rightarrow (M^{\omega})^m]$$

where the elements of V occur as additional parameter. For every state $v \in V$ we obtain a function

$$f : (M^{\omega})^n \rightarrow (M^{\omega})^m$$

by partial application:

$$f.x = (h.v).x$$

As a first simple example let us consider a simple store of elements from D

$$h: D \rightarrow [(D \cup \{?\})^{\omega} \rightarrow D]$$

specified by (let $d' \in D$):

$$(h.d)(d' \ \& \ x) = (h.d').x,$$

$$(h.d)(? \ \& \ x) = d \ \& \ (h.d).x \ .$$

As a further example we consider an interactive stack, again. It can be modelled by a function

$$f: (D \cup\{?\})^{\omega} \rightarrow D^{\omega}.$$

We specify f by using the set D* consisting of the finite sequences over D as concrete state:

$$f.x = (h.\langle\rangle).x$$

where the auxiliary function

$$h: D^{*} \rightarrow ((D \cup\{?\})^{\omega} \rightarrow D^{\omega})$$

is specified by the equations

$$(h.v).x = \textbf{if } ft.x = ? \quad \textbf{then } ft.v \ \& \ (h.rt.v).rt.x$$
$$\textbf{else } h(\langle ft.x\rangle\hat{}v).rt.x \qquad \textbf{fi} \ .$$

Note that the specification of h and therefore of f is in operational form. If we use the equation

$$(h.v).x = \textbf{if } ft.x = ? \quad \textbf{then } ft.v \ \& \ (h.rt.v).rt.x$$
$$\textbf{else } h(v\hat{}\langle ft.x\rangle).rt.x \qquad \textbf{fi}$$

instead of the one given above we obtain an interactive queue.

Two concrete states are (behavioural) equivalent, if they are mapped onto the same stream processing function. So far we have shown how concrete states can be used to specify deterministic systems. Nevertheless we obtain immediately nondeterministic systems, if we consider sets of concrete states. A set of states defines a set of possible behaviours of an agent.

With the help of states we may specify nondeterministic agents by techniques of so-called *oracles*. For our stack example above by

$$\exists \ v \in D^*: f.x = (h.v).x$$

we specify the set of functions f that behave like an interactive stack initialized by an arbitrary stack.

Generally given a function h with auxiliary state parameter ranging over set V we can specify an agent by the predicate Q with

$$Q.f \equiv \forall\ x: \exists\ v \in V: f.x = (h.v).x.$$

Note the similarity to the concept of time abstractions.

## 10.2.2 Abstract States

A stream processing function f can be easily understood as a state machine with input and output (also called I/O-automaton). This will be explained in detail in this section.

Generally *a deterministic state machine with input and output* is given by the quintuple

$$(S, I, O, \delta, \sigma_0)$$

the components of which denote the following items:

- a set of states S,
- a set of input elements I,
- a set of output elements O,
- a transition function with input and output:

$$\delta: I \times S \to O \times S,$$

- an initial state $\sigma_0$.

Every stream processing function (where for simplicity we assume $f.\langle\rangle = \langle\rangle$)

$$f: M^\omega \to N^\omega$$

can be seen as a state machine

$$([M^\omega \to N^\omega], M, N^\omega, \delta, f)$$

with

$$\delta(m, f) = (f.\langle m\rangle, g) \qquad \text{where for all } x: g.x = rt^{\#f.\langle m\rangle}(f(m\&x)) .$$

Note that g in the definition above denotes the resumption which characterizes the "state" of the considered machine after we have observed all the output caused by the input m. This shows that stream processing functions are a very concise and elegant representation of state machines with input and output (cf. also the association of traces to such functions). For explaining our notions let us consider a simple example:

**Example:** A simple store

A simple store that may contain exactly one data element at a time from a given set D of data can be defined as follows: let the set of "input messages" M be defined by

$$M = D \cup \{?\} .$$

We define a stream processing function f

$$f: M^\omega \to D^\omega$$

by the equations

$$f(? \& s) = \langle\rangle,$$

$$f(d \ \& \ d' \ \& \ s) = f(d' \ \& \ s),$$

$$f(d \ \& \ ? \ \& \ s) = d \ \& \ f(d \ \& \ s) .$$

f denotes the initial state of the store where no value is stored and therefore any attempt to read the value by sending the message "?" leads to the empty output. If the first input is the message d then the value d is stored and given as an answer to the signal "?" as long as no other message d' arrives. The resumption $f_{\langle d \rangle}$ denotes the state of the store where d is stored. With the definitions in the first example in the section above we obtain

$$(h.d).x = f(d \ \& \ x) .$$

This equation is easily proved from the specifying equations by induction on the length of x. $\quad\quad \lozenge$

Assume we send the input m to an agent represented by the function f. Then we obtain a finite or infinite stream y of output messages from f as a response to m. Formally we get

$$y = f.\langle m \rangle$$

Assume y contains $k \in \mathbb{N} \cup \{\infty\}$ elements, i.e. $k = \#y$. We specify the behaviour of the module after we have observed the input m and all the output y by the stream processing function g where for all streams x:

$$g.x = rt^k(f(m \ \& \ x)) = rt^k(f_{\langle m \rangle}(x))$$

g represents the abstract state the agent takes after having observed input $\langle m \rangle$ and the output $f.\langle m \rangle$.

Working with stream processing functions instead of state machines with input and output has a number of advantages. In particular we can use the functional calculus for reasoning about the behaviour of communicating systems.

Given a stream processing function (representing an abstract state as explained above) a stream processing function g is called (from f) *reachable* (abstract state), if there is a finite stream z such that f.z is finite and

$$f(z\hat{\ }x) = (f.z)\hat{\ }g.x .$$

Thus a stream processing function corresponds to a finite (control) state machine if the set of reachable functions is finite.

The state machines considered so far are deterministic. Every input corresponds to exactly one output stream and a unique abstract successor state. Now we show a functional model for nondeterministic state machines.

A *nondeterministic state machine* is given by the quintuple $(S, I, O, \delta, \sigma_0)$ the components of which denote the following items:

- a set of states S,
- a set of input elements I,
- a set of output elements O,
- a transition function with input and output:

$$\delta: I \times S \rightarrow \wp(O \times S),$$

- an initial state $\sigma_0$.

We consider predicates on stream processing functions (let $\mathbb{B}$ denote the set of truth values):

$$Q: [M^\omega \rightarrow N^\omega] \rightarrow \mathbb{B}$$

Every such predicate Q can be seen as a nondeterministic state machine

$$([M^\omega \to N^\omega] \to \mathbb{B}, M, N^\omega, \delta, Q)$$

with the transition relation defined by

$$\delta(m, R) = \{(f.\langle m\rangle, H): R.f \,\wedge$$

$$\forall h: H.h \equiv \exists g: R.g \wedge f.\langle m\rangle = g.\langle m\rangle \wedge \forall x: h.x = rt^{\#g.\langle m\rangle}(g(m\&x)) \}$$

Note that H in the definition above denotes the "state" of the considered machine after we have observed all the output caused by the input m. This shows that predicates on (or equivalently sets of) stream processing functions are a very concise and elegant representation of nondeterministic state machines with input and output.

Also input choice specifications R can be seen as nondeterministic automaton by using the following definition for the transition relation $\delta$:

$$\delta(m, R) = \{(y, H): \exists f: \quad R(f, \langle m\rangle) \wedge y \sqsubseteq f.\langle m\rangle \wedge \#y < \infty \,\wedge$$

$$\forall h, x: H(h, x) \equiv \exists g: R(g, m \& x) \wedge y \sqsubseteq g.\langle m\rangle \,\wedge$$

$$\forall z: h.z = rt.^{\#y} (g(m \& z))\}$$

Note that here we only produce finite output in every step. In an abstract state (reached by a step of the machine) now output without further input is possible, since we do not produce all the output that can be produced by the chosen function f an input $\langle m\rangle$. Note furthermore the difference between the state machines described by predicates on stream processing functions and state machines described by input choice specifications. Considering input choice specifications as state machines in the way introduced above liveness properties are not properly represented.

With the understanding of functional system specifications as state machines we may also specify agents by state transitions systems. However, not all input choice specifications lead to proper state machines in the sense that $\delta(m, H)$ is never empty.


## 11. Proper Specifications and Least Fixpoints

As in every powerful specification mechanism it is possible to write inconsistent input choice specifications. But apart from consistency there are additional properties we require for an input choice specification to make sure that we always get consistent specifications when we combine it with other proper specifications. Given an input choice specification

$$R: [(M^\omega)^n \to (M^\omega)^m] \times (M^\omega)^n \to \mathbb{B}$$

we want to formalize such properties in the following.

An input choice specification R is called *consistent*, if

$$\forall x: \exists f: R(f, x) .$$

From the viewpoint of a time abstraction consistency just means that for a nearly time insensitive agent T from which R is the time abstraction there is at least a timed stream-processing function f with T.f. So the definition coincides with in classical notion of consistency.

The consistency of R and R' implies the consistency of R∘R' and R‖R'. But the consistency of R does not imply the consistency of $\mu^k R$. Therefore we are especially interested in the question under which conditions an input choice specification (with $n = m$) has a least fixpoint, i.e. if for an input choice specification R we have:

$$\exists \ f \colon R(f, \textbf{fix} \ f) \ .$$

A stream z is called *least fixpoint* of an input choice specification, if there exists a function f with $z = \textbf{fix} \ f$, i.e. z is least fixpoint of f and $R(f, z)$.

In addition to this straightforward notion of consistency we are going to add the following requirement for input choice specifications which is a generalisation of the property of monotonicity. If we have $R(f, x)$ then for every output finite function $g \sqsubseteq f$ and every input $x'$ with $x \sqsubseteq x'$ we expect that there exists a behaviour function $f'$ such that $R(f', x')$ and $g \sqsubseteq f'$. Accordingly R is called *weakly monotonic*, if for every output finite prefix continuous function g we have

$$R(f, x) \wedge g \sqsubseteq f \wedge x \sqsubseteq x' \Rightarrow \exists \ f' \colon R(f', x') \wedge g \sqsubseteq f'.$$

With this assumption we may think about computations of agents as follows: let $\{x.i \colon i \in \mathbb{N}\}$ be chain (i.e. $x.i \sqsubseteq x.(i+1)$ for all i) of input streams. The output of the agent specified by R is computed for the input stream lub $\{x.i \colon i \in \mathbb{N}\}$ by the following strategy: for every i we choose functions f.i and output finite approximations g.i for f.i such that f.i is a correct choice for the input x.i, the output $(g.i).x.i$ is an approximation (a prefix) for the output generated by f.i on x.i and the function $f(i+1)$ is approximated by f.i . Formally we have (for all $i \in \mathbb{N}$):

$$R(f.i, \ x.i),$$

$$g.i \sqsubseteq f.i,$$

$$g.i \sqsubseteq g(i+1) \sqsubseteq f(i+1) \ .$$

For being able to guarantee that the limit of the chain $\{g.i \colon i \in \mathbb{N}\}$ fulfils required liveness properties, however, we need additional assumptions. The idea is to choose the chain $\{g.i \colon i \in \mathbb{N}\}$ such that (for all $i \in \mathbb{N}$):

$$R(\text{lub} \ \{g.i \colon i \in \mathbb{N}\}, \ \text{lub} \ \{x.i \colon i \in \mathbb{N}\}).$$

What we need is the existence of a function $\delta$ that indicates for given x.i and g.i how large we should choose $g(i+1)$ to guarantee both liveness and safety conditions and therefore allows to construct such a chain of $(g.i).x.i$ as approximations for the output.

Operationally one may think about a computation of an agent specified by the input choice specification R as follows: the agent gets step by step input and produces step by step output. The input can be modelled by a chain $\{x.i \colon i \in \mathbb{N}\}$ of finite elements, the corresponding output is also modelled by a chain $\{(g.i).x.i \colon i \in \mathbb{N}\}$ of finite elements. Every time the agent gets additional input, which can be modelled by going from x.i to $x(i+1)$, it may add messages to its output streams by going from $(g.i).x.i$ to $(g.i).x(i+1)$. However, the strategy has to be chosen carefully such that the stepwise produced output is large enough, such that the least upper bound of $\{(g.i).x.i \colon i \in \mathbb{N}\}$ is correct in the sense of liveness requirements, but it has to be small enough such that the output produced so far is still partially correct for any possible additional input.

A function

$$\delta \colon [(M^*)^n \to (M^*)^m] \times [(M^\omega)^n \to (M^\omega)^m] \to [(M^*)^n \to (M^*)^m]$$

is called *computation strategy*, if we have:

(1)    $g \sqsubseteq f \Rightarrow g \sqsubseteq \delta(g, f) \sqsubseteq f$,

(2)    $g.0 \sqsubseteq f \wedge \forall i: g(i+1) = \delta(g.i, f) \Rightarrow lub \{g.i: i \in \mathbb{N}\} = f$.

An input choice specification R is called *weakly continuous,* if there exists a computation strategy $\delta$ that fulfils the following requirements: it allows for some input x and some safety correct output y to produce a chain $\{g.i: i \in \mathbb{N}\}$ of approximations with the least upper bound f such that for every i we have: if we increase the input x to x' (i.e. $x \sqsubseteq x'$), then $g(i+1)$ produces safe output for the input x'.

More formally expressed we require for R the following properties: For all chains $\{x.i: i \in \mathbb{N}\} \subseteq (M^{\omega})^n$ we have:

$$(\forall i: \exists f: R(f, x.i) \wedge g.i+1 = \delta(g.i, f)) \Rightarrow R(lub \{g.i: i \in \mathbb{N}\}, lub \{x.i: i \in \mathbb{N}\}).$$

Intuitively speaking the function $\delta$ gives for every input x and every approximation g for a function f with $g \sqsubseteq f$ a better approximation $\delta(g, f)$ such that for every continuation of the input stream x the approximation is safe and can be continued to a live output by successively applying $\delta$.

By the assumptions above we may prove the existence of least fixpoints. An input choice specification is called *proper*, if it is consistent, weakly monotonic and weakly continuous.

**Theorem:** A proper input choice specification has a least fixpoint.

**Proof:** We define chains of tuples of streams $\{x.i: i \in \mathbb{N}\}$ and functions $\{g.i: i \in \mathbb{N}\}$ such that

$$x.0 = \diamond^n,$$

$$(g.0).x = \diamond^m, \qquad \text{for all x.}$$

g.0 is safe according to consistency of R. Given x.i, g.i we define

$$x(i+1) =_{df} (g.i).x.i,$$

$$g(i+1) =_{df} \delta(g.i, f.i), \text{ where f.i is arbitrary, such that } R(f.i, x.i) \wedge g.i \sqsubseteq f.i.$$

Such a function f.i always exists since R is consistent and weakly monotonic and $\delta$ is the computation strategy that exists according to weak continuity. According to weak continuity we have:

$$R(lub \{g.i: i \in \mathbb{N}\}, lub \{x.i: i \in \mathbb{N}\}).$$

By the construction of the x.i and f.i we obtain

$$lub \{g.i: i \in \mathbb{N}\}.lub\{x.i: i \in \mathbb{N}\} =$$

$$lub \{(g.i).x.i: i \in \mathbb{N}\} =$$

$$lub \{x(i+1): i \in \mathbb{N}\} =$$

$$lub \{x.i: i \in \mathbb{N}\}.$$

So lub $\{x.i: i \in \mathbb{N}\}$ is a fixpoint of lub $\{g.i: i \in \mathbb{N}\}$. According to the construction we have

$$lub \{g.i: i \in \mathbb{N}\}.x(i+1) \sqsupseteq (g.i).x.i.$$

So it is the least fixpoint. We obtain:

$$\textbf{fix } lub \{g.i: i \in \mathbb{N}\} = lub\{x.i: i \in \mathbb{N}\} \qquad \Diamond$$

The existence of least fixpoints is essential for proofs that rely on fixpoint arguments. An example is the alternating bit protocol again.

What is also important is the invariance of the properties used in the theorem above.

**Theorem:** The sequential and parallel composition and feedback applied to proper input choice specifications yields proper input choice specifications.

**Proof**: For sequential and parallel composition the proof of the theorem is rather straightforward. Therefore we concentrate only on feedback. Let R be a proper specification. According to the theory on the existence of least fixpoints $\mu^k R$ is consistent. Weak monotonicity and weak continuity follow from the continuity of the fixpoint operator. $\Diamond$

The notion of proper input choice specifications replaces the classical monotonicity and continuity requirements for functions.

## 12. Correctness Proof of the Alternating Bit Protocol

We return to the alternating bit protocol and give a proof of its correctness w.r.t. transmission. This proof uses just straightforward reasoning based on fixpoint properties and does not use induction. The proof is due to Birgit Schieder (cf. [Schieder 91]). It has to be shown that the function specified by the alternating bit agent is the identity, i.e. that the following equation holds for every input stream x:

$$(\mu \, (fs \circ f \circ fr \circ (id \parallel g)) \circ pr1).x = x$$

$$\textbf{where} \quad CS \, (fs, \, (x, \, (\mu \, (fs \circ f \circ fr \circ (id \parallel g)) \circ pr2).x)) \wedge CR.fr \wedge CT.f \wedge CT.g \wedge ID.id \wedge$$

$$CP.pr1 \wedge (\forall s1,s2: pr2(s1,s2) = s2)$$

We need some properties of the involved functions for our proof:

$$(0) \quad \forall y: \#alt.y \geq \#alt.g.y \qquad \qquad \text{(by CT)}$$

$$(1) \quad \forall y: \#alt.ack.y \geq \#alt.pr2.(id \parallel g).fr.f.y \qquad \text{(by (0) and CR)}$$

$$(2) \quad \forall y: \#res.y \geq \#alt.ack.y \qquad \qquad \text{(by res)}$$

$$(3) \quad \forall y: \#y = \infty \Rightarrow \#pr2.(id \parallel g).fr.f.y = \infty \qquad \text{(by CT and CR)}$$

$$(4) \quad \forall y: (\#res.y < \infty \wedge \#res.y = \#alt.ack.y \wedge \#alt.ack.y = \#alt.ack.f.y) \Rightarrow res.y = res.f.y$$

$$\text{(by CT)}$$

We abbreviate the second component of the least fixed point by v:

$$v =_{def} (\mu \, (fs \circ f \circ fr \circ (id \parallel g)) \circ pr2).x$$

From this definition it follows immediately that in particular v is a fixed point:

$$(5) \quad v = pr2.(id \parallel g).fr.f.fs(x,v)$$

At first we show that

$$\#x = \#alt.v \, .$$

We know that

$$\#alt.v \leq \#alt.ack.fs(x,v) \qquad \qquad \text{(by (1) and (5))}$$

$$\leq \text{\#res.fs}(x,v) \qquad\qquad \text{(by (2))}$$

$$\leq \#x \qquad\qquad \text{(by the 5th axiom of CS)}$$

In the other direction it holds that

$$\text{\#res.fs}(x,v) = \min(\#x, 1+\#alt.v) \qquad\qquad \text{(by the 5th axiom of CS)}$$

$$\Rightarrow \quad \#x < 1+\#alt.v \ \lor \ \text{\#res.fs}(x,v) = 1+\#alt.v$$

$$\Rightarrow \quad \#x \leq \#alt.v \ \lor \ (\#alt.v = \infty \ \lor \ \#v < \infty) \qquad\qquad \text{(by the 3rd axiom of CS)}$$

$$\Rightarrow \quad \#x \leq \#alt.v \ \lor \ \#alt.v = \infty \ \lor \ \#x \leq \#alt.v \qquad\qquad \text{(by the 4th axiom of CS and (3))}$$

$$\Rightarrow \quad \#x \leq \#alt.v$$

So we have shown the above equality.

From this equality we may conclude the following equations on lengths:

$$(6) \quad \#x = \text{\#res.fs}(x,v)$$

$$(7) \quad \text{\#res.fs}(x,v) = \text{\#alt.ack.fs}(x,v)$$

$$(8) \quad \text{\#alt.ack.fs}(x,v) = \text{\#alt.ack.f.fs}(x,v)$$

Let $\#x$ be finite now. Then we get

$$\text{pr1.fr.f.fs}(x,v) = \text{res.f.fs}(x,v) \qquad\qquad \text{(by CR)}$$

$$= \text{res.fs}(x,v) \qquad\qquad \text{(by } \#x < \infty, \text{ (6), (7), (8) and (4))}$$

$$= x \qquad\qquad \text{(by (6) and the 1st axiom of CS)}$$

Hence the correctness is proved for finite input streams. For infinite input it follows by continuity.


## 13. Full Abstractness

The three combining forms of sequential and parallel composition as well as feedback define contexts for agent specifications. We are interested in a relation $\approx$ on input choice specifications such that we may derive relational equivalence from it (remember that $\sim$ denotes relational equivalence):

$$R1 \approx R2 \Rightarrow R1 \sim R2$$

and moreover the relation $\approx$ forms a congruence w.r.t. the combining forms of parallel and sequential compositions as well as feedback:

$$R1 \approx R2 \Rightarrow R\|R1 \approx R\|R2 \land R1\|R \approx R2\|R,$$

$$R1 \approx R2 \Rightarrow R \circ R1 \approx R \circ R2 \land R1 \circ R' \approx R2 \circ R',$$

$$R1 \approx R2 \Rightarrow \mu^k R1 \approx \mu^k R2.$$

A relation $\approx$ with such properties is called *compositional*.

The concept for the specification of distributed system components as outlined in the previous chapters has been carefully chosen to be compositional. However, our specifications carry too much information, in general. In this section we answer the question under which circumstances two logically distinct input choice specifications are equivalent in all contexts. This leads to the concept of full abstractness. A compositional relation is called *fully abstract*, if there is no weaker compositional relation (cf. also [Kok 87], [Jonsson 88]).

For defining a fully abstract relation we introduce the notion of computation. A pair of chains $(\{x.i: i \in \mathbb{N}\}, \{y.i: i \in \mathbb{N}\})$ is called a *computation* for an input choice specification R, if there exists a function f with

$R(f, \text{lub } \{x.i: i \in \mathbb{N}\}),$

$\forall\, i \in \mathbb{N}: y.i \sqsubseteq f.x.i,$

$f.\text{lub } \{x.i: i \in \mathbb{N}\} = \text{lub } \{y.i: i \in \mathbb{N}\}.$

We define a relation $\approx$ on input choice specifications as follows:

$R \approx R'$          iff the sets of computations of R and R' coincide.

Now we prove that the relation $\approx$ is fully abstract.

**Theorem**: The relation $\approx$ is fully abstract.

**Proof**: We start by proving that the relation cannot be weakened without losing compositionality: R and R' are not relational equivalent for some context, if they have different sets of computations: Assume there is a function

$f: [(M^{\omega})^n \to (M^{\omega})^m]$

and a chain $\{x.i: i \in \mathbb{N}\}$ with $x.0 = (\diamond)^n$ and

$R(f, \text{lub } \{x.i: i \in \mathbb{N}\})$

and there does not exist a function f' with

$f.\text{lub } \{x.i: i \in \mathbb{N}\} = f'.\text{lub } \{x.i: i \in \mathbb{N}\}$

and $f.x.i \sqsubseteq f'.x.i$ for all $i \in \mathbb{N}$ and $R'(f', \text{lub } \{x.i: i \in \mathbb{N}\})$. W.l.o.g. we may assume $n = m$ (add dummy arguments or dummy results) and that either $f.x.i \neq f.x(i+1)$ for all i or there exists an index j with $f.x.i \neq f.x(i+1)$ for all $i < j$ and $f.x.i = f.x(i+1)$ for all $i \geq j$.

We define a specification R'' for (m, n)-ary functions g by

$R''(g, y) = \forall\, z: g.z = \text{lub } \{x.i: f.x.i \sqsubseteq z\}$

We consider the least fixpoint y of $f \circ g$ where $R''(g, f.y)$. This corresponds to an element of $\mu^n(R \circ R'')$. We show that y is not totally correct w.r.t. $\mu^n(R' \circ R'')$. We define the chain $\{z.i: i \in \mathbb{N}\}$ by

$z.0 = (\diamond)^n$

$z(i+1) = g.f.z.i$

We prove that $\text{lub } \{z.i: i \in \mathbb{N}\} = \text{lub } \{x.i: i \in \mathbb{N}\}$. If there exists an index j with $f.x.i \neq f.x(i+1)$ for all $i < j$ and $f.x.i = f.x(i+1)$ for all $i \geq j$, then this is obvious; otherwise for all $k \in \mathbb{N}: x.k = z(k+1)$:

$z(k+1) =$

$g.f.x.k =$

$\text{lub } \{x.i: f.x.i \sqsubseteq f.x.k\} =$

$x.k$

Thus the lubs of $\{z.i: i \in \mathbb{N}\}$ and of $\{x.i: i \in \mathbb{N}\}$ coincide (and $y = \text{lub } \{z.i: i \in \mathbb{N}\}$).

For every function f' with not $f.x.k \sqsubseteq f'.x.k$ for some k we obtain for the chain z'.i defined by

$z'.0 = (\langle\rangle)^n$

$z'.i{+}1 = g.f'.z.i$

the equation

$z'.i = z.k$ for all $i \geq \min \{j: \neg(f.x.j \sqsubseteq f'.x.j)\}$

since with $k = \min \{j: \neg(f.x.j \sqsubseteq f'.x.j)\}$ we obtain

$z'.k{+}1 =$

$g.f'.x.k =$

$\text{lub } \{x.i: f.x.i \sqsubseteq f'.x.k\} =$

$z'.k$

Thus y cannot be a behaviour of $\mu^n(R'{\circ}R'')$. This concludes the first part of the proof. Next we prove that the relation $\approx$ is a congruence, i.e. it is invariant under our compositional forms. For every continuous operator

$$\tau: [(M^\omega)^n \to (M^\omega)^m] \times (M^\omega)^{n'} \to [(M^\omega)^{n'} \to (M^\omega)^{m'}] \times (M^\omega)^n$$

we have if $R \approx R'$, then $Q \approx Q'$ where

$Q(f,x) = \exists\ f',\ x': R(f',\ x') \wedge (f,\ x') = \tau(f',\ x)$

(and in analogy Q' is defined based on R'). Since parallel and sequential composition and feedback operator correspond to such continuous operators, the condition is invariant under the considered compositional forms. $\Diamond$

Based on this theorem we may prove equivalences of the form $R \approx R'$ for given specifications R and R'. Moreover we may derive a fully abstract model.

In terms of input choice specifications R and R' we have

$R \approx R'$  iff      $(R \approx> R' \wedge R' \approx> R)$

where

$R \approx> R'$    iff      $\forall\ f, x: R(f, x) \Rightarrow \exists\ g: R'(g, x) \wedge f|_{(f.x)}{\downarrow} = g|_{(f.x)}{\downarrow}$

This leads to a unique normal form for proper specifications. An input choice specification R is normal form, if for every function f that is partially correct w.r.t. R we have

$R(g, x) \wedge f|_{(f.x)}{\downarrow} = g|_{(f.x)}{\downarrow} \Rightarrow R(f, x)$ .

For proper input choice specifications R and R' in this normal form we have

$R \approx R'$ iff $\forall\ f, x: R(f, x) \Leftrightarrow R'(f, x)$

So on proper input choice specifications in normal form the logical identity is a fully abstract relation.

## 14. Timed Agents Revisited

Even for timed agents that are not time insensitive we may find ways to abstract away the timing information and still retain a sufficient representation of their behaviour. For making this more precise we first introduce the notion of finite time transformation.

A *finite time transformation* is a continuous function

$$t: [((M^{\sqrt{}})^\omega)^n \to ((M^{\sqrt{}})^\omega)^n]$$

that causes only finite time-shifts, i.e. we require

(i) $\quad \varphi.t.x = \varphi.x,$

(ii) $\quad \#(\sqrt{}©x).i = \infty \Leftrightarrow \#(\sqrt{}©t.x).i = \infty \qquad$ for all i, $1 \le i \le n.$

We write FTT.t, iff a function t fulfils the requirements (i) and (ii). Finite time transformations formalize the concept of independent timing (or speed) of system components. This is often used informally when speaking about time dependent systems and programs when it is stated that "nothing is assumed about the relative speed of two units running in parallel".

An (n,m)-ary timed agent specification T is called *timing stable,* if for all finite time transformations t

$$T.f \Rightarrow T(t \circ f)$$

Otherwise the agent specification is called *time critical.* In a specification of a timing stable agent there are no sharp timing requirements. The time information in the input is not relevant for the output or not exploited effectively due to the nondeterminism of the agent.

Having time information available it is rather simple to specify certain behaviours. For instance using timed agents it is rather simple to specify timed nondeterministic "strict" fair merge:

$$\text{TNFM.f} \equiv_{df} \forall\, x, y: \exists\, r, t: \text{FTT.t} \wedge f(x, y) = t.\text{scd}(x, y, r) \wedge \#(1©r) = \infty \wedge \#(0©r) = \infty$$

$$\textbf{where } \forall\, x, y, r: \ \text{scd}(x, y, 1\&r) = \text{ft.x} \ \& \ \text{scd}(rt.x, y, r) \wedge$$

$$\text{scd}(x, y, 0\&r) = \text{ft.y} \ \& \ \text{scd}(x, rt.y, r)$$

The agent specified by TNFM is timing stable. The timing influences the output only in a nondeterministic way such that finite time shifts do not essentially change the behaviour according to the nondeterminism contained in T.

However, this agent behaves essentially differently on finite timed streams and infinite timed streams. If the input on both of the input lines is infinite, then all the input on both lines is guaranteed to appear as output. So for infinite input the merge is fair. This is not true for finite input. This indicates that the specification TFNM is not time insensitive as defined so far. Otherwise there would exist a nontimed function g such that (for certain i and j):

$$g(M©(\sqrt{}^{i}{}^{\wedge}\langle 1\rangle{}^{\wedge}\sqrt{}^{\infty}), M©(\sqrt{}^{j}{}^{\wedge}\langle 2\rangle{}^{\wedge}\sqrt{}^{\infty})) = \langle 1\rangle{}^{\wedge}\langle 2\rangle,$$

and

$$g(\langle\rangle, M©(\sqrt{}^{j}{}^{\wedge}\langle 2\rangle{}^{\wedge}\sqrt{}^{\infty})) = \langle 2\rangle.$$

This, however, is in conflict with our monotonicity assumptions for g.

The inclusion of explicit time considerations into the specification of communicating systems leads to a trade off. Certain properties are easier and more explicitly expressible, however, the detailed consideration of timing may lead to problems with overspecification. This trade off can be partly solved by techniques that allow us to specify parts of a system with explicit timing and other parts without explicit timing. For those techniques, however, we need interfaces between timed and nontimed parts of system specifications. The interfaces can be again described by appropriately specified agents. Basically we may use agents that filter out time informations as well as agents that (re-)introduce time information. The first is simple. The later is more difficult.

A function

$$t: (M^\omega)^n \to ((M^{\sqrt{}})^\omega)^n$$

is called *complete timing function*, if

(i)  $M©t.x = x$,

(ii)  $\#(t.x).i = \infty$ $\qquad$ for all i, $1 \le i \le n$.

Of course a complete timing function t is not prefix monotonic as can be seen by the fact that all images of t are maximal. Consider for instance the following equations

$$t.\langle\rangle = \sqrt{}^\infty \quad \text{and} \quad t.\langle a\rangle = \sqrt{}^k {}^\wedge \langle a\rangle^\wedge \sqrt{}^\infty.$$

Nevertheless we may give an input choice specification $\tau$ for complete timing functions

$$\tau(t, x) \equiv_{df} (\forall\, z \in (M^\omega)^n: \varphi.t.z \sqsubseteq z) \wedge \#t.x = \infty \wedge \varphi.t.x = x.$$

The input choice specification $\tau$ fulfils all the requirements necessary for guaranteeing the existence of least fixpoints. This shows that by the technique of input choice specification we may combine timed and nontimed models of communicating systems.

**Lemma**:  The input choice specification

$$\tau \circ \varphi$$

represents the identity.

**Proof**:  For f.x = y with $(\tau \circ \varphi).(f, x)$ we have

$$\exists\, t: f = t \circ \varphi \wedge \tau(t, x) \wedge f.x = y$$

which is by definition of $\tau$ equivalent to

$$\exists\, t: f = t \circ \varphi \wedge (\forall\, z \in (M^\omega)^n: \varphi.t.z \sqsubseteq z) \wedge \#t.x = \infty \wedge \varphi.t.x = x \wedge f.x = y$$

From this we obtain:

$$\varphi.t.x = x \wedge \varphi.t.x = y$$

which yields

$$x = y. \hspace{10cm} \Diamond$$

Note that the behaviours specified by $\tau$ correspond to so-called arbiters.

Moreover assuming the agent specified by T we may easily program fair nonstrict merge by timed merge:

$$\tau \circ TNFM \circ \varphi.$$

We may now specify the sender by the predicate CS by the timed sender specification:

$$CS = \tau \circ TCS \circ \varphi.$$

where we include complete time information in the stream of acknowledgements and later filter out the time information again.

General a specification

$$T: [((M^{\surd})^{\omega})^n \to ((M^{\surd})^{\omega})^m] \to \mathbb{B}$$

T is called *timing stable* if there exists an agent specification

$$Q: [(M^{\omega})^n \to (M^{\omega})^m] \to \mathbb{B}$$

such that

$$Q = \tau \circ T \circ \varphi.$$

**Lemma**: Every time insensitive agent is timing stable.

**Proof**: For a time insensitive agent T we have by definition

$$T \circ \varphi = \varphi \circ Q$$

This allows as to derive

$$\tau \circ T \circ \varphi = \tau \circ \varphi \circ Q$$

and since is the identity we have

$$\tau \circ T \circ \varphi = Q \qquad \lozenge$$

If a stream processing function is not time insensitive, then abstracting away time can be seen as introducing some kind of nondeterminism.

Nevertheless, if parts of a system are time sensitive, such that the correct behaviour of one part strongly depends on the particular timing of the other parts time abstraction does not work for proving certain properties.


## 15. Conclusion

Functional techniques comprising fixpoint theory with concepts like monotonicity and continuity together with logical techniques can be seen as the backbone of formal methods for program construction. When considering communicating systems a straightforward application of such techniques seems difficult. This is due to the very particular abstractions by information hiding (such as hiding time information) and the nondeterminism introduced thereby. Fortunately these difficulties can be mastered by a careful choice of the specification formalism. This way a relatively simple and powerful formal framework for the functional treatment of communicating systems is obtained.

For many applications time aspects are important. According to the principle of abstraction computing scientists are interested in techniques that allow to talk about timing wherever this seems appropriate and to avoid time consideration wherever possible. A rigorous formal foundation for time sensitive communicating systems therefore is of high importance.

After all we have identified the following classifications of timed functions and specifications of time functions. This classification has to do with the following two aspects of timed behaviour:

-   does the timing of the input streams influence the output streams only with respect to their timing,

- is it possible to capture the timed behaviour by nontimed behaviour using the concept of nondeterminism, i.e. can every time abstract output be seen as an output for a particular timing of the input.

Intuitively speaking regarding the first aspect a timed function or a timed specification T is called

- *time insensitive*, if the time information on the input streams does not influence the produced message streams, i.e. if there exists a nontimed specification Q such that

$$T \circ \varphi = \varphi \circ Q$$

where $\varphi$ is the time abstraction function on timed streams. In particular T is time insensitive if

$$T = \varphi \circ Q \circ \tau$$

- *timing stable*, if the timing does influence the output, but there is a nondeterministic abstraction, i.e. if

$$Q = \tau \circ T \circ \varphi$$

Note that these two formal characterisations just represent notions of refinements for communicating agents. This way introducing explicit timing and time abstraction can just be seen as particular concepts of refinement.

## Acknowledgement

Discussions with Leslie Lamport have motivated part of this work. I am grateful to Frank Dederichs, Thomas Streicher, and Rainer Weber for helpful remarks on draft versions of this paper. I thank Birgit Schieder for careful reading and for the equational proof for the alternating bit protocol.

## References

[Brock, Ackermann 81]
J.D. Brock, W.B. Ackermann: Scenarios: A model of nondeterminate computation. In: J. Diaz, I. Ramos (eds): Lecture Notes in Computer Science 107, Springer 1981, 225-259

[Broy 83]
M. Broy: Applicative real time programming. Information Processing 83, IFIP World Congress, Paris 1983, North Holland Publ. Company 1983, 259-264

[Broy 85]
M. Broy: Specification and top down design of distributed systems. In: H. Ehrig et al. (eds.): Formal Methods and Software Development. Lecture Notes in Computer Science 186, Springer 1985, 4-28, Revised version in JCSS 34:2/3, 1987, 236-264

[Broy 86]
M. Broy: A theory for nondeterminism, parallelism, communication and concurrency. Habilitation, Fakultät für Mathematik und Informatik der Technischen Universität München, 1982, Revised version in: Theoretical Computer Science 45 (1986) 1-61

[Broy 87a]
M. Broy: Semantics of finite or infinite networks of communicating agents. Distributed Computing 2 (1987), 13-31

[Broy 87b]
M. Broy: Predicative specification for functional programs describing communicating networks. Information Processing Letters 25 (1987) 93-101

[Dybier, Sander 88]
P. Dybier, H. Sander: A functional programming approach to the specification and verification of concurrent systems. Chalmers University of Technology and University of Göteborg, Department of Computer Sciences 1988

[Jonsson 88]
B. Jonsson: A fully abstract trace model for dataflow networks. 16th POPL 1989, 155 - 165

[Kahn, MacQueen 77]
G. Kahn and D. MacQueen, Coroutines and networks of processes, Proc. IFIP World Congress 1977, 993-998

[Kok 87]
J.N. Kok: A fully abstract semantics for data flow networks. Proc. PARLE, Lecture Notes in Computer Science 259, Berlin-Heidelberg-New York: Springer 1987, 214-219

[Lamport 83]
L. Lamport: Specifying concurrent program modules. ACM Toplas 5:2, April 1983, 190-222

[Park 80]
D. Park: On the semantics of fair parallelism. In: D. Björner (ed.): Abstract Software Specification. Lecture Notes in Computer Science 86, Berlin-Heidelberg-New York: Springer 1980, 504-526

[Schieder 91]
B. Schieder: Private Communications, 1991

[Schneider 87]
F.B. Schneider: Decomposing properties into safety and liveness using predicate logic. Cornell University, Department of Computer Science, Technical Report 87-874, October 1987

# SYSTEMATIC DERIVATION OF COMMUNICATING PROGRAMS


Ernst-Rüdiger Olderog

FB Informatik
Universität Oldenburg, Postfach 2503
2900 Oldenburg, Fed. Rep. Germany



ABSTRACT. We present an approach to the top-down derivation of communicating, concurrent programs from their specification. As a programming language we use a blend of Milner's CCS, Hoare's CSP and Lauer's COSY, and as a specification language we use a version of Zwiers' trace logic. The link between communicating programs and trace specifications is given by a notion of program correctness which deals with both safety and liveness properties. The top-down derivation proceeds by application of compositional transformation rules which refine the given trace specification stepwise into a communicating program satisfying the safety and liveness requirements of the specification.

Key words: Concurrency, communication, nondeterminism, CCS, CSP, COSY, Petri nets, specification, many-sorted first order logic, trace logic, program correctness, safety, liveness, externally deterministic, readiness semantics, top-down derivation, transformations, mixed terms.


## CONTENTS



## 1. INTRODUCTION

We present a simple approach to the top-down derivation of communicating, concurrent programs from specifications. Such a program interacts with its user or environment by communication. In between two subsequent communications the program may engage in internal actions. These are not visible for the user, but as a result of such internal actions the program behaviour may appear nondeterministic to the user. Concurrency arises firstly because there can be more than one user and secondly because the program can generate more than one active subprocess. For the user(s), however, only the communication behaviour in of interest.

In our approach the communication behaviour is defined as a set of finite communication sequences that are possible between user and program. Such sequences are known as *histories* or *traces* [Ho 78]. Since traces are insensitive to intervening internal actions and concurrent process activities, this definition achieves abstraction from both internal activity and concurrency. Note that our notion of trace differs from the one used by Mazurkiewicz [Ma 77]; there a trace is a certain partial order expressing concurrency.

As a specification language for trace sets we use a *many-sorted first-order predicate logic*. Since its main sort is "trace", it is called *trace logic* and its formulas are called *trace formulas* or *trace specifications*. Informal use of trace logic appears in a number of papers (e.g. [CHo 81, MC 81, Sn 85, Rm 87, WGS 87]). Precise syntax and semantics, however, is given only in [Zw 89]. We shall adopt Zwiers' proposal, but we need only a simplified version of it because we deal here with atomic communications instead of messages sent along channels.

As a programming language we use a blend of CCS, CSP and COSY [Mi 80, Ho 85, LTS 79]. An intuitively clear operational semantics of communicating programs is given by labelled Petri nets. This enables us to exhibit all details about the internal process activity and the possible concurrency. We refer here to the operational net semantics of process terms defined and discussed in [Ol 88/89, Ol 89a].

The link between programs and specifications is given by notion of *program correctness*. It is a relationship

$$P \ sat \ S$$

stating when a communicating program P *satisfies* or is *correct with respect to* a trace specification S. In most papers [CHo 81, MC 81, ZRE 85, Zw 89] trace formulas express only *safety properties* (cf. [OL 82]). In other words, P *sat* S if every trace of P satisfies the formula S. As a consequence, there exists a single process term which satisfies every trace specification with the same alphabet. Such a process term is called a *miracle* after Dijkstra [Di 76].

However, with miracles the task of *program derivation*, i.e. given a trace formula S derive a program P with P *sat* S, becomes trivial: always choose the miracle. Therefore we shall be more demanding and use trace formulas to express also a simple type of *liveness property* (cf. [OL 82]). Essentially, P *sat* S requires the following:

* Safety:               P may only engage in traces satisfying S.
* Liveness:           P must engage in every trace satisfying S.

The terminology of "may" and "must" originates from [DH 88] but the details are different here. The liveness condition is due to [OH 86] and related to the idea of Misra and Chandy to use so-called *quiescent* infinite trace specifications to express liveness in the setting of asynchronous communication (see [Jo 87]). It implies that every program P satisfying a trace formula S is divergence free and *externally deterministic*. That is: in every run of the program the user has exactly the same possibilities of communication, no matter which actions the program has pursued internally. Thus in our approach trace formulas can specify only a subset of programs. We are interested in this subset because it has many applications and yields simple compositional transformation rules for top-down derivation.

The concepts of "may" and "must" are clear intuitively, but not helpful for discovering such rules. To simplify this task, we develop a second, more abstract semantics for communicating programs and trace specifications. It is a *modified* version $\mathfrak{R}^*$ of the *readiness semantics* $\mathfrak{R}$ introduced in [OH 86]. For programs, $\mathfrak{R}^*$ is defined by filtering out certain informations from the operational net semantics. Among these informations are pairs consisting of a trace and a so-called *ready set* [Ho 81, FLP 84]. Ready sets are used to explain the liveness requirement of P *sat* S.

In $\mathfrak{R}^*$ program correctness boils down to a *semantic equation* between P and S:

$$P \; sat \; S \quad \text{iff} \quad P \equiv S.$$

Moreover, $\mathfrak{R}^*$ has an equivalent *denotational* definition in the sense of Scott and Strachey [Sc 70, St 77, Ba 80]. This implies that $\mathfrak{R}^*$ is compositional with respect to the program operators and that recursion is dealt with by fixed point techniques. Therefore $\mathfrak{R}^*$ serves well as a stepping stone for developing compositional transformation rules for program derivation

Such a construction is presented as a sequence

$$
\begin{aligned}
S &\equiv Q_1 \\
&\; ||| \\
&\; \vdots \\
&\; ||| \\
Q_n &\equiv P
\end{aligned}
$$

of semantic equations where $Q_1$ is the given trace specification and $Q_n$ is the derived program P. By the transitivity of $\equiv$, it follows $P \equiv S$ which means P *sat* S. Thus by construction, P satisfies the safety and liveness requirements of S. The terms $Q_i$ in between are *mixed terms*, i.e. syntactic constructs mixing programming notation with specification parts. This mixture formalises ideas form the work of Dijkstra and Wirth on program development by stepwise refinement [Di 76, Wi 71].

The sequence of equations in the top-down construction is obtained by applying the principle of *transformational programming* as e.g. advocated in the Munich Project CIP [Bau 85, Bau 87]. Thus each equation $Q_i \equiv Q_{i+1}$ is justified by applying a transformation rule on mixed terms. We present a system of such transformation rules based on the modified readiness semantics $\mathfrak{R}^*$. Most rules are extremely simple. For example, parallel composition is reflected by logical conjunction of trace formulas. Only two programming operators need transformation rules which are difficult to apply: *hiding* and *renaming with aliasing*. The reason is that these operators can completely restructure the communication behaviour of programs.

We apply our transformation rules to derive several communicating programs from their trace specification.

## 2. TRACE LOGIC

We start from an infinite set Comm of unstructured *communications* with typical elements a, b. By a *communication alphabet* or simply *alphabet* we mean a finite subset of Comm. We let letters A, B range over alphabets. Syntax and semantics of trace logic we adopt from Zwiers [Zw 89]. It is a many-sorted predicate logic with the following sorts:

> *trace*     (finite communication sequences)
> *nat*     (natural numbers)
> *comm*     (communications)
> *log*     (logical values)

*Trace logic* then consists of sorted expressions built up from sorted constants, variables and operator symbols. For notational convenience, trace formulas count here as expressions of sort *log*.

All communications appear as constants of sort *trace* and *comm*, and all natural numbers $k \geq 0$ appear as constants of sort *nat*. The set Var of variables is partitioned into a set Var:*trace* of variables t of sort *trace* and a set Var:*nat* of variables n of sort *nat*. Among the trace variables there is a *distinguished trace variable* called h; it will be used in the definiton of trace *specification*. For all communication alphabets A and all communications a, b there are unary operator symbols $\cdot \upharpoonright A$ and $\cdot [b/a]$ of sort *trace* $\longrightarrow$ *trace*. Further on, there are binary operator symbols $\cdot . \cdot$ of sort *trace* $\times$ *trace* $\longrightarrow$ *trace* and $\cdot [ \cdot ]$ of sort *trace* $\times$ *nat* $\longrightarrow$ *comm*, and a unary operator symbol $| \cdot |$ of sort *trace* $\longrightarrow$ *nat*. The remaining symbols used in trace logic are all standard.

**Definition.** The syntax of trace logic is given by a set

$$\text{Exp} = \text{Exp:}\textit{trace} \cup \text{Exp:}\textit{nat} \cup \text{Exp:}\textit{comm} \cup \text{Exp:}\textit{log}$$

of *expressions* ranged over by xe. The constituents of Exp are defined as follows.

(1) The set Exp:*trace* of *trace expressions* consists of all expressions te of the form

$$\text{te:: } = \varepsilon \mid a \mid t \mid te_1 . te_2 \mid te \upharpoonright A \mid te[b/a]$$

where every trace variable t in te occurs within a subexpression of the form $te_0 \upharpoonright A$.

(2) The set Exp:*nat* of *natural number* consists of the following expressions ne:

$$\text{ne:: } = k \mid n \mid ne_1 + ne_2 \mid ne_1 * ne_2 \mid |te|$$

(3) The set Exp:*comm* of *communication expressions* consists of the following expressions ce:

$$\text{ce:: } = a \mid te[ne]$$

(4) The set Exp: *log* of *trace formulas* or *logical expressions* consists of the following expressions le:

$$\text{le:: } = \textit{true} \mid te_1 \leq te_2 \mid ne_1 \leq ne_2 \mid ce_1 = ce_2$$
$$\mid \neg \text{ le} \mid le_1 \wedge le_2 \mid \exists t. \text{ le} \mid \exists n. \text{ le} \qquad \square$$

Let xe{te/t} denote the result of *substituting* the trace expression te for every free occurrence of the trace variable t in xe. Furthermore, let xe{b/a} denote the result of literally replacing every occurrence of the communication a in xe by b.

The *standard semantics* or *interpretation of trace logic* is introduced along the lines of Tarski's semantic definition for predicate logic. It is a mapping

$$\mathfrak{I} : \text{Exp} \longrightarrow (\text{Env}_{\mathfrak{I}} \longrightarrow \text{DOM}_{\mathfrak{I}})$$

assigning a value to every expression with the help of so-called *environments*. These are mappings

$$\rho \in Env_{\mathcal{J}} = Var \longrightarrow DOM_{\mathcal{J}}$$

assigning values to the free variables in expressions. The semantic domain of $\mathcal{J}$ is

$$DOM_{\mathcal{J}} = Comm^* \cup \mathbb{N}_0 \cup Comm \cup \{\bot\} \cup \{true, false\},$$

and the environments $\rho$ respect sorts, i.e. trace variables t get values in $Comm^*$ and natural number variables n get values in $\mathbb{N}_0$.

**Definition.** With the above conventions the standard semantics $\mathcal{J}$ of trace logic is defined as follows.

(1) *Semantics of trace expressions* yielding values in $Comm^*$:

$\mathcal{J}[\![\,\varepsilon\,]\!](\rho) = \varepsilon$ , the empty trace

$\mathcal{J}[\![\,a\,]\!](\rho) = a$

$\mathcal{J}[\![\,t\,]\!](\rho) = \rho(t)$

$\mathcal{J}[\![\,te_1 . te_2\,]\!](\rho) = \mathcal{J}[\![\,te_1\,]\!](\rho) ._{\mathcal{J}} \mathcal{J}[\![\,te_2\,]\!](\rho)$ , the *concatenation* of the traces

$\mathcal{J}[\![\,te \upharpoonright A\,]\!](\rho) = \mathcal{J}[\![\,te\,]\!](\rho) \upharpoonright_{\mathcal{J}} A$ , the *projection* onto A, i.e. with all communications outside A removed

$\mathcal{J}[\![\,te[b/a]\,]\!](\rho) = \mathcal{J}[\![\,te\,]\!](\rho) \{b/a\}$ , i.e. every occurrence of a is *renamed* into b. Brackets [...] denote an unevaluated renaming operator and brackets {...} its evaluation.

(2) *Semantics of natural number expressions* yielding values in $\mathbb{N}_0$:

$\mathcal{J}[\![\,k\,]\!](\rho) = k$ for $k \in \mathbb{N}_0$

$\mathcal{J}[\![\,n\,]\!](\rho) = \rho(n)$

$\mathcal{J}[\![\,|te|\,]\!](\rho) = |\mathcal{J}[\![\,te\,]\!](\rho)|_{\mathcal{J}}$, the *length* of the trace

Expressions $ne_1 + ne_2$ and $ne_1 * ne_2$ are interpreted as addition and multiplication.

(3) *Semantics of communication expressions* yielding values in $Comm \cup \{\bot\}$:

$\mathcal{J}[\![\,a\,]\!](\rho) = a$

$\mathcal{J}[\![\,te[ne]\,]\!](\rho) = \mathcal{J}[\![\,te\,]\!](\rho)[\mathcal{J}[\![\,ne\,]\!](\rho)]_{\mathcal{J}}$ , the *selection* of the $\mathcal{J}[\![\,ne\,]\!](\rho)$-th element of the trace $\mathcal{J}[\![\,ne\,]\!](\rho)$ if it exists and $\bot$ otherwise

(4) *Semantics of trace formulas* yielding values in $\{true, false\}$:

$\mathcal{J}[\![\,true\,]\!](\rho) = true$

$\mathcal{J}[\![\,te_1 \le te_2\,]\!](\rho) = (\mathcal{J}[\![\,te_1\,]\!](\rho) \le_{\mathcal{J}} \mathcal{J}[\![\,te_2\,]\!](\rho))$, the *prefix* relation on $Comm^*$

$\mathcal{J}[\![\,ne_1 \le ne_2\,]\!](\rho) = (\mathcal{J}[\![\,ne_1\,]\!](\rho) \le_{\mathcal{J}} \mathcal{J}[\![\,ne_2\,]\!](\rho))$, the standard ordering relation on $\mathbb{N}_0$

$\mathcal{J}[\![\,ce_1 = ce_2\,]\!](\rho) = (\mathcal{J}[\![\,ce_1\,]\!](\rho) =_{\mathcal{J}} \mathcal{J}[\![\,ce_2\,]\!](\rho))$, the *strong, non-strict equality* on $DOM_{\mathcal{J}}$. Thus a value $\bot$, which is possible for a communication expression, does not propagate to the logical level.

Formulas $\neg le$, $le_1 \wedge le_2$, $\exists t.le$, $\exists n. le$ are interpreted as negation, conjunction and existential quantification over $Comm$ and $\mathbb{N}_0$, respectively.

(5) A trace formula le is called *valid*, abbreviated $\models le$, if $\mathcal{J}[\![\,le\,]\!](\rho) = true$ for all environments $\rho$. $\qquad\qquad\square$

To specify trace sets we use a subset of trace formulas.

**Definition.** The set Spec of *trace specifications* ranged over by S, T, U consists of all trace formulas where at most the distinguished variables h of sort *trace* is free. $\qquad\square$

Thus the logical value $\mathfrak{I}[S](\rho)$ of a trace specification $S$ depends only on the trace value $\rho(h)$. We say that a trace $\mathfrak{h} \in \text{Comm}^*$ *satisfies* $S$ and write $\mathfrak{h} \vDash S$ if $\mathfrak{I}[S](\rho) = \text{true}$ for $\rho(h) = \mathfrak{h}$. Note the following relationship between satisfaction and validity:

$$\mathfrak{h} \vDash S \quad \text{iff} \quad \vDash S \{ \mathfrak{h}/h \}$$

A trace specification $S$ specifies the set of all traces satisfying $S$. In fact, wether or not a trace satisfies a trace specification $S$ depends only on the trace value within the *projection alphabet* $\alpha(S)$. This is the smallest set of communications such that h is accessed only via trace projections within $\alpha(S)$. Its definition can be found in [Ol 88/89, Ol 89b]. We also consider the *extended alphabet* $\alpha\alpha(S)$. This is the set of all communications appearing somewhere in S. For example, for

$$S = (\text{lk.h}){\restriction}\{\text{dn}\} \le (\text{lk.h}){\restriction}\{\text{up}\}$$

we have $\alpha(S) = \{\text{dn, up}\}$ and $\alpha\alpha(S) = \{\text{lk, dn, up}\}$.

**Projection Lemma.** Let $S$ be a trace specification. Then

$$\mathfrak{h} \vDash S \quad \text{iff} \quad \mathfrak{h}{\restriction}\alpha(S) \vDash S$$

for all traces $\mathfrak{h} \in \text{Comm}^*$. $\qquad\qquad\square$

Since trace logic includes the standard interpretation of Peano arithmetic, viz. the model $(\mathbb{N}_0, 0, 1, +_{\mathfrak{Z}}, *_{\mathfrak{Z}}, =_{\mathfrak{Z}})$, trace specifications are very expressive.

**Expressiveness Theorem.** [Zw 89] Let $\mathfrak{L} \subseteq A^*$ be a recursively enumerable set of traces over the alphabet A. Then there exists a trace specification $\text{TRACE}(\mathfrak{L})$ with projection alphabet $\alpha(\text{TRACE}(\mathfrak{L})) = A$ such that

$$\mathfrak{h} \in \mathfrak{L} \quad \text{iff} \quad \mathfrak{h} \vDash \text{TRACE}(\mathfrak{L})$$

for all traces $\mathfrak{h} \in A^*$. The same is true for sets $\mathfrak{L} \subseteq A^*$ whose complement in $A^*$ is recursively enumerable. $\qquad\qquad\square$

For practical specification, such a general expressiveness result is not very helpful. Then a concise and clear notation is important. We use the following:

* Natural number expressions *counting* the number of communications in a trace:

$$a \# te =_{df} | te{\restriction}\{a\} |$$

* Communication expressions *selecting* specific elements of a trace: e.g.

$$last\ te =_{df} te[|te|]$$

* Extended syntax for logical expressions: e.g. for $k \ge 3$

$$ne_1 \le \ldots \le ne_k =_{df} \bigwedge_{j=1}^{k-1} ne_j \le ne_{j+1}$$

* *Regular expressions* denoting sets of traces.

## 3. COMMUNICATING PROGRAMS

Communicating programs are defined as recursive terms over a certain signature of operator symbols taken from Lauer's COSY [LTS 79, Be 87], Milner's CCS [Mi 80] and Hoare's CSP as in [Ho 85]. More specifically, we take the parallel composition $\|$ from COSY, prefix a., choice + and renaming $[a_1/b_1]$ from CCS, and deadlock $stop$ : A, divergence $div$ : A, hiding $\setminus$ B and the idea of using communication alphabets to express certain context-sensitive restrictions on from CSP.

To the set Comm of communication we add an element $\tau \in$ Comm yielding the set Act = Comm $\cup$ $\{\tau\}$ of $actions$. The element $\tau$ is called $internal$ action and the communications are also called $external$ actions. We let u,v range over Act. As before let a,b range over Comm and A,B over communication alphabets. The set of $(process)$ $identifiers$ is denoted by Idf; it is partitioned into sets Idf:A $\subseteq$ Idf of $identifiers$ $with$ $alphabet$ A, one for each communication alphabet A. We let X,Y,Z range over Idf.

**Definition**. The set Rec of $(recursive)$ $terms$, with typical elements P,Q,R, consists of all terms generated by the following context-free production rules:

$$
\begin{array}{lll}
P ::= & stop : A & (\text{ deadlock }) \\
| & div : A & (\text{ divergence }) \\
| & a.P & (\text{ prefix }) \\
| & P + Q & (\text{ choice }) \\
| & P \| Q & (\text{ parallelism }) \\
| & P[b_1/a_1] & (\text{ renaming }) \\
| & P \setminus B & (\text{ hiding }) \\
| & X & (\text{ identifier }) \\
| & \mu X.P & (\text{ recursion })
\end{array}
$$

An occurrence of an identifier X in a term P is said to be $bound$ if it occurs in P within a subterm of the form $\mu$X.Q. Otherwise the occurence is said to be $free$. A term P $\in$ Rec without free occurences of identifiers is called $closed$. P$\{$Q$/$X$\}$ denotes the result of $substituting$ Q for every free occurence of X in P.

A term P is called $action$-$guarded$ if in every recursive subterm $\mu$X.Q of P every free occurence of X in Q occurs within a subterm of the form a.R of Q. E.g. $\mu$X . a. X is action-guarded, but a.$\mu$X. X is not.

To every term P we assign a communication alphabet $\alpha(P)$ defined inductively as follows :

$$
\begin{aligned}
&\alpha(stop : A) = \alpha(div : A) = A , \\
&\alpha(a.P) = \{a\} \cup \alpha(P) , \\
&\alpha(P+Q) = \alpha(P \| Q) = \alpha(P) \cup \alpha(Q) , \\
&\alpha(P[b_1/a_1]) = (\alpha(P)) - \{a_1\}) \cup \{b_1\} , \\
&\alpha(P \setminus B) = \alpha(P) - B , \\
&\alpha(X) = A \text{ if } X \in Idf(A) , \\
&\alpha(\mu X.P) = \alpha(X) \cup \alpha(P) .
\end{aligned}
$$

Thus for every n-ary operator symbol op of Proc there is a set-theoretic operator $op_\alpha$ satisfying $\alpha(op(P_1,...,P_n)) = op_\alpha(\alpha(P_1),...,\alpha(P_n))$.

**Definition.** A *(communicating) program* is a term $P \in \text{Rec}$ which satisfies the following context-sensitive restrictions:

(1)  P is action-guarded,
(2)  every subterm a.Q of P satisfies $a \in \alpha(Q)$,
(3)  every subterm Q+R of P satisfies $\alpha(Q) = \alpha(R)$,
(4)  every subterm $\mu X.Q$ of P satisfies $\alpha(X) = \alpha(P)$.

Let Prog denote the set of all programs and CProg the set of all closed programs.    □

The operational semantics of communicating programs can be defined using Petri nets. We consider here *labelled place/transition nets* with arc weight 1 and place capacity ω [Re 85] but we will mainly work in the subclass of safe Petri nets. We deviate slightly from the standard definition and use the following one which is inspired by [Go 88].

**Definition.** A *Petri net* or simply *net* is a structure $\mathfrak{N} = (\text{A, Pl, } \longrightarrow\text{, } M_0)$  where
(1)  A is a communication alphabet,
(2)  Pl is a possibly infinite set of *places*,
(3)  $\longrightarrow \subseteq \mathfrak{P}_{nf}(Pl) \times (A \cup \{\tau\}) \times \mathfrak{P}_{nf}(Pl)$ is the *transition relation*,
(4)  $M_0 \in \mathfrak{P}_{nf}(Pl)$ is the *initial marking*.    □

Here $\mathfrak{P}_{nf}(Pl)$ denotes the set of all non-empty, finite subsets of Pl. An element $(I, u, O) \in \longrightarrow$ is called a *transition (labelled with the action u)* and will usually be written as

$$ I \xrightarrow{\text{u}} O \ . $$

For a transition $t = I \xrightarrow{\text{u}} O$ its *preset* or *input* is given by pre(t) = I, its *postset* or *output* by post(t) = O and its action by act(t) = u.

The graphical representation of a net $\mathfrak{N} = (A, Pl, \longrightarrow, M_0)$ is as follows. We draw a rectangular box divided into an upper part displaying the alphabet A and a lower part displaying the remaining components Pl, $\longrightarrow$ and $M_0$ in the usual way. Thus places $p \in Pl$ are represented as *circles* with the name "p" outside and transitions

$$ t = \{ p_1, ..., p_m \} \xrightarrow{\text{u}} \{ q_1, ..., q_n \} $$

as *boxes* carrying the label "u" inside and connected via directed arcs to the places in pre(t) and post(t). Since pre(t) and post(t) need not be disjoint, some of the outgoing arcs of u actually point back to places in pre(t) and thus introduce *cycles*. The initial marking is represented by putting a token into the circle of each $p \in M_0$.

As usual the dynamic behaviour of a Petri net is defined by its *token game*; it describes which transitions are concurrently enabled at a given marking and what the result of their concurrent execution is. Though the initial marking of a net is defined to be a set of places, the token game can result in more general markings, viz. multisets. A net $\mathfrak{N}$ is *safe* if in every reachable marking is a set, i.e. each place contains at most one token.

Consider a net $\mathfrak{N} = (A, Pl, \longrightarrow, M_0)$. The set place($\mathfrak{N}$) of *statically reachable places* of $\mathfrak{N}$ is the smallest subset of Pl satisfying

(1)  $M \subseteq \text{place}(\mathfrak{N})$ ,

(2) If $I \subseteq \text{place}(\mathfrak{N})$ and $I \xrightarrow{u} O$ for some $u \in A \cup \{ \tau \}$ and $O \subseteq Pl$ then also $O \subseteq \text{place}(\mathfrak{N})$.

The term "statical" emphasises that, by (2), the set $\text{place}(\mathfrak{N})$ is closed under the execution of any transition $t = I \xrightarrow{u} O$ independently of whether $t$ is ever enabled in the token game of $\mathfrak{N}$.

When considering nets, we wish to ignore the identity of places and forget about places that are not statically reachable. We do this by introducing suitable notions of isomorphism and abstract net.

**Definition.** Two nets $\mathfrak{N}_1 = ( A_1, Pl_1, \longrightarrow_1, M_{01} )$, i=1,2, are *weakly isomorphic*, abbreviated

$$\mathfrak{N}_1 =_{isom} \mathfrak{N}_2,$$

if $A_1 = A_2$ and there exists a bijection $\beta : \text{place}(\mathfrak{N}) \longrightarrow \text{place}(\mathfrak{N})$ such that $\beta(M_{01}) = M_{02}$ and for all $I, O \subseteq \text{place}(\mathfrak{N}_1)$ and all $u \in A \cup \{ \tau \}$

$$I \xrightarrow{u}_1 O \quad \text{iff} \quad \beta(I) \xrightarrow{u}_2 \beta(O)$$

where $\beta(M_{01})$, $\beta(I)$, $\beta(O)$ are understood elementwise. The bijection $\beta$ is called an *weak isomorphism between $\mathfrak{N}_1$ and $\mathfrak{N}_2$* .  □

Clearly, $=_{isom}$ is an equivalence relation. An *abstract net* is defined as the isomorphism class

$$[\mathfrak{N}]_{=_{isom}} = \{ \mathfrak{N}' \mid \mathfrak{N} =_{isom} \mathfrak{N}' \}$$

of a net $\mathfrak{N}$. It will be written shorter as $[\mathfrak{N}]$. For abstract nets, we use the same graphical representation as for nets; we only have to make sure that all places are statically reachable and eliminate their names. Most concepts for nets can be lifted in a straightforward way to abstract nets. For example, we shall call an abstract net $[\mathfrak{N}]$ safe, if $\mathfrak{N}$ is safe. Let Net denote the set of nets and ANet the set of abstract nets.

Formally, the semantics of communicating programs is a mapping $\mathfrak{N}[\![ \cdot ]\!] : \text{CProg} \longrightarrow$ ANet which assigns to every proces term $P \in \text{CProg}$ an safe abstract net of the form

$$\mathfrak{N}[\![ P ]\!] = [ ( \alpha(P), Pl, \longrightarrow , M_0) ].$$

For the definition of the components $Pl$, $\longrightarrow$ and $M_0$ we refer to operational net semantics given in $\lfloor Ol\ 89a \rfloor$. Here we have space only for an example.

**Example.** Consider the program $SEM = \mu X. (p_1. v_1 . X + p_2 .v_2. X)$ modelling a semaphore and $CYCLE = \mu Y.n.p.b.e.v.Y$ modelling a user cycle with non-critical action n and critical section b.e sheltered by semaphore communications p and v. Applying the renaming operator we generate two copies of CYCLE:

$$CYCLE1 = CYCLE [ n_1, p_1, b_1, e_1, v_1 / n, p, b, e, v ] ,$$
$$CYCLE2 = CYCLE [ n_2, p_2, b_2, e_2, v_2 / n, p, b, e, v ] .$$

Applying parallel composition to CYCLE1, SEM and CYCLE2 we obtain the program

$$MUTEX = CYCLE1 \parallel SEM \parallel CYCLE2$$

modelling the mutual exclusion of the critical sections $b_1 . e_1$ and $b_2 . e_2$. Then $\mathfrak{N}[\![ \text{MUTEX} ]\!]$ is the following abstract net:

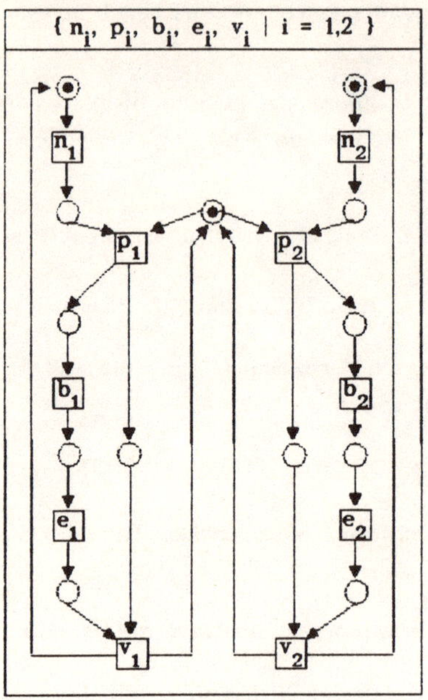


## 4. PROGRAM CORRECTNESS

We begin with an informal description. Consider a program P and a communication trace $\mathfrak{h} = a_1 ... a_n$ over $\alpha(P)$. We say that P *may engage* in $\mathfrak{h}$ if there exists a transition sequence of P where the user is able to communicate $a_1 ... a_n$ in that order. We say that P *must engage* in $\mathfrak{h}$ if the following holds. When started P eventually becomes stable. Then it is possible for the user to communicate $a_1$. Now P may engage in some internal activity, but eventually it becomes stable again. Then it is ready for the next communication $a_2$ with the user, etc. for $a_3, ..., a_n$. Also after the last communication $a_n$ the program P eventually becomes stable again. Summarising, in every transition sequence of P the user is able to communicate $a_1, ..., a_n$ in that order after which the program eventually becomes stable. Stability can be viewed as an acknowledgement of P for a successful communication with the user. We say that P is *stable immediately* if initially P cannot engage in any internal action.

**Definition.** Consider a closed program P and a trace specification S. Then

$$P \ sat \ S$$

if $\alpha(P) = \alpha(S)$ and the following conditions hold:

(1) *Safety.*     For every trace $\mathfrak{h} \in \alpha(P)^*$ whenever P may engage in $\mathfrak{h}$ then $\mathfrak{h} \vDash S$.

(2) *Liveness.*   For every trace $\mathfrak{h} \in \alpha(S)^*$ whenever *pref* $\mathfrak{h} \vDash S$ then P must engage in $\mathfrak{h}$.
     The notation *pref* $\mathfrak{h} \vDash S$ means that $\mathfrak{h}$ and all its prefixes satisfy S.

(3) *Stability.*  P is stable immediately.

The distinction between safety and liveness properties is due to Lamport (see e.g. [OL 82] and [AS 85]). Note that the notion of safety is different from safeness defined for nets in Section 3: safeness can be viewed as a specific safety property of the token game of a net. Stability is also a safety property, but it is singled out here because its rôle is more technical. Its presence allows a more powerful verification rule for the choice operator in Section 6.

In the following we give formal definitions of the notions of "may" and "must engage" and of initial stability by looking at the Petri net denoted by P. The intuition behind these definitions is as follows. Whereas transitions labelled by a communication occur only if the user participates in them, transitions labelled by $\tau$ occur autonomously at an unknown, but positive speed. Thus $\tau$-transitions give rise to unstability and divergence.

**Definition.** Consider a net $\mathfrak{N} = (A, Pl, \longrightarrow, M_0)$, reachable markings M, M' of $\mathfrak{N}$ and a trace $\mathfrak{h} \in \text{Comm}^*$.

(1) *Progess properties.* The set of *next possible actions* at M is given by

$$\text{next}(M) = \{u \in \text{Act} \mid \exists t \in \longrightarrow : \text{pre}(t) \subseteq M \text{ and } \text{act}(t) = u\}.$$

M is called *stable* if $\tau \notin \text{next}(M)$ otherwise it is called *unstable*. M is *ready* for a communication b if M is stable and $b \in \text{next}(M)$. M is ready for the communication set A if M is stable and $\text{next}(M) = A$. $\mathfrak{N}$ is *stable immediately* if $M_0$ is stable. We write

$$M \overset{\mathfrak{h}}{\Longrightarrow} M'$$

if there exists a finite transition sequence $M \xrightarrow{t_1} M_1 \ldots M_{n-1} \xrightarrow{t_n} M_n = M'$ such that $\mathfrak{h} = (\text{act}(t_1) \ldots \text{act}(t_n)) \setminus \tau$, i.e. $\mathfrak{h}$ results from the sequence of actions $\text{act}(t_1) \ldots \text{act}(t_n)$ by deleting all internal actions $\tau$.

(2) *Divergence properties.* $\mathfrak{N}$ *can diverge from* M if there exists an infinite transition sequence

$$M \xrightarrow{t_1} M_1 \xrightarrow{t_2} M_2 \xrightarrow{t_3} \ldots$$

such that $\tau = \text{act}(t_1) = \text{act}(t_2) = \text{act}(t_3) = \ldots$ $\mathfrak{N}$ *can diverge immediately* if $\mathfrak{N}$ can diverge from $M_0$. $\mathfrak{N}$ *can diverge after* $\mathfrak{h}$ if there exists a marking M with

$$M_0 \overset{\mathfrak{h}}{\Longrightarrow} M$$

such that $\mathfrak{N}$ can diverge from M. $\mathfrak{N}$ *can diverge only after* $\mathfrak{h}$ if whenever $\mathfrak{N}$ can diverge after some trace $\mathfrak{h}'$ then $\mathfrak{h} \leq \mathfrak{h}'$. $\mathfrak{N}$ *can diverge* if there is a reachable marking M of $\mathfrak{N}$ from which $\mathfrak{N}$ can diverge. $\mathfrak{N}$ is *divergence free* if $\mathfrak{N}$ cannot diverge.

(3) *Deadlock properties.* $\mathfrak{N}$ *deadlocks* at M if $\text{next}(M) = \Phi$. $\mathfrak{N}$ *deadlocks immediately* if $\mathfrak{N}$ deadlocks at $M_0$. $\mathfrak{N}$ *can deadlock after* $\mathfrak{h}$ if there exists a marking M with

$$M_0 \overset{\mathfrak{h}}{\Longrightarrow} M$$

such that $\mathfrak{N}$ deadlocks at M. $\mathfrak{N}$ *can deadlock only after* $\mathfrak{h}$ if whenever $\mathfrak{N}$ can deadlock after some trace $\mathfrak{h}'$ then $\mathfrak{h} \leq \mathfrak{h}'$. $\mathfrak{N}$ *can deadlock* if there is a reachable marking M of $\mathfrak{N}$ at which $\mathfrak{N}$ deadlocks. $\mathfrak{N}$ is *deadlock free* if $\mathfrak{N}$ cannot deadlock. $\qquad\square$

We now turn to communicating programs.

**Definition.** Consider a closed program P, a representative $\mathfrak{N}_0 = (\alpha(P), Pl, \longrightarrow, M_0)$ of the abstract net $\mathfrak{N}[\![ P ]\!]$, and a trace $\mathfrak{h} \in \text{Comm}^*$.

(1) P is *stable immediately* if $\mathfrak{N}_0$ is so.

(2) P *can diverge (immediately* or *after* $\mathfrak{h}$ or *only after* $\mathfrak{h}$) if $\mathfrak{N}_0$ can do so. P is *divergence free* if $\mathfrak{N}_0$ is so.

(3) P *deadlocks immediately* if $\mathfrak{N}_0$ does so. P *can deadlock (after* $\mathfrak{h}$ or *only after* $\mathfrak{h}$) if $\mathfrak{N}_0$ can do so. P is *deadlock free* if $\mathfrak{N}$ is so.

(4) P *may engage* in $\mathfrak{h}$ if there exists a marking M with $M_0 \xoverset{\mathfrak{h}}{\Longrightarrow} M$.

(5) P *must engage* in $\mathfrak{h} = a_1 \ldots a_n$ if the process term $P \parallel a_1 \ldots a_n . \; stop: \alpha(P)$ is divergence free and can deadlock only after $\mathfrak{h}$. $\qquad\qquad\square$

Clearly, these definitions are independent of the choice of the representative $\mathfrak{N}_0$. The formalisations of immediate stability and "may engage" capture the above intuitions, but the formalisation of "must engage" requires some explanation. The process term $a_1 \ldots a_n$. *stop*: $\alpha(P)$ models a user wishing to communicate the trace $a_1 \ldots a_n$ to P and stop afterwards. Communication is enforced by making the alphabet of user and process identical. Thus the parallel composition $P \parallel a_1 \ldots a_n$. *stop*: $\alpha(P)$ can behave only as follows: it can engage in some prefix $a_1 \ldots a_k$ of $\mathfrak{h}$ with $0 \le k \le n$ and then either diverge (i.e. never become stable again) or deadlock (i.e. become stable, but unable to engage in any further communication). The user's wish to communicate $\mathfrak{h}$ is realised if and only if $P \parallel a_1 \ldots a_n$. *stop*: $\alpha(P)$ never diverges and if it deadlocks only after $\mathfrak{h}$. A final deadlock is unavoidable because the user wishes to stop. This is how we formalise the notion of "must engage".

The terminology of "may" and "must engage" originates from DeNicola and Hennessy's work on testing of processes [DH 84]. There it is used to define several so-called testing equivalences on processes, among them one for the "may" case and one for the "must" case. Here the definition of "must engage" is stronger than in [DH 84] because we require stability after each communication, also after the last one.

**Proposition.** Consider a closed program P and a trace specification S. Then P *sat* S implies the following:

(1) "May" is equivalent to "must", i.e. for every trace $\mathfrak{h}$ the program P may engage in $\mathfrak{h}$ if and only if P must engage in $\mathfrak{h}$.

(2) P is divergence free.

(3) P is externally deterministic. $\qquad\qquad\square$

Intuitively, a program is externally deterministic if the user cannot detect any nondeterminism by communicating with it. Formally, we define this notion as follows:

**Definition.** Consider a closed program P and some representative $\mathfrak{N}_0 = (\alpha(P), Pl, \longrightarrow, M_0)$ of $\mathfrak{N}[P]$. Then P is called *externally deterministic* if for all traces $\mathfrak{h} \in \text{Comm}^*$ and all markings $M_1, M_2$ of $\mathfrak{N}_0$ whenever

$$M_0 \xoverset{\mathfrak{h}}{\Longrightarrow} M_1 \text{ and } M_0 \xoverset{\mathfrak{h}}{\Longrightarrow} M_2$$

such that $M_1$ and $M_2$ are stable then $\text{next}(M_1) = \text{next}(M_2)$. That is: every communication trace $\mathfrak{h}$ uniquely determines the next stable set of communications. $\qquad\qquad\square$

Thus trace formulas specify only divergence free and exernally deterministic programs. This is a clear restriction of our approach, but it is an interesting class of programs with many applications and simplest verification rules (see Sections 6 and 7).

**Example.** Let us consider the trace specification

$$S = 0 \le up\#h - dn\#h \le 2$$

which is an abbreviation for $dn\#h \le up\#h \le 2+dn\#h$, and examine how a program P satisfying S should behave. Since P *sat* S implies $\alpha(P) = \alpha(S) = \{ up, dn \}$, P should engage only in the communications up and dn. By the safety condition, in every communication trace $\mathfrak{h}$ that P may engage in, the difference of the number of up's and the number of dn's is between 0 and 2. If P has engaged in such a trace $\mathfrak{h}$ and the extension $\mathfrak{h}.dn$ still satisfies S, the liveness condition of P *sat* S requires that after $\mathfrak{h}$ the program P must engage in the communication dn. The same is true for up.

Thus S specifies that P should behave like *2-counter* which can internally store a natural number n with $0 \le n \le 2$. After a communication trace $\mathfrak{h}$, the number stored is $n = up\#\mathfrak{h} - dn\#\mathfrak{h}$. Thus initially, when $\mathfrak{h}$ is empty, n is 0. Communicating up increments n and communicating dn decrements n provided these changes of n do not exceed the bounds 2 and 0.

A program satisfying S is

$$P = \mu X. \, up. \, \mu Y. \, ( \, dn. \, X \, + \, up. \, dn. \, Y \, )$$

denoting the following abstract net

$$\mathfrak{N}[\![ P ]\!] =$$

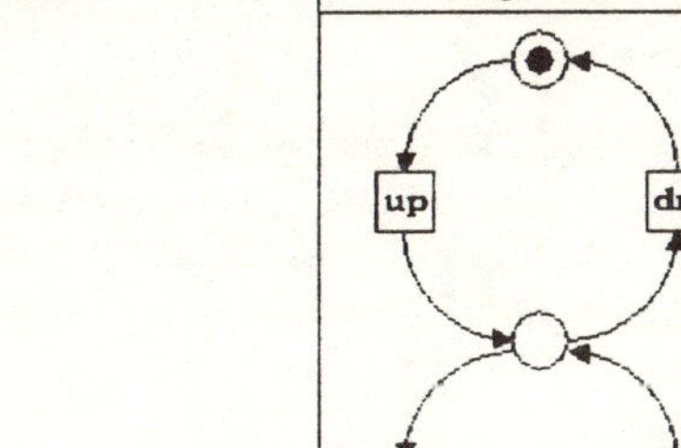


This net is purely sequential, i.e. every reachable marking contains at most one token, and there are no internal actions involved. Another program satisfying S is

$$Q = (( \, \mu X. \, up. \, dn. \, X) \, [ \, lk/dn \, ] \, \| \, (\mu X. \, up. \, dn. \, X) \, [ \, lk/up \, ]) \setminus lk$$

denoting the following abstract net.

$$\mathfrak{N}[\![\, Q \,]\!] = \quad \{\ up,\ dn\ \}$$

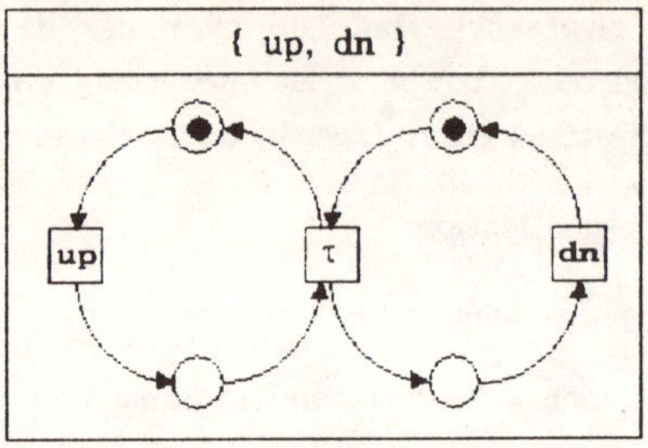

Here, after each up-transition the net has to engage in an internal action $\tau$ before it is ready for the corresponding dn-transition. Since $\tau$-actions occur autonomously, readiness for the next dn is guaranteed, as required by the specification S. This leads in fact to a marking where up and dn are concurrently enabled.

The examples of P and Q demonstrate that presence or absence of concurrency or intervening internal activity are treated here as properties of the implementation (program and and net), not of the specification. $\square$


## 5. READINESS SEMANTICS


The liveness condition of the satisfaction relation P *sat* S is difficult to check when only the net semantics of P is available. To simplify matters, we introduce now a second, more abstract semantics for both programs and specifications. It is a modification of the readiness semantics $\mathfrak{R}$ introduced in [OH 86]. To avoid confusion, we call this modification $\mathfrak{R}^*$. Formally, it is a mapping

$$\mathfrak{R}^*[\![\, \cdot \,]\!] : CProg \cup Spec \longrightarrow DOM_{\mathfrak{R}}$$

which assigns to every program $P \in CProg$ an element $\mathfrak{R}^*[\![\, P \,]\!]$ and to every trace specification $S \in Spec$ an element $\mathfrak{R}^*[\![\, S \,]\!]$ in the readiness domain $DOM_{\mathfrak{R}}$. This domain consists of pairs $(A, \Gamma)$ where A is a communication alphabet and $\Gamma$ is a set of *process informations*. We consider three types of process information:

(1) The element $\tau$ indicating initial *unstability*.
(2) *Ready pairs* $(\mathfrak{h}, \mathfrak{F})$ consisting of a trace $\mathfrak{h} \in A^*$ and a *ready set* $\mathfrak{F} \subseteq A$.
(3) *Divergence points* $(\mathfrak{h}, \uparrow)$ consisting of a trace $\mathfrak{h} \in A^*$ and a special symbol $\uparrow$ standing for divergence.

The set of these process informations can be expressed as follows:

$$Info_{\mathfrak{R}}\!:\!A = \{\ \tau\ \} \cup A^* \times \mathfrak{P}(A) \cup A^* \times \{\ \uparrow\ \}.$$

Define

$$DOM_{\mathfrak{R}}\!:\!A = \{\ (A, \Gamma) \mid \Gamma \subseteq Info\!:\!A\ \}.$$

The readiness domain is then given by

$$DOM_{\mathfrak{R}} = \bigcup DOM_{\mathfrak{R}}\!:\!A$$

where the union is taken over all communication alphabets A. We adopt the following notational conventions: letters $\gamma, \delta$ range over $Info_{\mathfrak{R}}\!:\!A$, letters $\Gamma, \Delta$ over subsets of $Info_{\mathfrak{R}}\!:\!A$ and hence pairs $(A, \Gamma), (B, \Delta)$ over $DOM_{\mathfrak{R}}$, letters $\mathfrak{F}, \mathfrak{G}$ range over ready sets and the letter $\mathfrak{X}$ can either be a ready set or the symbol $\uparrow$.

The mapping $\mathfrak{R}^*$ retrieves the relevant process information from the operational Petri net semantics and enforces three closures: the "chaotic closure" due to [BHR 84], the "acceptance closure" due to [DH 84], and a new "radiation closure" by which a divergence after a trace $\mathfrak{h}.a$ "radiates up" by affecting the ready sets after the immediate prefix $\mathfrak{h}$.

**Definition.** For a program $P \in CProc$ with $\mathfrak{N}[\![\,P\,]\!] = [\,(\alpha(P), \mathrm{Pl}, \longrightarrow, M_0)\,]$ the *operational readiness semantics* is given by

$$\mathfrak{R}^*(\mathfrak{N}) = \mathrm{close}(\,A,\ \{\quad \tau \quad |\ M_0 \text{ is unstable }\}$$
$$\cup \{\,(\mathfrak{h}, \mathfrak{F})\ |\ \exists \text{ marking } M \text{ of } \mathfrak{N}:$$
$$M_0 \overset{\mathfrak{h}}{=\!=}\!\!> M \text{ and } M \text{ is stable and } \mathfrak{F} = \mathrm{next}(M)\,\}$$
$$\cup \{\,(\mathfrak{h}, \uparrow)\ |\ \exists \text{ marking } M \text{ of } \mathfrak{N}:$$
$$M_0 \overset{\mathfrak{h}}{=\!=}\!\!> M \text{ and } \mathfrak{N} \text{ can diverge from } M\,\}\qquad)$$

where the *closure operator* close: $\mathrm{DOM}_\mathfrak{R} \longrightarrow \mathrm{DOM}_\mathfrak{R}$ is defined as follows:

$$\mathrm{close}(A, \Gamma) =$$

$$(A,\ \Gamma \cup \{\,(\mathfrak{h}, \mathfrak{G})\ |\ \exists\,\mathfrak{F}:\ (\mathfrak{h}, \mathfrak{F}) \in \Gamma \text{ and } \mathfrak{F} \subseteq \mathfrak{G} \subseteq \mathrm{succ}(\mathfrak{h}, \Gamma)\,\} \qquad \text{"acceptance closure"}$$
$$\cup \{\,(\mathfrak{h}', \mathfrak{X})\ |\ \exists\,\mathfrak{h} \leq \mathfrak{h}': (\mathfrak{h}, \uparrow) \in \Gamma \text{ and } \mathfrak{h}' \in A^*$$
$$\text{and } (\mathfrak{X} \subseteq A \text{ or } \mathfrak{X} = \uparrow)\,\} \qquad \text{"chaotic closure"}$$
$$\cup \{\,(\mathfrak{h}, \mathfrak{G})\ |\ \exists\,a:\ (\mathfrak{h}.a, \uparrow) \in \Gamma \text{ and } \mathfrak{G} \subseteq \mathrm{succ}(\mathfrak{h}, \Gamma)\,\}\,) \qquad \text{"radiation closure"}$$

Here $\mathrm{succ}(\mathfrak{h}, \Gamma)$ denotes the set of all *successor communications* of $\mathfrak{h}$ in $\Gamma$:

$$\mathrm{succ}(\mathfrak{h}, \Gamma) = \{\,a\ |\ \exists\,\mathfrak{G}: (\mathfrak{h}.a, \mathfrak{G}) \in \Gamma\,\}\ .$$

For a trace specification $S \in Spec$ the *readiness semantics*

$$\mathfrak{R}^*[\![\,S\,]\!] = (\,\alpha(S),\ \{\,(\mathfrak{h}, \mathfrak{F})\ |\ \mathfrak{h} \in \alpha(S)^* \text{ and } \textit{pref } \mathfrak{h} \models S$$
$$\text{and } \mathfrak{F} = \{\,a \in \alpha(S)\ |\ \mathfrak{h}.a \models S\,\}\,\})$$

where, as before, *pref* $\mathfrak{h} \models S$ means that $\mathfrak{h}$ and all its prefixes satisfy $S$. $\qquad \square$

The readiness semantics $\mathfrak{R}^*$ is an interleaving semantics because it is insensitive to concurrency. This is demonstrated by the law

$$\mathfrak{R}^*[\![\,a.\ \textit{stop}:\{a\} \parallel b.\ \textit{stop}:\{b\}\,]\!] \ =\ \mathfrak{R}^*[\![\,a.\,b.\ \textit{stop}:\{a, b\} + b.\,a.\ \textit{stop}:\{a, b\}\,]\!]$$

which is easily established by retrieving the readiness information from the corresponding nets.

Note that the three closures add ready sets and divergence points to $\mathfrak{R}^*[\![\,P\,]\!]$ which are not justified by the token game of $\mathfrak{N}[\![\,P\,]\!]$. These additions make the semantics $\mathfrak{R}^*$ more abstract so that less programs can be distinguished under $\mathfrak{R}^*$. The resulting level of abstraction is in perfect match with the distinctions that we can make among programs under the satisfaction relation $P$ *sat* $S$. Technically speaking, $\mathfrak{R}^*$ is *fully abstract* with respect to this relation (see [Ol 88/89, Ol 89 b]).

**Correctness Theorem.** [Ol 88/89] For every closed program $P$ and trace specification $S$

$$P \textit{ sat } S \qquad \text{iff} \qquad \mathfrak{R}^*[\![\,P\,]\!] = \mathfrak{R}^*[\![\,S\,]\!],$$

i.e. in the readiness semantics program correctness reduces to semantics equality. $\qquad \square$

The Correctness Theorem simplifies, at least conceptually, the task of proving that a program P satisfies a trace specification S.

**Example.** In Section 4 we considered the trace specification $S = 0 \le up*h - dn*h \le 2$ and argued informally that the programs

$$P = \mu X.\ up.\ \mu Y.\ (\ dn.\ X + up.\ dn.\ Y\ )$$

and

$$Q = (\ (\ \mu X.\ up.\ dn.\ X\ )[\ lk/dn\ ]\ \|\ (\ \mu X.\ up.\ dn.\ X\ )[\ lk/up\ ]\ )\ \backslash lk$$

both satisfy S. We can now prove this claim by comparing the readiness semantics of S with that of P and Q:

$$\mathcal{R}^*[\![S]\!] = (\ \{\ up, dn\ \},\ \{\ (\mathfrak{h},\ \mathfrak{F})\ |\ \forall\ \mathfrak{h}' \le \mathfrak{h}:\ 0 \le up*\mathfrak{h}' - dn*\mathfrak{h}' \le 2$$

$$\text{and } (\text{ if } 0 = up*\mathfrak{h} - dn*\mathfrak{h} \quad\text{ then } \mathfrak{F} = \{\ up\ \}\quad\ )$$

$$\text{and } (\text{ if } 0 < up*\mathfrak{h} - dn*\mathfrak{h} < 2\ \text{ then } \mathfrak{F} = \{\ up, dn\ \})$$

$$\text{and } (\text{ if }\quad up*\mathfrak{h} - dn*\mathfrak{h} = 2\ \text{ then } \mathfrak{F} = \{\ dn\ \}\quad\ )\ )$$

By an exhaustive analysis of the reachable markings of the nets $\mathfrak{N}[\![P]\!]$ and $\mathfrak{N}[\![Q]\!]$ shown in Section 4 we see that $\mathcal{R}^*[\![P]\!] = \mathcal{R}^*[\![S]\!] = \mathcal{R}^*[\![Q]\!]$. Thus indeed P *sat* S and Q *sat* S. $\square$

Above we referred to the obvious way of determining the readiness semantics of a programs P, viz. by analysis of the underlying Petri net $\mathfrak{N}[\![P]\!]$. The drawback of this method is that it is very inflexible: whenever a part of P is exchanged, the analysis of $\mathfrak{N}[\![P]\!]$ has to start again. Instead we would like to reuse the results established for those parts of P which are left unchanged. To this end, we wish to determine the readiness semantics $\mathcal{R}^*[\![P]\!]$ by induction on the structure of P.

Technically speaking, we provide an alternative, denotational definition $\mathcal{R}^{**}$ of the readiness semantics and show that on closed programs it coincides with the previous operational definition $\mathcal{R}^*$. In Section 6 this denotational definition will be the basis for developing verification rules for program correctness.

Following the approach of Scott and Strachey [Sc 70, St 77, Ba 80], a semantics of a programming language is called *denotational* if its definition is compositional and employs fixed point techniques when dealing with recursion. These techniques stipulate a certain structure of the composition operators and the semantic domains. We will use here the standard set-up and work with monotonic or (chain-) continuous operators over complete partial orders (cpo's). For the definition and basic properties of these notions we refer to the literature, e.g. [Ba 80]. However, the specific format of our programming language, viz. recursive terms with alphabets as types, requires a few extra details which we explain now.

**Definition.** The *denotational readiness semantics for process terms* is a mapping

$$\mathcal{R}^{**}:\ Proc \longrightarrow (\ Env_{\mathcal{R}} \longrightarrow DOM_{\mathcal{R}}\ )$$

which satisfies the following conditions:

(1) The semantic domain $DOM_{\mathcal{R}}$ is equipped with the following partial ordering $\sqsubseteq$:

$$(A,\ \Gamma) \sqsubseteq (B,\ \Delta)\quad \text{if}\quad A = B \text{ and } \Gamma \supseteq \Delta.$$

Intuitively, the reverse set inclusion $\Gamma \supseteq \Delta$ expresses that the process generating $\Delta$ is *more controllable*, i.e. *more deterministic*, *less divergent* and *more stable* than the process generating $\Gamma$ [Ho 85]. For each alphabet $A \subseteq$ Comm the subdomain $DOM_{\mathfrak{R}} : A \subseteq DOM_{\mathfrak{R}}$ is a complete partial order under $\subseteq$.

(2) The set $Env_{\mathfrak{R}}$ of environments consists of mappings $\rho \in Env_{\mathfrak{R}} = Idf \longrightarrow DOM_{\mathfrak{R}}$ which respect alphabets, i.e. $\alpha(X) = A$ implies $\rho(X) \in DOM_{\mathfrak{R}} : A$.

(3) For every n-ary operator symbol op of Proc there exists a corresponding semantic operator

$$op_{\mathfrak{R}} : \underbrace{DOM_{\mathfrak{R}} \times ... \times DOM_{\mathfrak{R}}}_{n \text{ times}} \longrightarrow DOM_{\mathfrak{R}}$$

defined in [Ol 88/89]. Each of these operators is $\subseteq$-monotonic and satisfies $op_{\mathfrak{R}}((A_1, \Gamma_1), ..., (A_n, \Gamma_n)) \in DOM_{\mathfrak{R}} : op_\alpha(A_1, ..., A_n)$. Here $op_\alpha$ is the set-theoretic operator on alphabets that corresponds to op (c.f. Section 3).

(4) The definition of $\mathfrak{R}^{**}$ proceeds inductively and obeys the following principles:

(4.1) *Environment technique*.

$$\mathfrak{R}^{**}[\![X]\!](\rho) = \rho(X)$$

(4.2) *Compositionality*.

$$\mathfrak{R}^{**}[\![op(P_1, ..., P_n)]\!](\rho) = op_{\mathfrak{R}}(\mathfrak{R}^{**}[\![P_1]\!](\rho), ..., \mathfrak{R}^{**}[\![P_n]\!](\rho))$$

(4.3) *Fixed point technique*.

$$\mathfrak{R}^{**}[\![\mu X.P]\!](\rho) = \text{fix } \Phi_{P,\rho}$$

Here fix $\Phi_{P,\rho}$ denotes the least fixed point of the mapping

$$\Phi_{P,\rho} : DOM_{\mathfrak{R}} : \alpha(X) \longrightarrow DOM_{\mathfrak{R}} : \alpha(X)$$

defined by

$$\Phi_{P,\rho}((A, \Gamma)) = \mathfrak{R}^{**}[\![P]\!](\rho[(A, \Gamma)/X])$$

where $\rho[(A, \Gamma)/X]$ is a modified environment which agrees with $\rho$, except for the identifier $X$ where its value is $(A, \Gamma)$. $\qquad\qquad\square$

By (1)-(3), the inductive definition (4) of $\mathfrak{R}^{**}$ is well-defined and yields a value

$$\mathfrak{R}^{**}[\![P]\!](\rho) \in DOM_{\mathfrak{R}} : \alpha(P)$$

for every program P. In clause (4.3) we apply Knaster and Tarski's fixed point theorem: if $DOM_{\mathfrak{R}} : A$ is a complete partial order and if all operators $op_{\mathfrak{R}}$ and thus $\Phi_{P,\rho}$ are monotonic then $\Phi_{P,\rho}$ has a least fixed point.

Environments $\rho$ assign values to the free identifiers in programs, analogously to Tarski's semantic definition for predicate logic in Section 2. Thus for closed programs the environment parameter $\rho$ of $\mathfrak{R}^{**}$ is not needed. We will therefore write $\mathfrak{R}^{**}[\![P]\!]$ instead of $\mathfrak{R}^{**}[\![P]\!](\rho)$.

**Equivalence Theorem.** [Ol 88/89] For every closed programs P the operational and denotational readiness semantics coincide:

$$\mathfrak{R}^{*}\llbracket\, P\,\rrbracket \;=\; \mathfrak{R}^{**}\llbracket\, P\,\rrbracket. \qquad\qquad \square$$

As a consequence of the Equivalence Theorem we shall henceforth write $\mathfrak{R}^{*}$ for both the operational and the denotational version of readiness semantics.

## 6. TRANSFORMATIONS

Based on the above notion of program correctness we present now an approach to systematic derivation of communicating, concurrent programs from specifications. We proceed by *stepwise refinement* as advocated by Dijkstra and Wirth [Di 76, Wi 71]. Thus at each step of the derivation we replace a part of the specification by a piece of process syntax. Thus at the intermediate steps of the derivation we use so-called *mixed terms*. These are syntactic constructs mixing programming notation with trace specifications. An example of a mixed term is

$$a.(\,S \parallel T\,).$$

It describes a program where some parts are already constructed (prefix and parallel composition) whereas other parts are only specified (S and T) and remain to be constructed in future refinement steps.

From now on letters P, Q, R will range over mixed terms whereas letters S, T, U continue to range over the set Spec of trace specifications.

**Definition.** The set Prog + Spec of *mixed terms* consists of all terms P that are generated by the context-free production rules

$$
\begin{array}{lll}
P ::= & S & (\text{ specification }) \\
 & |\;\; stop : A & (\text{ deadlock }) \\
 & |\;\; div : A & (\text{ divergence }) \\
 & |\;\; a.P & (\text{ prefix }) \\
 & |\;\; P + Q & (\text{ choice }) \\
 & |\;\; P \parallel Q & (\text{ parallelism }) \\
 & |\;\; P\,[a_{i}/b_{i}] & (\text{ renaming }) \\
 & |\;\; P \setminus B & (\text{ hiding }) \\
 & |\;\; X & (\text{ identifier }) \\
 & |\;\; \mu X.P & (\text{ recursion })
\end{array}
$$

and that satisfy the context-sensitive restrictions stated in Section 3. Mixed terms without free process identifiers are called *closed*. The set of all closed mixed terms is denoted by CProg + Spec. $\qquad \square$

Semantically, mixed terms denote elements of the readiness domain $DOM_{\mathfrak{R}}$. In fact, since trace specifications S have a semantics $\mathfrak{R}^{*}\llbracket\, S\,\rrbracket \in DOM_{\mathfrak{R}}$ and since for every operator symbol op of Prog there exists a corresponding semantic operator $op_{\mathfrak{R}}$ on $DOM_{\mathfrak{R}}$, we immediately obtain a denotational readiness semantics

$$\mathfrak{R}^{*}\llbracket\,\cdot\,\rrbracket : Prog + Spec \longrightarrow (\,Env_{\mathfrak{R}} \longrightarrow DOM_{\mathfrak{R}}\,)$$

for mixed terms which is defined as in Section 5. For closed mixed terms the environment parameter $\rho$ of $\mathfrak{R}^*$ is not needed. Thus we write $\mathfrak{R}^*[\![P]\!]$ instead of $\mathfrak{R}^*[\![P]\!](\rho)$.

Under the readiness semantics $\mathfrak{R}^*$, programs, trace specifications and mixed terms are treated on equal footing so that they are easy to compare. For $P, Q \in CProg + Spec$ we write

$$P \equiv Q \quad \text{if} \quad \mathfrak{R}^*[\![P]\!] = \mathfrak{R}^*[\![Q]\!]$$

and call $P \equiv Q$ a *semantic equation* in the readiness semantics.

By the Correctness Theorem, program correctness boils down to a semantic equation, i.e.

$$P \ sat \ S \quad \text{iff} \quad P \equiv S$$

for all $P \in CProc$ and $S \in Spec$. Therefore a top-down derivation of a program $P$ from a trace specification $S$ will be presented as a sequence

$$S_1 \equiv Q$$
$$|||$$
$$\vdots$$
$$|||$$
$$Q_n \equiv P$$

of semantics equations between mixed terms $Q_1, \dots, Q_n$ where $Q_1$ is the given trace specification $S$ and $Q_n$ is the derived program $P$. The transitivity of $\equiv$ ensures $P \equiv S$, the desired result.

The sequence of equations is generated by applying the principles of *transformational programming* as e.g. advocated in the Munich Project CIP [Bau 85, Bau 87 ]. Thus each equation

$$Q_i \equiv Q_{i+1}$$

in the sequence is obtained by applying a *transformation rule* to a subterm of $Q_i$, usually a trace specification, say $S_i$. Most of the transformation rules to be introduced below state that under certain conditions a semantic equation of the form

$$S_i \equiv op( S_{i_1}, \dots, S_{i_n} )$$

holds, i.e. the trace specification $S_i$ can be refined into a mixed term consisting of an n-ary process operator op applied to trace specifications $S_{i_1}, \dots, S_{i_n}$ which are related to $S_i$ by some logical operator. Then $Q_{i+1}$ results from $Q_i$ by replacing the subterm $S_i$ with $op( S_{i_1}, \dots, S_{i_n} )$. Since the underlying readiness semantics is denotational, $S_i \equiv op( S_{i_1}, \dots, S_{i_n} )$ implies $Q_i \equiv Q_{i+1}$. In this way, the original trace specification $S$ is gradually transformed into a program $P$.

Transformation rules are theorems about the readiness semantics of mixed terms that can be expressed as deduction rules of the form

$$( \mathfrak{D} ) \qquad \frac{\models S_1 , \dots , \models S_m \quad P_1 \equiv Q_1, \dots , P_n \equiv Q_n}{P \equiv Q} \quad \text{where} \dots$$

where m, n $\geq$ 0 and "..." is a condition on the syntax of the trace specifications $S_1$, ... , $S_m$ and the mixed terms $P_1$, $Q_1$, ... , $P_n$, $Q_n$, P, Q. The rule $\mathcal{D}$ states that if the condition "..." is satisfied, the trace specifications $S_1$, ... , $S_m$ are valid and the semantic equations $P_1 \equiv Q_1$, ... , $P_n \equiv Q_n$ are true then also the semantic equation $P \equiv Q$ holds. If m = n = 0, the notation simplies to

$$P \equiv Q \quad \text{where} \dots$$

We say that a deduction rule $\mathcal{D}$ is *sound* if the theorem denoted by $\mathcal{D}$ is true.

The condition "..." will always be a simple decidable property, e.g. concerning alphabets. By contrast, the validity of trace specifications is in general not decidable due to the Expressiveness Theorem in Section 2. Thus in general the transformation rules $\mathcal{D}$ are effectively applicable only *relative to* the logical theory of trace specifications, i.e. the set

$$\text{Th(Spec)} = \{\ S \in \text{Spec} \mid\ \vDash S\ \}\ .$$

To enhance the readability of transformation rules, we will sometimes deviate from the form $\mathcal{D}$ above and use equivalent ones. For example, we will often require that certain traces $\mathfrak{h}$ satisfy a trace specification S, but this is of course equivalent to a validity condition:

$$\mathfrak{h} \vDash S \quad \text{iff} \quad \vDash S\{\mathfrak{h}/h\}\ .$$

Depending on the syntactic form of the mixed terms P and Q in the conclusion of $\mathcal{D}$ we distinguish among several types of transformation rules:

(1) *Equation Rules*: P, Q are arbitrary mixed terms.
(2) *Specification Rules*: P, Q are trace specifications.
(3) *Construction Rules*: P is a mixed term involving one process operator op or one recursion symbol $\mu$ and Q is a trace specification.
(4) *Algebraic Rules*: P, Q are mixed terms which both involve certain process operators or recursion symbols.

Some rules of type (1) - (4) will be classified as *derived rules* because they can be deduced from other rules. Their purpose is to organise the presentation of process constructions more clearly. We shall consider here only transformation rules of type (1) - (3); rules of type (4) are not used.

To state the semantic and syntactic conditions of our transformation rules, we need the following notions. The *prefix kernel* of a trace specification S is given by the formula

$$\text{kern}(S) =_{df} \forall\, t.\ t \leq h \upharpoonright \alpha(S) \longrightarrow S\,\{\,t\,/\,h\,\}$$

Thus kern(S) is a trace specification which denotes the largest prefix closed set of traces satisfying S (cf. [Zw 89]). The set of *initial communications* of a trace specification S is given by

$$\text{init}(S) = \{\ a \in \alpha(S) \mid\ \textit{pref}\ a \vDash S\ \}$$

Recall that *pref* a $\vDash$ S is equivalent to requiring $\varepsilon \vDash$ S and a $\vDash$ S.

By definition, every mixed term P is action-guarded. We now introduce a stronger notion of communication-guardedness whereby the guarding communication may not be hidden. The purpose of this notion is a sufficent syntactic condition implying that every recursion in P has a unique fixed point in the readiness semantics. In particular, if this fixed point is given by a trace specification, the recursion is divergence free. This observation gives rise to a corresponding recursion rule. Since uniqueness of fixed points is an undecidable property for the class of terms considered here, no syntactic condition for it will be a necessary one.

**Definition.** Let A be a set of communications. A mixed term $P \in$ Prog+Spec is called *communication-guarded by A* if for every recursive subterm $\mu X.\ Q$ of P every free occurrence of X in Q occurs within a subterm of Q which is of the form a. R with $a \in A$, but not within a subterm of Q which is of the form R [ b/a ] or R\a with $a \in A$ and $b \notin A$. P is called *communication-guarded* if there exists a set A of communications such that P is communication-guarded by A.  $\square$

For example, the following terms are all communication-guarded:

$$P_1 = \mu X.\ \text{up.}\ \mu Y.\ (\ \text{dn.}\ X + \text{up. dn.}\ Y\ )\ ,$$

$$P_2 = ((\ \mu X.\ \text{up. dn.}\ X)\ [\ \text{lk/dn}\ ]\ \|\ 0 \le \text{lk}*\text{h} - \text{dn}*\text{h} \le 1)\ \backslash \text{lk}\ ,$$

$$P_3 = \mu X.\ \text{up.}\ (X[\ \text{lk/dn}\ ]\ \|\ \text{dn.}\ (\ \text{dn}*\text{h} \le \text{lk}*\text{h}\ ))\ \backslash\ \text{lk}.$$

By contrast,

$$Q_1 = \mu X.\ \text{up.}\ (X\ [\text{lk/up}\ ])\ \backslash \text{lk}$$

$$Q_2 = \mu X.\ \text{up.}\ (X\ [\text{lk/dn}\ ]\ \|\ \text{dn.}\ X[\ \text{lk/up}\ ])\ \backslash \text{lk}$$

are action-guarded, but not communication-guarded.

We can now introduce the announced transformation rules for mixed terms.

**Definition.** In the following transformation rules, letters P, Q, R range over CProg+Spec unless stated otherwise. As usual, letters S, T; X; a, b; A range over Spec, Idf, Comm, and alphabets, respectively.

(1) Equation Rules:

   (Reflexivity)
$$P \equiv P$$

   (Symmetry)
$$\frac{P \equiv Q}{Q \equiv P}$$

   (Transitivity)
$$\frac{P \equiv Q\ ,\ Q \equiv R}{P \equiv R}$$

   (Context)
$$\frac{Q \equiv R}{P \{ Q/X \} \equiv P \{ R/X \}}$$

where $P \in$ Proc+Spec and $P \{ Q/X \}, P \{ R/X \} \in$ CProg+Spec

390

(2) Specification Rules:

(Kernel)

$$S \equiv \mathrm{kern}(S)$$

(Logic)

$$\frac{\models S \leftrightarrow T}{S \equiv T} \qquad \text{where } \alpha(S) = \alpha(T)$$

(3) Construction Rules:

(Deadlock)

$$stop: A \equiv h \upharpoonright A \leq \varepsilon$$

(Prefix)

$$\frac{\mathrm{init}(S) = \{\, a \,\}}{a.\, S \{\, a.\, h\, /h \,\} \equiv S}$$

(Choice)

$$\frac{\varepsilon \models S \wedge T \,,\; \mathrm{init}(S \wedge T) = \Phi}{S + T \;\equiv\; S \vee T} \qquad \text{where } \alpha(S) = \alpha(T)$$

(Parallelism)

$$S \parallel T \;\equiv\; S \wedge T$$

(Renaming)

- *same traces*:

  $\forall\, \hbar' \in \alpha(T)^{*}:$

  $\hbar' \models T$ iff $\exists\, \hbar \in \alpha(S)^{*}: \hbar \models S$ and $\hbar\{\, b_i/a_i \,\} = \hbar'$

- *same liveness*:

  $\forall\, \hbar_1, \hbar_2 \in \alpha(S)^{*} \quad \forall\, c \in \alpha(S) :$

  $\hbar_1 \models S$ and $\hbar_2 \models S$ and $\hbar_1\{\, b_i/a_i \,\} = \hbar_2\{\, b_i/a_i \,\}$

  and $\hbar_1.\, c \models S$ imply

  $\exists\, d \in \alpha(S): \hbar_2.\, d \models S$ and $c\{\, b_i/a_i \,\} = d\{\, b_i/a_i \,\}$

$$\frac{}{S\,[\, b_i/a_i \,] \;=\; T}$$

where $\alpha(T) = \alpha(S)\{\, b_i/a_i \,\}$
and $b_i/a_i$ stands for a renaming
$b_1 , \ldots, b_n /a_1, \ldots, a_n$ with $n \geq 1$

(Hiding)

- *empty trace*:        $\varepsilon \models S$
- *logical implication*:    $\models S \longrightarrow T$
- *stability*:             $\forall\, b \in B: b \models \neg\, S$
- *divergence freedom*:   $\forall\, \hbar \in \alpha(S)^{*}\; \exists\, b_1, \ldots, b_n \in B:$

                   $\hbar.\, b_1 \ldots b_n \models \neg\, S$
- *same liveness*:      $\forall\, \hbar \in \alpha(S)^{*}\; \forall\, c \in \alpha(T) :$

                   $\hbar \models S$ and $\forall\, b \in B: \hbar.\, b \models \neg\, S$

                   and $\hbar.\, c \models T$ imply $\hbar.\, c \models S$

$$\frac{}{S \backslash B \equiv T} \qquad \text{where } \alpha(T) = \alpha(S) - B$$

(Recursion)

$$\frac{P \{\, S/X \,\} \equiv S}{\mu X.\, P \qquad \equiv S}$$

where $\mu X.\, P \in \mathrm{CProg+Spec}$ is
communication-guarded and $\alpha(S) = \alpha(X)$

(Expansion: derived)

$$\frac{\text{init}(S) = \{ a_1, \ldots, a_n \}}{\sum_{i=1}^{n} a_i \cdot S \{ a_i \cdot h / h \} \equiv S}$$

where $a_1, \ldots, a_n$ are $n \geq 2$ distinct communications

(Disjoint renaming: derived)

$$S [ b_i/a_i ] \equiv S \{ b_i/a_i \}$$

where $b_i/a_i$ stands for a renaming
$b_1, \ldots, b_n / a_1, \ldots, a_n$
with $n \geq 1$ and $b_1, \ldots, b_n \notin \alpha\alpha(S)$

□

The equation rules are standard for any type of equational reasoning. The kernel rule reflects a property of the process domain, i.e. that the set of traces in which a program may engage in is prefix closed. Hence only the prefix kernel of a trace specification is relevant. The need for such a rule in trace-based reasoning has been observed in [Jo 87, WGS 87, Zw 89]. The logic rule provides the interface between trace logic and equality in the readiness domain.

Of the construction rules those for renaming and hiding are surprisingly complicated. The reason is that renaming and hiding can completely restructure the communication behaviour of a process. In particular, both operators can introduce externally observable nondeterminism and additionally hiding can introduce divergence and initial unstability. Programs with such properties fail to satisfy the liveness conditions of any trace specification. To exclude such failure the rules for renaming and hiding have various semantic premises.

The difficult case of renaming is *aliasing*, i.e. when two originally different communications get the same name. For example, if

$$S \equiv a \# h \leq 1 \wedge b \# h \leq 2$$

then

$$S [ b/a ] \equiv b \# h \leq 3$$

but if

$$S \equiv a \# h \leq 1 \vee b \# h \leq 2$$

then $S[ b/a]$ is externally nondeterministic ( it either deadlocks after the first b or accepts a second one ) and hence not equivalent to any other trace specification. Fortunately, in most cases we are only interested in renamings $S[ b_i/a_i ]$ without aliasing. Then we can assume that the communications $b_i$ are not in the extended alphabet $\alpha\alpha( S )$ introduced in Section 2 and apply the extremely simple rule for disjoint renaming. It states that the semantics of $S[ b_i/a_i ]$ is captured by a simple syntactic substitution $S \{ b_i/a_i \}$.

For hiding we have not found a special case that would allow an equally simple treatment. Thus we always have to check all the premises of the hiding rule. In practical applications we found that the first four premises are easily established. Only the premise "same liveness" is more difficult: it is proven by a case analysis for all communications $c \in \alpha( T )$.

The remaining rules are all extremely simple. Parallel composition is just logical conjunction, choice is essentially logical disjunction, prefix is treated by logical substitution, the expansion rule combines prefix with ( multiple ) choice, and the recursion rule asks for a recursive semantic equation in its premise. Thus when applying the above transformation rules difficulties arise only when they are unavoidable: in the cases of renaming with aliasing and hiding.

We remark that the simple conjunction rule for parallelism was one of our aims when selecting the operators for programs. That is why we have chosen COSY's version of parallel composition. By contrast, the parallel composition of CCS would be extremely difficult to deal with in our approach because it combines in a single operator the effect of COSY's parallelism, renaming with aliasing ( to model *autoconcurrency*, i.e. concurrent communications with the same name ) and hiding. Here we follow CSP and treat all these issues separately.

A simple disjunction rule for choice was another aim. It led us to require that processes satisfying a trace specification should be stable initially. Indeed, if S and T would allow programs to have an initial instability then S + T would exhibit externally observable nondeterminism and hence not satisfy the liveness condition of any other trace specification.

**Soundness Theorem.** [Ol 88/89]   The above transformation rules are sound.   □


## 7. PROGRAM DERIVATION

**Example: Milner's Scheduling Problem.** In his books on CCS, Milner considers a small scheduling problem [Mi 80, Mi 89]. Specification and implementation are described by CCS terms: the proof that the implementation meets the specification is purely algebraic, i.e. by application of the algebraic rules of CCS. Interestingly, in [Mi 80] Milner also gives an informal Petri net semantics of his implementation. Milner's example provides a very good illustration for our approach: we start from a trace specification and systematically derive two alternative programs, one of them corresponding to Milner's solution in CCS. For each of the derived programs we examine its Petri net semantics. Since we use COSY's parallel operator, our derivation and the resulting Petri net solution turns out to be simpler than Milner's CCS solution.

In trace logic Milner's scheduling problem can be stated as follows:

$$\text{SPEC}_k =_{df} \bigwedge_{i=1}^{k} h \upharpoonright \{a_i, b_i\} \in \mathit{pref}\, (a_i . b_i)^*$$

$$\wedge\ h \upharpoonright \{a_1, \ldots, a_k\} \in \mathit{pref}\, (a_1 \ldots a_k)^*$$

where $k \geq 1$. Thus we use here regular expressions and the prefix operator *pref* to specify a process that can engage in communications $a_1, \ldots, a_k$ and $b_1, \ldots, b_k$. When projected onto $\{a_i, b_i\}$ the process engages in a small cycle $a_i b_i$ and when projected onto $\{a_1, \ldots, a_k\}$ it engages in a large cycle $a_1 \ldots a_k$. There are no further restrictions imposed on the communication behaviour of the process.

For distinct communications $c_1, \ldots, c_n$ consider the trace specifications

$$S =_{df} h \upharpoonright \{c_1, \ldots, c_n\} \in pref\,(c_1, \ldots, c_n)^*$$

and

$$T =_{df} c_n \# h \leq \ldots \leq c_1 \# h \leq 1 + c_n \# h.$$

Since $\alpha(S) = \alpha(T)$ and $\models S \longleftrightarrow kern\,(T)$, the kernel and logic rule yield

(A)
$$S \equiv T.$$

Applying this observation to $SPEC_k$, we obtain

$$SPEC_k \equiv \bigwedge_{i=1}^{k} b_i \# h \leq a_i \# h \leq 1 + b_i \# h$$

$$\wedge \; a_k \# h \leq \ldots \leq a_1 \# h \leq 1 + a_k \# h.$$

We find regular expressions easier to read, but inequalities between the number of occurrences of certain communications easier to calculate with. We now present two derivations of programs from $SPEC_k$.

*First Derivation.* Since $SPEC_k$ is given as a conjunction of several conditions, the application of the parallelism rule is straightforward. Define

$$PROC_i = \quad h \upharpoonright \{a_i, b_i\} \in pref\,(a_i . b_i)^*$$

for $i = 1, \ldots, k$ and

$$SCH_k =_{df} h \upharpoonright \{a_1, \ldots, a_k\} \in pref(a_1 \ldots a_k)^*,$$

standing for "process i" and "scheduler of size k". Then by successive applications of the parallelism rule, we obtain:

(C)
$$SPEC_k$$
$$|||$$
$$(\, \|_{i=1}^{k} PROC_i) \,\| \, SCH_k$$

Programs satisfying the specifications $PROC_i$ and $SCH_k$ are easily constructed using the rules for prefix and recursion. We record only the results of these constructions:

(D)
$$PROC_i \qquad\qquad SCH_k$$
$$||| \qquad\quad and \qquad |||$$
$$\mu XX_i . a_i . b_i . XX_i \qquad \mu YY_k . a_1 \ldots a_k . YY_k$$

where $\alpha(XX_i) = \{a_i, b_i\}$ and $\alpha(YY_k) = \{a_1, \ldots, a_k\}$. Combining (C) and (D) completes the first derivation:

$$SPEC_k$$
$$|||$$
$$(\, \|_{i=1}^{k} PROC_i) \,\| \, SCH_k$$
$$|||$$
$$(\, \|_{i=1}^{k} \mu XX_i . a_i . b_i . XX_i) \,\| \, \mu YY_k . a_1 \ldots a_k . YY_k$$

For k=3, the derived program denotes the following abstract net:

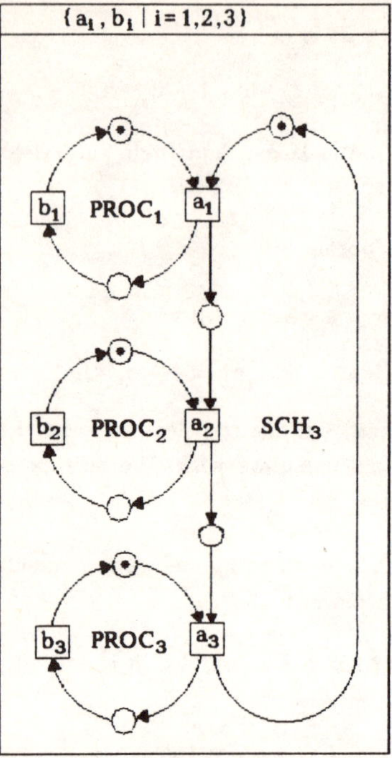


The annotation indicates which cycles in the net implement the specifications PROC$_i$ and SCH$_3$. Note that a concurrent activation of the transitions labeled by $b_i$ is possible. For arbitrary k, the net has 3k places. The drawback of this implementation is that for each k a new scheduler SCH$_k$ is needed. To overcome this drawback, we present now a derivation corresponding to Milner's original solution.

*Second derivation.* Milner's aim was a *modular* scheduler which for arbitrary problem size could be built as a "ring of elementary identical components" [Mi 80]. This idea can be realised very neathy using transformations on mixed terms. The point is to break down the large cycle $a_1 \ldots a_k$ of SCH$_k$ into k+1 cycles of size 2, viz. $a_1 a_2$, $a_2 a_3$, $\ldots, a_{k-1} a_k$ and $a_1 a_k$.

Formally, this idea is justified by calculations with inequalities based on the representation (A) of cyclic regular expressions. Represent the large cycle $a_1 \ldots a_k$ by

$$CYC_k =_{df} a_k \# h \leq \ldots \leq a_1 \# h \leq 1 + a_k \# h$$

and for $1 \leq i < j \leq k$ represent a cycle $a_i a_j$ by

$$SCH_{i,j} =_{df} a_j \# h \leq a_i \# h \leq 1 + a_j \# h.$$

By simple calculations with inequalities, we prove the logical equivalence

$$\text{(E)} \qquad \models CYC_k \longleftrightarrow ( \bigwedge_{i=1}^{k-1} SCH_{i,i+1} ) \wedge SCH_{1,k} .$$

By (A) above, we have $SCH_k \equiv CYC_k$ and hence by (E), the logic rule yields

$$SCH_k \equiv (\bigwedge_{i=1}^{k-1} SCH_{i,i+1}) \wedge SCH_{1,k}$$

Thus successive applications of the parallelism rule lead to the following decomposition of the large cycle $SCH_k$ into small cycles $SCH_{i,j}$ :

(F)
$$SCH_k$$
$$|||$$
$$(\underset{i=1}{\overset{k}{\|}} SCH_{i,i+1}) \| SCH_{1,k}$$

Combining (F) with (C) of the previous derivation we obtain:

(G)
$$SPEC_k$$
$$|||$$
$$(\underset{i=1}{\overset{k}{\|}} PROC_i) \| SCH_k$$
$$|||$$
$$(\underset{i=1}{\overset{k}{\|}} PROC_i) \| (\underset{i=1}{\overset{k-1}{\|}} SCH_{i,i+1}) \| SCH_{1,k}$$

Note that each of the specifications $SCH_{i,j}$ is equivalent to

$$0 \le a_i * h - a_j * h \le 1.$$

By (A), also each of the specifications $PROC_i$ is equivalent to

$$0 \le a_i * h - b_i * h \le 1$$

Thus $SCH_{i,j}$ and $PROC_i$ are all renamed copies of a 1-counter specified by

$$S_1 =_{df} 0 \le up * h - dn * h \le 1.$$

Formally,

(H)
$$PROC_i \qquad\qquad\qquad SCH_{i,j}$$
$$||| \qquad\qquad and \qquad\qquad |||$$
$$S_1[a_i,b_i / up,dn] \qquad\qquad S_1[a_i,a_j / up,dn]$$

by the logic and disjoint renaming rule. By the rules for prefix and recursion,

(I)
$$\mu X.up.dn.X \equiv S_1 .$$

Thus combining (G)-(I) we obtain the following program satisfying $SPEC_k$ :

$$(\underset{i=1}{\overset{k}{\|}} (\mu X.up.dn.X)[a_i,b_i / up,dn])$$
$$\| (\underset{i=1}{\overset{k-1}{\|}} (\mu X.up.dn.X)[a_i,a_{i+1} / up,dn])$$
$$\| (\mu X.up.dn.X)[a_1,a_k / up,dn]$$

It consists of the parallel composition of 2k copies of the 1-counter. For k=4 this term denotes the following abstract net:

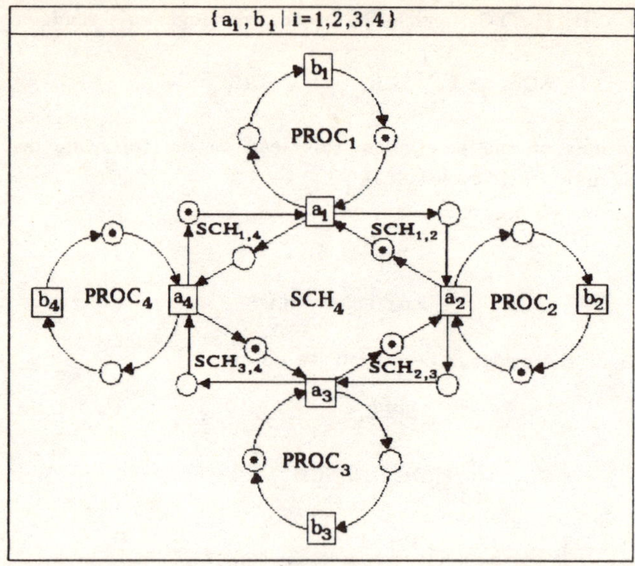


The annotation indicates the cycles that implement the specifications $PROC_i$. The rectangular part in the centre is the implementation of $SCH_4$ consisting of an appropriate synchronisation of the cycles $SCH_{1,2}$ , $SCII_{2,3}$ , $SCII_{3,4}$ and $SCH_{1,4}$. Note that the asymmetry in the initial token distribution is due to the cycle $SCH_{1,4}$. For arbitrary k the net has 4k places. (The net given in [Mi 80] needs 5k places.) Since only the part implementing the scheduler $SCH_k$ differs from the previous net implementation, concurrency of the $b_i$-transitions is still possible. We believe that this is an example where it is fairly difficult to get the net implementation right without the systematic guidance of formulas and terms.

**Example: Binary Variable.** A binary variable stores at any moment of time one of the values 0 or 1, We assume that its initial value is 0. If its value is 0 or 1, the variable allows a read instruction r0 or r1, respectively. The value of the variable can be changed only by one of its write instructions w0 or w1 which set the value to 0 or 1, respectively.

Viewing the instructions r0, r1, w0 and w1 as comunications, and using the abbreviations RW = {r0,r1,w0,w1} and W = {w0,w1}, we can specify a binary variable in trace logic:

$$\text{BINO} =_{df} \quad (\text{\textit{last} h} \upharpoonright RW = r0 \longrightarrow \text{\textit{last}}(w0.h \upharpoonright W) = w0) \tag{1}$$
$$\wedge \ (\text{\textit{last} h} \upharpoonright RW = r1 \longrightarrow \text{\textit{last} h} \upharpoonright W = w1) \tag{2}$$

Thus if the last read is r0 then the last write must be w0. Note that the initialisation with 0 is represented in line (1) by the extra write w0 preceding $h \upharpoonright W$. If the last read is r1 then the last write must be w1. The write communications w0 and w1 are always possible. This specification is essentially due to Hoare [Ho 85b]. We derive now two different communicating programs from BINO, a sequential one and a concurrent one.

*First derivation.* A sequential program can be obtained from BIN0 by applying the rules for expansion and recursion. Let

$$\text{BIN1} =_{df} \quad (\textit{last } h\!\restriction RW = r0 \longrightarrow \textit{last } h\!\restriction W = w0)$$
$$\wedge \ (\textit{last } h\!\restriction RW = r1 \longrightarrow \textit{last } (w1.h\!\restriction W) = w1)$$

by the trace specification of a binary variable with initial value 1. Then we obtain

$$\text{BIN1}$$
$$|||\ \{\text{expansion, logic and context}\}$$
$$r1.\text{BIN1} + w1.\text{BIN1} + w0.\text{BIN0}$$
$$|||\ \{\text{recursion and transitivity}\}$$
$$\mu Y.\,(r1.Y + w1.Y + w0.\text{BIN0})$$

and hence

$$\text{BIN0}$$
$$|||\ \{\text{expansion, logic and context}\}$$
$$r0.\text{BIN0} + w0.\text{BIN0} + w1.\text{BIN1}$$
$$|||\ \{\text{context and transitivity}\}$$
$$r0.\text{BIN0} + w0.\text{BIN0}$$
$$+ \ w1.\mu Y.(r1.Y + w1.Y + w0.X)$$
$$|||\ \{\text{recursion and transitivity}\}$$
$$\mu X.(\ r0.X + w0.X$$
$$+ \ w1.\mu Y.(r1.Y + w1.Y + w0.X))$$

where $\alpha(X) = \alpha(Y) = RW$. Thus we have constructed

$$\mu X.(\ r0.X + w0.X$$
$$+ \ w1.\mu Y.(r1.Y + w1.Y + w0.X))$$

as a process term satisfying BIN0. This term denotes the following net:

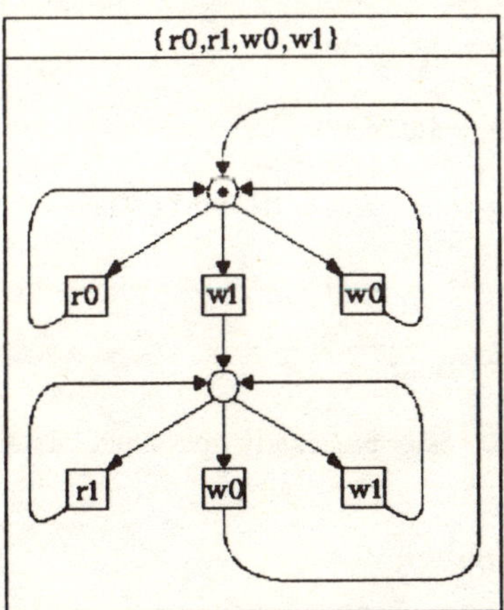


The marked place replesents the value 0 and the other one the value 1. The dynamic behaviour of the net is purely sequential because every reachable marking has exactly one token.

*Second derivation.* We derive now a concurrent program from BIN0. First note that a direct application of the parallelism rule to BIN0 is possible and yields

$$(1) \parallel (2) \equiv \text{BIN0}$$

where (1) and (2) are the two conditions of BIN0. However, since (1) and (2) are trace specifications with the same alphabet, the definition of parallel composition enforces full synchrony and hence does not lead to any concurrency in the resulting net.

Nevertheless we can obtain a concurrent implementation for BIN0 by a somewhat more elaborate construction using the general renaming rule. The idea is to distinguish between two different kinds of write communications: those which *change* the value stored by the variable and those which *preserve* the existing value. Then preserving writes can occur concurrently with read communications.

Thus we shall describe the internal behaviour of a binary variable in more detail. To this end, we introduce two new communications $c0$ and $c1$ for writes that change the value of the variable to 0 and 1, respectively. We abbreviate $C = \{c0, c1\}$, $CR = \{c0, c1, r0, r1\}$ and $CW = \{c0, c1, w0, w1\}$. The communications $w0$ and $w1$ are only used for preserving writes. Since changes occur alternatively, this leads to the following specification of the internal behaviour of a binary variable initialised with 0:

$$\text{INTBIN0} =_{df} \quad 0 \leq c1\#h - c0\#h \leq 1 \tag{3}$$
$$\wedge \ (last \ h{\upharpoonright} CR \ = r0 \longrightarrow last \ (c0.h{\upharpoonright} C) = c0) \tag{4}$$
$$\wedge \ (last \ h{\upharpoonright} CR \ = r1 \longrightarrow last \ h{\upharpoonright} C = c1) \tag{5}$$
$$\wedge \ (last \ h{\upharpoonright} CW = w0 \longrightarrow last \ (c0.h{\upharpoonright} C) = c0) \tag{6}$$
$$\wedge \ (last \ h{\upharpoonright} CW = w1 \longrightarrow last \ h{\upharpoonright} C = c1) \tag{7}$$

The new specification INTBIN0 can be successfully decomposed by the parallelism rule. Let

$$\text{READ0} =_{df} (3) \wedge (4) \wedge (5)$$

and

$$\text{PRES0} =_{df} (3) \wedge (6) \wedge (7)$$

with (3)-(7) referring to the corresponding conditions in INTBIN0. Then we obtain

$$\text{(A)} \qquad\qquad\qquad \text{INTBIN0}$$
$$|||$$
$$\text{READ0} \parallel \text{PRES0} \ .$$

Programs satisfying READ0 and PRES0 can be easily obtained using the rules for expansion and recursion:

$$\text{(B)} \qquad\qquad\qquad \text{READ0}$$
$$|||$$
$$\mu XR.(r0.XR + c1.\mu YR.(r1.YR + c0.XR))$$

and

$$\text{(C)} \qquad\qquad\qquad \text{PRES0}$$
$$|||$$
$$\mu XW.(w0.XW + c1.\mu YW.(w1.YW + c0.XW))$$

where $\alpha(XR) = \alpha(YR) = CR$ and $\alpha(XW) = \alpha(YW) = CW$. Since the alphabets of READ0 and PRES0 intersect only on C, the parallel composition of READ0 and PRES0 allows asynchrony between w0, w1 and r0, r1 and thus concurrency in the resulting net. In this way we obtain a concurrent implementation for INTBIN0, but not for the original specification BIN0.

However, BIN0 can be obtained from INTBIN0 by just "forgetting" that some writes are called c0 and c1. Formally, this is done by renaming. We claim that

(D)
$$\text{BIN0}$$
$$|||$$
$$\text{INTBIN0}\,[\,w0,w1\,/\,c0,c1\,]$$

Since this renaming is not disjoint, we have to check the more difficult premises of the general renaming rule. Fortunately, this is easily done once we have shown that the renaming is a bijection. To this end, we introduce the trace sets

$$\Gamma = \{\,\mathfrak{h} \in (CR \cup CW)^{*} \mid \mathfrak{h} \models \text{INTBIN0}\,\}$$

and

$$\Delta = \{\,\mathfrak{h}' \in RW^{*} \mid \mathfrak{h}' \models \text{BIN0}\,\}$$

and the mapping $\beta : \Gamma \longrightarrow \Delta$ by

$$\beta(\mathfrak{h}) = \mathfrak{h}\,\{\,w0,w1\,/\,c0,c1\,\}.$$

We show that $\beta$ is a bijection.

(i) *$\beta$ is well defined* : For every $\mathfrak{h} \in \Gamma$ the conditions (3) – (7) of INTBIN0 imply that $\beta(\mathfrak{h})$ satisfies the conditions (1) – (2) of BIN0.

(ii) *$\beta$ is surjective* : Given a trace $\mathfrak{h}' \in \Delta$ construct $\mathfrak{h}$ by renaming every first write w0 or w1 of a new value in the extended trace $w0.\mathfrak{h}'$ into c0 or c1, respectively, and then deleting the leading communication c0. For example,

$$\mathfrak{h}' = r0\,.\,w0\,.\,w1\,.\,r1\,.\,w1\,.\,w1\,.\,w0\,.\,w0$$

yields

$$\mathfrak{h} = r0\,.\,w0\,.\,c1\,.\,r1\,.\,w1\,.\,w1\,.\,c0\,.\,w0\,.$$

Then $\mathfrak{h} \in \Gamma$ and $\beta(\mathfrak{h}) = \mathfrak{h}'$.

(iii) *$\beta$ is injective*: This follows from the fact that the pre-image $\mathfrak{h}$ of a trace $\mathfrak{h}'$ under $\beta$ is uniquely determined by $\mathfrak{h}'$, viz. as explained in (ii).

Now the premises of the renaming rule are easily established. The premise "same traces" follows from the fact that $\beta$ is well-defined and surjective, and the premise "same liveness" follows from the fact that $\beta$ is injective. This verifies the chained relationship (D) between BIN0 and INTBIN0.

Putting (A) – (D) together by the context rule yields our second derivation of a program from BIN0:

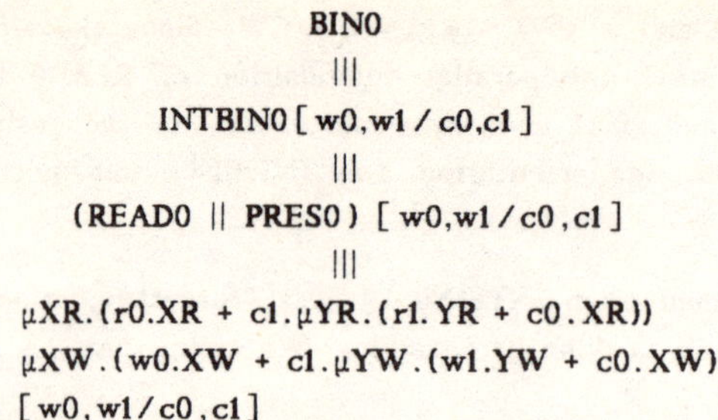

$$\text{BIN0}$$
$$|||$$
$$\text{INTBIN0}\,[\,w0,w1\,/\,c0,c1\,]$$
$$|||$$
$$(\,\text{READ0}\;\|\;\text{PRES0}\,)\;[\,w0,w1\,/\,c0\,,c1\,]$$
$$|||$$
$$(\;\mu XR.(r0.XR\,+\,c1.\mu YR.(r1.YR\,+\,c0.XR))$$
$$\|\;\mu XW.(w0.XW\,+\,c1.\mu YW.(w1.YW\,+\,c0.XW))$$
$$)\;[\,w0,w1\,/\,c0\,,c1\,]$$

This term denotes the following abstract net:

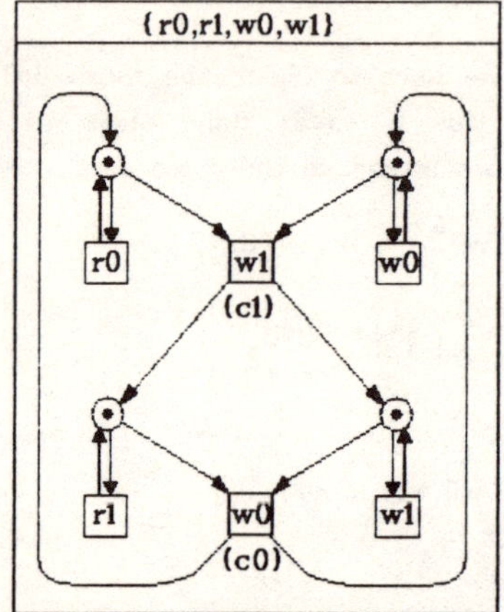


The transitions whose labels are introduced by the renaming are indicated by (c1) and (c0). This implementation allows the concurrent activation of reads and preserving writes.

**Example: Access Control.** Consider a system consisting of readers and writers accessing a shared data structure. We wish to enforce certain access disciplines in this system. Verjus et al. have proposed to solve this task in two steps [AHV 85].

*Step 1*: Introduction of FIFO disciplines for readers and writers separately.

*Step 2*: Introduction of scheduling disciplines which define when the first reader in the FIFO queue is allowed to access the shared data structure.

Hence the whole system can be specified as a conjunction

$$\text{SYS = READERS} \wedge \text{WRITERS} \wedge \text{FIFO} \wedge \text{SCH}$$

of trace formulas describing the readers, the writers, the FIFO disciplines and the desired scheduling discipline separately. By the parallelism rule, a program satisfying SYS can then be constructed as the parallel composition of subprograms satisfying the individual conjuncts of SYS. We shall treat here Step 2 dealing with the scheduler part SCH.

Readers and writers are instrances of a generic concept of a user. As far as the scheduling task is concerned, a user engages in the following three communications:

$$\left.\begin{array}{ll} r & \text{request} \\ a & \text{access} \\ t & \text{terminate} \end{array}\right\} \quad \text{the use of the shared data}$$

Whenever a user wants to access the shared data, it engages in the comunications r,a and t in that order. If the user is a reader, we write rR, aR and tR instead of r,a and t, and if it is a writer, we write rW,aW and tW. According to Step 1 above we imagine now that the names of readers and writers that have requested but not yet accessed the shared data are kept in two separate FIFO queues. For Step 2 we consider now three different scheduling disciplines for the case of $m \geq 1$ readers and $n \geq 1$ writers.

*First scheduler* : The first reader in the reader queue is allowed to access the shared data if no writer is active, and the first writer in the writer queue is allowed to access the shared data if no reader or writer is active. Thus any number of readers may access the shared data at a time, but only one writer.

In trace logic we express the number of active readers and writers by

$$\text{activeR} =_{df} \text{aR} \# h - \text{tR} \# h$$

and

$$\text{activeW} =_{df} \text{aW} \# h - \text{tW} \# h$$

and then consider the following specification:

$$
\begin{array}{lll}
\text{SCH1}_m =_{df} & 0 \leq \text{activeR} \leq m & (1) \\
& \wedge \ 0 \leq \text{activeW} \leq 1 & (2) \\
& \wedge \ (\textit{last } h \upharpoonright \{aW, tW, aR\} = aR \longrightarrow \text{activeW} = 0) & (3) \\
& \wedge \ (\textit{last } h \upharpoonright \{aR, tR, aW\} = aW \longrightarrow \text{activeR} = 0) & (4)
\end{array}
$$

Using the parallelism rule we decompose $\text{SCH1}_m$ as follows:

$$(A) \qquad\qquad \text{SCH1}_m$$
$$|||$$
$$\text{A--READ} \ || \ \text{A--WRITE}$$

where $\text{A--READ} =_{df} (1) \wedge (4)$ deals with the active readers and $\text{A--WRITE} =_{df} (2) \wedge (3)$ deals with the active writers.

A--READ and A--WRITE are renamed copies of bounded counters with an additional test for zero. Using up,dn and zero as communications and $k \geq 1$ as capacities, such a counter can be specified as follows:

$$
\begin{array}{ll}
\text{COUNT}_k =_{df} & 0 \leq \text{up} \# h - \text{dn} \# h \leq k \\
& \wedge \ (\textit{last } h \upharpoonright \{up, dn, zero\} = zero \longrightarrow \text{up} \# h = \text{dn} \# h).
\end{array}
$$

Thus by the logic and disjoint renaming rule, we obtain:

$$(B) \qquad\qquad \text{A--READ}$$
$$|||$$
$$\text{COUNT}_m [\text{aR}, \text{tR}, \text{aW} / \text{up}, \text{dn}, \text{zero}]$$

and

(C)
$$\text{A-WRITE}$$
$$|||$$
$$\text{COUNT}_1 \, [\, aW, tW, aR \,/\, up, dn, zero \,].$$

From $\text{COUNT}_k$ one can derive communicating programs inductively. For $k=1$

$$Q_1 \, =_{df} \, \mu Z.(\, zero.Z \,+\, up.dn.Z\,)$$

satisfies $\text{COUNT}_1$, and for $k>1$

(D)
$$Q_k \, =_{df} \, (Q_{k-1}\,[\,lk\,/\,dn\,] \,\|\, Q_1\,[\,lk\,/\,up\,]) \setminus lk$$

satisfies $\text{COUNT}_k$. We skip the details of these derivations. Thus

(E)
$$Q_k \, \equiv \, \text{COUNT}_k$$

for each $k \geq 1$. Combining steps (A)-(C) and (D) we thus obtain

$$\text{SCH1}_m$$
$$|||$$
$$Q_m\,[\, aR, tW, aW \,/\, up, dn, zero\,]$$
$$\| \; Q_1\,[\, aW, tW, aR \,/\, up, dn, zero\,].$$

Thus using (D) the scheduler $\text{SCH1}_m$ is implemented in a modular fashion by $m+1$ copies of a 1-counter with zero-test.

For $m=2$ the net view of this implementation is as follows:

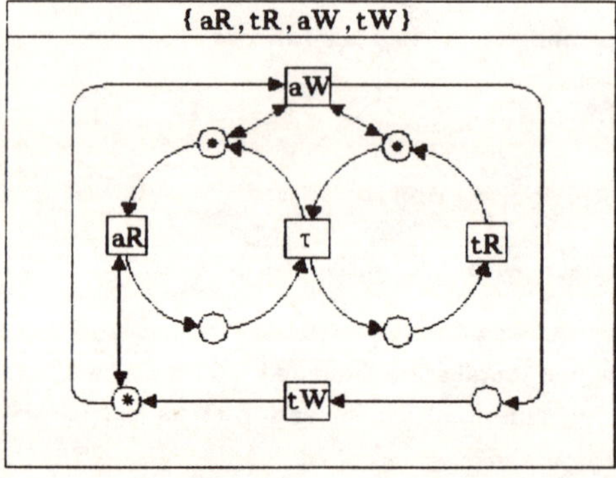


In this net double arrows $\longleftrightarrow$ abbreviate cyclic transitions

Note that the transitions aR and tR (of different readers) can be activated concurrently. For arbitrary $m$ the net consists of $2(m+1)$ places and $m+3$ transitions.

*Second Scheduler*. Now we wish to schedule readers and writers such that *priority* is given to the *readers*. Thus the first writer in the writer queue is allowed to access the shared data only if no reader is waiting in the reader queue.

Therefore we record the number of readers that have requested but not yet accessed the shared data by

$$\text{requestR} =_{df} rR*h - aR*h$$

and start from the following trace specification:

$$
\begin{aligned}
\text{SCH2}_m =_{df}\quad & 0 \le \text{requestR} \le m & (1')\\
\wedge\ & 0 \le \text{activeR} \le m & (2')\\
\wedge\ & 0 \le \text{activeW} \le 1 & (3')\\
\wedge\ & (\ last\ h\upharpoonright\{aW,tW,aR\} = aR \longrightarrow \text{activeW} = 0\ ) & (4')\\
\wedge\ & (\ last\ h\upharpoonright\{rR,aR,tR,aW\} = aW \longrightarrow & \\
& \qquad\qquad\qquad \text{requestR} = 0 \wedge \text{activeR} = 0\ ) & (5')
\end{aligned}
$$

Note that condition (5') is logically equivalent to the conjunction of

$$last\ h\upharpoonright\{rR,aR,aW\} = aW \longrightarrow \text{requestR} = 0 \qquad (6')$$

and

$$last\ h\upharpoonright\{aR,tR,aW\} = aW \longrightarrow \text{activeR} = 0, \qquad (7')$$

each one with smaller projection alphabet than (5'). Hence by using the logic and the parallelism rule we decompose $\text{SCH2}_m$ as follows:

(A')
$$\text{SCH2}_m$$
$$|||$$
$$\text{R-READ} \parallel \text{SCH1}_m$$

where $\text{R-READ} =_{df}$ (1') $\wedge$ (6'). R-READ is another copy of a bounded counter with zero-test. Formally, we have

(B')
$$\text{R-READ}$$
$$|||$$
$$\text{COUNT}_m\,[\,rR,aR,aW\,/\,up,dn,zero\,].$$

Combining (A') and (B') with the construction steps used for $\text{SCH1}_m$ we finally obtain

$$\text{SCH2}_m$$
$$|||$$
$$Q_m\,[\,rR,aR,aW\,/\,up,dn,zero\,]$$
$$\parallel\ Q_m\,[\,aR,tR,aW\,/\,up,dn,zero\,]$$
$$\parallel\ Q_1\,[\,aW,tW,aR\,/\,up,dn,zero\,].$$

Thus using (D) above, $\text{SCH2}_m$ is implemented in a modular fashion by $2m+1$ copies of a 1-counter with zero-test.

For m=2 the net view of this implementation is as follows:

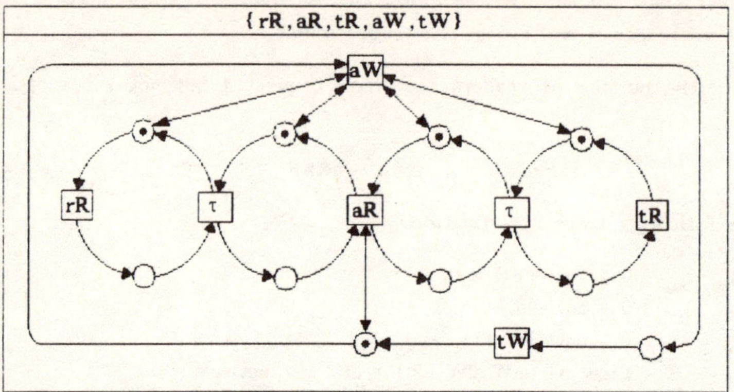


The transitions rR, aR and tR (of different writers) can be activated concurrently. For arbitrary m the net consists of $4m+2$ places and $2m+3$ transitions.

*Third scheduler*. Finally, we wish to schedule readers and writers such that *priority* is given to the *writers*. Thus the first reader in the reader queue is allowed to access the shared data only if no writer is waiting in the writer queue.

Specification and derivation of this scheduler is similar to that of the previous scheduler, but the resulting nets are quite different. We record the number of writers that have requested but not yet accessed the shared data by

$$requestW =_{df} rW * h - aW * h$$

and start from the following trace specification:

$$SCH3_{m,n} =_{df} \quad 0 \leq activeR \quad \leq m \tag{1''}$$
$$\wedge\ 0 \leq requestW \leq n \tag{2''}$$
$$\wedge\ 0 \leq activeW \leq 1 \tag{3''}$$
$$\wedge\ (\textit{last } h \upharpoonright \{rW, aW, tW, aR\} = aR \longrightarrow$$
$$requestW = 0 \wedge activeW = 0) \tag{4''}$$
$$\wedge\ (\textit{last } h \upharpoonright \{aR, tR, aW\} = aW \longrightarrow activeR = 0) \tag{5''}$$

Since condition (4") is logically equivalent to the conjunction of

$$\textit{last } h \upharpoonright \{rW, aW, aR\} = aR \longrightarrow requestW = 0 \tag{6''}$$

and

$$\textit{last } h \upharpoonright \{aW, tW, aR\} = aR \longrightarrow activeW = 0 , \tag{7''}$$

the logic and parallelism rule yield

$$SCH3_{m,n}$$
$$|||$$
$$R\text{-WRITE} \parallel SCH1_m$$

where R-WRITE $=_{df}$ (2") $\wedge$ (6"). Proceeding as in the previous derivation, we obtain

$$\text{SCH3}_{m,n}$$
$$|||$$
$$Q_n[\ rW,aW,aR/up,dn,zero\ ]$$
$$||\ Q_m[aR,tR,aW\ /up,dn,zero\ ]$$
$$||\ Q_1\ [aW,tW,aR\ /up,dn,zero\ ].$$

By referring to (D) above, $\text{SCH3}_m$ is thus implemented by $m+n+1$ copies of a 1-counter with zero-test.

For $m=n=2$ the net view of this implementation is as follows:

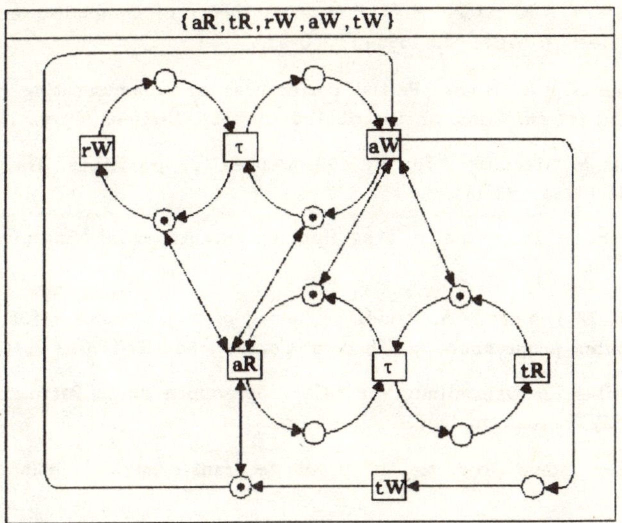


The transitions rW can be activated concurrently to any other transition. As before, the transitions aR and tR (of different readers) can also be activated concurrently. For arbitrary m and n the net consists of $2(m+n+1)$ places and $m+n+4$ transitions. This example illustrates that by performing the same type of derivation steps rather different communicating programs can be obtained. By construction, these these programs satisfy the safety and liveness properties required by the original specification. The Petri nets visualise subtle implementation details such as internal actions or the possible concurrency.

## 8. REFERENCES


[AHV 85]   F. André, D. Herman, J.-P. Verjus, Synchronization of Parallel Programs (MIT Press, Cambridge, Mass., 1985).

[AS 85]   B. Alpern, F.B. Schneider, Defining liveness, Inform. Proc. Letters 21 (1985) 181-185.

[Ba 80]   J.W. de Bakker, Mathematical Theory of Program Correctness (Prentice-Hall, London, 1989).

[Bau 85]    F.L. Bauer et al., The Munich Project CIP, Vol. I: The Wide Spectrum Language CIP-L, Lecture Notes in Comput. Sci. 183 (Springer-Verlag, 1985).

[Bau 87]    F.L. Bauer et al., The Munich Project CIP, Vol. II: The Program Transformation System CIP-S, Lecture Notes in Comput. Sci. 292 (Springer-Verlag, 1987).

[Be 87]    E. Best, COSY: its relation to nets and CSP, in: W. Brauer, W. Reisig, G. Rozenberg (Eds.), Petri Nets: Applications and Relationships to Other Models of Concurrency, Lecture Notes in Comput. Sci. 255 (Springer-Verlag, 1987) 416-440.

[BHR 84]    S.D. Brookes, C.A.R. Hoare, A.W. Roscoe, A theory of communicating sequential processes, J. ACM 31 (1984) 560-599.

[CHo 81]    Z. Chaochen, C.A.R. Hoare, Partial correctness of communicating processes, in: Proc. 2nd Intern. Conf. on Distributed Comput. Systems, Paris, 1981.

[DH 84]    R. DeNicola, M. Hennessy, Testing equivalences for processes, Theoret. Comput. Sci. 34 (1984) 83-134.

[Di 76]    E.W. Dijkstra, A Discipline of Programming (Prentice-Hall, Englewood Cliffs, NJ, 1976).

[FLP 84]    N. Francez, D. Lehmann, A. Pnueli, A linear history semantics for languages for distributed programmming, Theoret. Comput. Sci. 32 (1984) 25-46.

[Go 88]    U. Goltz, Über die Darstellung von CCS-Programmen durch Petrinetze, Doctoral Diss., RWTH Aachen, 1988.

[Ho 78]    C.A.R. Hoare, Some properties of predicate transformers, J. ACM 25 (1978) 461-480.

[Ho 81]    C.A.R. Hoare, A calculus of total correctness for communicating processes, Sci. Comput. Progr. 1 (1981) 44-72.

[Ho 85]    C.A.R. Hoare, Communicating Sequential Processes (Prentice-Hall, London, 1985).

[Jo 87]    B. Jonsson, Compositional Verification of Distributed Systems, Ph.D. Thesis, Dept. Comput. Sci., Uppsala Univ., 1987.

[LTS 79]    P.E. Lauer, P.R. Torrigiani, M.W. Shields, COSY - A system specification language based on paths and processes, Acta Inform. 12 (1979) 109-158.

[Mz 77]    A. Mazurkiewicz, Concurrent program schemes and their interpretations, Tech. Report DAIMI PB-78, Aarhus Univ., 1977.

[Mi 80]    R. Milner, A Calculus of Communicating Systems, Lecture Notes in Comput. Sci. 92 (Springer-Verlag, 1980).

[Mi 89]    R. Milner, Communication and Concurrency (Prentice-Hall, London, 1989).

[MC 81]    J. Misra, K.M. Chandy, Proofs of networks of processes, IEEE Trans. Software Eng. 7 (1981) 417-426.

[Ol 88/89]   E.-R. Olderog, Nets, Terms and Formulas: Three Views of Concurrent Processes and Their Relationship, Habilitationsschrift, Univ. Kiel, 1988/89.

[Ol 89a]   E.-R. Olderog, Strong bisimilarity on nets: a new concept for comparing net semantics, in: J.W. de Bakker, W.P. de Roever, G. Rozenberg (Eds.), Linear Time, Branching Time and Partial Order in Logics and Models of Concurrency, Lecture Notes in Comput. Sci. 354 (Springer-Verlag, 1989) 549-573.

[Ol 89b]   E.-R. Olderog, Correctness of concurrent processes, invited paper, in: A. Kreczmar, G. Mirkowska (Eds.), Math. Found. of Comput. Sci. 1989, Lecture Notes in Comput. Sci. 379 (Springer-Verlag, 1989) 107-132.

[OH 86]   E.-R. Olderog, C.A.R. Hoare, Specification-oriented semantics for communicating processes, Acta Inform. 23 (1986) 9-66.

[OL 82]   S. Owicki, L. Lamport, Proving liveness properties of concurrent programs, ACM TOPLAS 4 (1982) 199-223.

[Re 85]   W. Reisig, Petri Nets, An Introduction, EATCS Monographs on Theoret. Comput. Sci. (Springer-Verlag, 1985).

[Rm 87]   M. Rem, Trace theory and systolic computation, in: J.W. de Bakker, A.J. Nijman, P.C. Treleaven (Eds.), Proc. PARLE Conf., Eindhoven, Vol. I, Lecture Notes in Comput. Sci. 258, (Springer-Verlag, 1987) 14-33.

[Sc 70]   D.S. Scott, Outline of a mathematical theory of computation, Tech. Monograph PRG-2, Progr. Research Group, Oxford Univ., 1970.

[St 77]   J.E. Stoy, Denotational Semantics: The Scott-Strachey Approach to Programming Language Theory (MIT Press, Cambridge, Mass., 1977).

[Sn 85]   J.L.A. van de Snepscheut, Trace Theory and VLSI Design, Lecture Notes in Comput. Sci. 200 (Springer-Verlag, 1985).

[WGS 87]   J. Widom, D. Gries, F.B. Schneider, Completeness and incompleteness of trace-baced network proof systems, in: Proc. 14th ACM Symp. on Principles of Progr. Languages, München, 1987, 27-38.

[Wi 71]   N. Wirth, Program development by stepwise refinement, Comm. ACM 14 (1971) 221-227.

[Zw 89]   J. Zwiers, Compositionality, Concurrency and Partial Correctness, Lecture Notes in Comput. Sci. 321 (Springer-Verlag, 1989).

[ZRE 85]   J. Zwiers, W.P. de Roever, P. van Emde-Boas, Compositionality and concurrent networks, in: W. Brauer (Ed.), Proc. 12th Coll. Automata, Languages and Programming, Lecture Notes in Comput. Sci. 194 (Springer-Verlag, 1985) 509-519.

Printing: Druckerei Zechner, Speyer
Binding: Buchbinderei Schäffer, Grünstadt

# NATO ASI Series F

# NATO ASI Series F

# NATO ASI Series F

*Including Special Programmes on Sensory Systems for Robotic Control (ROB) and on Advanced Educational Technology (AET)*

# NATO ASI Series F